Lecture Notes in Computer Scien

Edited by G. Goos, J. Hartmanis, and J. van L

T0250549

Springer
Berlin
Heidelberg
New York
Barcelona
Hong Kong
London
Milan
Paris
Tokyo

José D. P. Rolim   Salil Vadhan (Eds.)

Randomization and Approximation Techniques in Computer Science

6th International Workshop, RANDOM 2002
Cambridge, MA, USA, September 13-15, 2002
Proceedings

 Springer

Series Editors

Gerhard Goos, Karlsruhe University, Germany
Juris Hartmanis, Cornell University, NY, USA
Jan van Leeuwen, Utrecht University, The Netherlands

Volume Editors

José D. P. Rolim
University of Geneva
Centre Universitaire d'Informatique
24, Rue Général Dufour
1211 Genève 4, Switzerland
E-mail: jose.rolim@cui.unige.ch

Salil Vadhan
Harvard University, DEAS
Maxwell Dworkin 337, 33 Oxford Street
Cambridge, MA 02138, USA
E-mail: salil@eecs.harvard.edu

Cataloging-in-Publication Data applied for

Die Deutsche Bibliothek - CIP-Einheitsaufnahme

Randomization and approximation techniques in computer science :
6th international workshop ; proceedings / RANDOM 2002, Camebridge,
MA, USA, September 13 - 15, 2002. José D. P. Rolim ; Salil Vadhan (ed.). -
Berlin ; Heidelberg ; New York ; Barcelona ; Hong Kong ; London ; Milan ;
Paris ; Tokyo : Springer, 2002
  (Lecture notes in computer science ; Vol. 2483)
  ISBN 3-540-44147-6

CR Subject Classification (1998): F.2, G.1, G.2

ISSN 0302-9743
ISBN 3-540-44147-6 Springer-Verlag Berlin Heidelberg New York

Springer-Verlag Berlin Heidelberg New York,
a member of BertelsmannSpringer Science+Business Media GmbH

http://www.springer.de

© Springer-Verlag Berlin Heidelberg 2002
Printed in Germany

Typesetting: Camera-ready by author, data conversion by PTP-Berlin, Stefan Sossna e.K.
Printed on acid-free paper   SPIN: 10870588   06/3142   5 4 3 2 1 0

Foreword

This volume contains the papers presented at the *6th International Workshop on Randomization and Approximation Techniques in Computer Science* (RANDOM 2002), which took place at Harvard University, Cambridge, Massachusetts, from September 13–15, 2002. RANDOM 2002 was concerned with applications of randomness to computational and combinatorial problems, and was the sixth workshop in the series following Bologna, Barcelona, Berkeley, Geneva, and Berkeley again.

The volume contains 21 contributed papers, selected by the program committee from 48 submissions received in response to the call for papers. We thank all of the authors who submitted papers, our invited speakers, the members of the program committee:

Dimitris Achlioptas, Microsoft Research
Martin Dyer, U. of Leeds
Uriel Feige, Weizmann Institute
Russell Impagliazzo, UC San Diego
Sampath Kannan, U. of Pennsylvania
David Karger, MIT
Nati Linial, Hebrew U.
Rafail Ostrovsky, Telcordia Technologies
Paul Spirakis, U. of Patras and CTI
Angelika Steger, TU Munich
Rüdiger Urbanke, Swiss Federal Inst. of Tech.
Salil Vadhan, Harvard U., chair,

and the external reviewers: N. Alon, R. Alur, A. Ambainis, T. Batu, J. Feigenbaum, S. Gerke, Y. Gertner, A. Goerdt, L. Goldberg, J. Hastad, C. Iliopoulos, Y. Ishai, V. Kabanets, S. Khot, L. Kirousis, S. Kontogiannis, M. Krivelevich, M. Mavronicolas, A. McGregor, F. McSherry, D. van Melkebeek, M. Molloy, E. Mossel, S. Nikoletseas, R. Raz, D. Ron, P. Tetali, L. Trevisan, E. Vigoda, J. Watrous, and P. Winkler.

We gratefully acknowledge support from the Division of Engineering and Applied Sciences at Harvard University and from the Département d'Informatique of the University of Geneva. We also would like to thank Hafssa Benaboud for her help in preparing this volume.

July 2002

José D.P. Rolim, Workshop Chair
Salil Vadhan, Program Chair

Table of Contents

Counting Distinct Elements in a Data Stream

Ziv Bar-Yossef[1]*, T.S. Jayram[2], Ravi Kumar[2], D. Sivakumar[2], and
Luca Trevisan[3]**

[1] Computer Science Division, Univ. of California at Berkeley, Berkeley, CA 94720.
`zivi@cs.berkeley.edu`
[2] IBM Almaden Research Center, 650 Harry Road, San Jose, CA 95120. {`jayram,`
`ravi, siva`}`@almaden.ibm.com`
[3] Computer Science Division, Univ. of California at Berkeley, Berkeley, CA 94720.
`luca@cs.berkeley.edu`

Abstract. We present three algorithms to count the number of distinct
elements in a data stream to within a factor of $1 \pm \epsilon$. Our algorithms
improve upon known algorithms for this problem, and offer a spectrum
of time/space tradeoffs.

1 Introduction

Let $\mathbf{a} = a_1, \ldots, a_n$ be a sequence of n elements from the domain $[m] =
\{1, \ldots, m\}$. The *zeroth-frequency moment* of this sequence is the number of
distinct elements that occur in the sequence and is denoted $F_0 = F_0(\mathbf{a})$. In this
paper we present three space- and time-efficient algorithms for approximating
F_0 in the data stream model.

In the data stream model, an algorithm is considered efficient if it makes
one (or a small number of) passes over the input sequence, uses very little
space, and processes each element of the input very quickly. In our context,
a data stream algorithm to approximate F_0 is considered efficient if it uses only
$\mathrm{poly}(1/\epsilon, \log n, \log m)$ bits of memory, where $1 \pm \epsilon$ is the factor within which F_0
must be approximated.

Let $\epsilon, \delta > 0$ be given. An algorithm $\mathcal{A}$ is said to be (ϵ, δ)-*approximate* F_0 if for
any sequence $\mathbf{a} = a_1, \ldots, a_n$, with each $a_i \in [m]$, it outputs a number $\tilde{F}_0$ such
that $\Pr[|F_0 - \tilde{F}_0| \leq \epsilon F_0] \geq 1 - \delta$, where the probability is taken over the internal
coin tosses of $\mathcal{A}$. Two main parameters of $\mathcal{A}$ are of interest: the workspace and
the time to process each item. We study these quantities as functions of the
domain size m, the number n of elements in the stream, the approximation
parameter ϵ, and the confidence parameter δ.

There are several reasons for designing algorithms for F_0 in the data stream
model. Counting the number of distinct elements in a (column of a relational)
table of data is a fairly fundamental problem in databases. This has applications

---

* Part of this work was done while the author was visiting IBM Almaden Research
Center. Supported by NSF Grant CCR-9820897.
** Work supported by a Sloan Research Fellowship and an NSF Career Award.

to estimating the selectivity of queries, designing good plans for executing a query, etc.—see, for instance, [WVT90,HNSS96]. Another application of counting distinct elements is in routing of Internet traffic. The router usually has very limited memory, but it is desirable to have the router gather various statistical properties (say, the number of distinct destination addresses) of the traffic flow. The number of distinct elements is also a natural quantity of interest in several large data set applications (eg., the number of distinct queries made to a search engine over a week).

Flajolet and Martin [FM85] designed the first algorithm for approximating F_0 in the data stream (or what was then thought of as a one-pass) model. Unfortunately, their algorithm assumed the existence of hash functions with some ideal properties; it is not known how to construct such functions with limited space. Alon, Matias, and Szegedy [AMS99] built on these ideas, but used random pairwise independent hash functions [CW77,WC79] and gave an (ϵ, δ)-approximation algorithm for $\epsilon > 1$; their algorithm uses $O(\log m)$ space. For arbitrarily small ϵ, Gibbons and Tirthapura [GT01] gave an algorithm that used $S = O(1/\epsilon^2 \cdot \log m)$ space and $O(S)$ processing time per element; Bar-Yossef et al. [BKS02] gave an algorithm that used $O(1/\epsilon^3 \cdot \log m)$ space (and time per element) but that had some other nice property required for their application. Cohen [Coh97] considered this problem in the context of graph-theoretic applications; her algorithm is similar in spirit to that of [FM85,AMS99]; specifically, it has a high-level viewpoint similar to the first algorithm in this paper. However, the implementation is very different, and does not yield a $o(m)$ space algorithm.

One of the drawbacks of the algorithms of [GT01,BKS02] is that the space and time are the *product* of $\mathrm{poly}(1/\epsilon)$ and $\log m$. Even with modestly small constants in the O notation and $\epsilon = 0.01$, the space required might be prohibitive in certain applications (eg., a router with very little memory or a database application where the frequency estimation is required to be piggy-backed on some other computation). This leads to the question of whether it is possible to obtain space/time upper bounds that are $\mathrm{poly}(1/\epsilon) + \log m$. In this paper we achieve this, modulo factors of the form $\log \log m$ and $\log(1/\epsilon)$.

Results. We give three algorithms with different space-time tradeoffs for approximating F_0. Each of our algorithms is an improvement over any of the existing algorithms in either space or processing time or both.

We will state the bounds for our algorithms in terms of ϵ and $\log m$ (suppressing the dependence on n). This is without loss of generality: If indeed $m < n$, it is clearly advantageous to have an algorithm whose bounds depend on $\log m$ and not on $\log n$. If, on the other hand, $m > n$, we can employ a simple hashing trick (with $O(\log(m + n))$ space and time per element) that reduces the description of each stream element to $O(\log n)$ bits. Thus we will assume for the rest of the paper that $\log m = O(\log n)$. We will also assume that there exists $\epsilon_0 < 1$ such that the accuracy parameter ϵ given to the algorithms is at most ϵ_0. (Note that this is, in fact, the interesting case. We make this assumption explicit only so that we may abbreviate $\max\{1/\epsilon, \epsilon_0\}$ by $1/\epsilon$.)

The following table summarizes our results. The $\tilde{O}$ notation suppresses $\log(1/\epsilon)$ and $\log \log m$ factors. For simplicity, the dependence on δ, which is a multiplicative factor of $\log(1/\delta)$ for both space and time, is also dropped.

Algorithm	Space	Time/element
1, Thm. 1	$O(1/\epsilon^2 \cdot \log m)$	$\tilde{O}(\log m)$
2, Thm. 2	$\tilde{O}(1/\epsilon^2 + \log m)$	$\tilde{O}(1/\epsilon^2 \cdot \log m)$
3, Thm. 3	$\tilde{O}(1/\epsilon^2 + \log m)$	$\tilde{O}(\log m)$ [amortized]

A lower bound of $\Omega(\log m)$ was shown in [AMS99]. It is also easy to show an $\Omega(1/\epsilon)$ lower bound by a reduction from the "indexing" problem in the one-way communication complexity model. An interesting open question is to obtain an algorithm with space bound $(1/\epsilon) \cdot$ polylog (m), or a lower bound of $\Omega(1/\epsilon^2)$.

2   The First Algorithm

Our first algorithm is a generalization of the algorithm of [FM85,AMS99] to work for any $\epsilon > 0$.

To make our exposition clearer, we first describe an intuitive way to look at the algorithm of [FM85,AMS99]. This algorithm first picks a random hash function $h : [m] \rightarrow [0, 1]$. It then applies $h(\cdot)$ to all the elements in $\mathbf{a}$ and maintains the value $v = \min_{j=1}^{n} h(a_j)$. In the end, the estimation is $\tilde{F}_0 = 1/v$. The algorithm has the right approximation (in the expectation sense) because if there are F_0 independent and uniform values in $[0, 1]$, then their expected minimum is around $1/F_0$. Of course, the technical argument in [AMS99] is to quantify this precisely, even when h is chosen from a pairwise independent family of hash functions.

In our algorithm, we also pick a hash function $h : [m] \rightarrow [0, 1]$, but we keep the $t = O(1/\epsilon^2)$ elements a_i on which h evaluates to the t smallest values. If we let v be the t-th smallest such value, we estimate $\tilde{F}_0 = t/v$. This is because when we look at F_0 uniformly distributed (and, say, pairwise independent) elements of $[0, 1]$, we expect about t of them to be smaller than t/F_0. The formal description is given below.

Theorem 1. *There is an algorithm that for any $\epsilon, \delta > 0$, (ϵ, δ)-approximates F_0 using $O(1/\epsilon^2 \cdot \log m \cdot \log(1/\delta))$ bits of memory and $O(\log(1/\epsilon) \cdot \log m \cdot \log(1/\delta))$ processing time per element.*

Proof. We pick at random a pairwise independent hash function $h : [m] \rightarrow [M]$, where $M = m^3$. Note that, with probability at least $1 - 1/m$, h is injective over the elements of $\mathbf{a}$.

Let $t = \lceil 96/\epsilon^2 \rceil$. Our algorithm maintains the t smallest distinct values of $h(a_i)$ seen so far. The algorithm updates (if necessary) this list each time a new element arrives. Let v be the value of the t-th smallest such value when the entire sequence has been processed. The algorithm outputs the estimation $\tilde{F}_0 = tM/v$.

The algorithm can be implemented in $O(1/\epsilon^2 \cdot \log m)$ space, since the hash function h requires $O(\log m)$ space, and each of the $t = O(1/\epsilon^2)$ values to be stored requires $O(\log m)$ space. The t values could be stored in a balanced binary search tree, so that each step can be implemented in $O(\log(1/\epsilon) \cdot \log m)$ time (rather than $O(1/\epsilon^2 \cdot \log m)$ which would be necessary if the elements are stored as a list).

We can assume that $1/M < \epsilon t/(4F_0)$. Since $F_0 \leq m$, $M = m^3$, and $t \geq 96/\epsilon^2$, the condition is satisfied as long as $m \geq \sqrt{\epsilon/24}$. Let $b_1, \ldots, b_{F_0}$ be the distinct elements of **a**.

Let us first consider the case $\tilde{F}_0 > (1+\epsilon)F_0$, i.e., the case when the algorithm outputs a value above $(1+\epsilon)F_0$. This means the sequence $h(b_1), \ldots, h(b_{F_0})$ contains at least t elements that are smaller than $tM/(F_0(1+\epsilon)) \leq (1-\epsilon/2)tM/F_0$ (using the fact $\epsilon \leq 1$). Each $h(b_i)$ has a probability at most $(1-\epsilon/2)t/F_0 + 1/M < (1-\epsilon/4)t/F_0$ (taking into account rounding errors) of being smaller than $(1-\epsilon/2)tM/F_0$. Thus, we are in a situation where we have F_0 pairwise independent events, each one occurring with probability at most $(1-\epsilon/4)t/F_0$, and at least t such events occur. Let $X_i, i = 1, \ldots, F_0$, be an indicator r.v. corresponding to the event "$h(b_i) < (1-\epsilon/2)tM/F_0$". Clearly, $E[X_i] \leq (1-\epsilon/4)t/F_0$. Let $Y = \sum_{i=1}^{F_0} X_i$. It follows $E[Y] \leq (1-\epsilon/4)t$ and by pairwise independence, $\mathrm{Var}(Y) \leq (1-\epsilon/4)t$. The event that the algorithm outputs a value above $(1+\epsilon)F_0$ occurs only if Y is more than t, and therefore the probability of error is:

$$\Pr[Y > t] \leq \Pr[|Y - E[Y]| > \epsilon t/4] \leq 16 \cdot \mathrm{Var}[Y]/(\epsilon^2 t^2) \leq 16/(\epsilon^2 t) \leq 1/6,$$

using Chebyshev's inequality.

Let us consider now the case in which the algorithm outputs $\tilde{F}_0$ which is below $(1-\epsilon)F_0$. This means the sequence $h(b_1), \ldots, h(b_{F_0})$ contains less than t elements that are smaller than $tM/(F_0(1-\epsilon)) \leq (1+\epsilon)tM/F_0$. Let X_i be an indicator r.v. corresponding to the event "$h(b_i) \leq (1+\epsilon)tM/F_0$", and let $Y = \sum_{i=1}^{F_0} X_i$. Taking into account rounding errors, $(1+\epsilon/2)t/F_0 \leq E[X_i] \leq (1+3\epsilon/2)t/F_0$, and therefore $E[Y] \geq t(1+\epsilon/2)$ and $\mathrm{Var}[Y] \leq E[Y] \leq t(1+3\epsilon/2)$, as before. Now, $\tilde{F}_0 < (1-\epsilon)F_0$ only if $Y < t$, and therefore the probability of error is:

$$\Pr[Y < t] \leq \Pr[|Y - E[Y]| \geq \epsilon t/2] \leq 4 \cdot \mathrm{Var}[Y]/(\epsilon^2 t^2) \leq 12/(\epsilon^2 t) < 1/6$$

Thus, the probability that the algorithm outputs $\tilde{F}_0$ which is not within $(1 \pm \epsilon)$ factor of F_0 is at most $1/3 + 1/m$. As usual, this probability can be amplified to $1 - \delta$ by running in parallel $O(\log(1/\delta))$ copies of the algorithm, and taking the median of the resulting approximations.

3   The Second Algorithm

The second algorithm is based on recasting the F_0 problem as estimating the probability of an appropriately defined event. The main idea is to define a quantity that can be approximated in the data stream model and that, in turn, can be used to approximate F_0.

Theorem 2. *There is an algorithm that for any $\epsilon, \delta > 0$, (ϵ, δ)-approximates F_0 using $\widetilde{O}((1/\epsilon^2 + \log m) \cdot \log(1/\delta))$ bits of memory. The processing time per data item is $\widetilde{O}(1/\epsilon^2 \cdot \log m \cdot \log(1/\delta))$. (The $\widetilde{O}$ notation suppresses $\log(1/\epsilon)$ and $\log \log m$ factors).*

Proof. We take advantage of the fact that the algorithm of [AMS99] can be used to provide a rough estimate of F_0; namely, it is possible to obtain an estimate R such that $2F_0 \leq R \leq 2cF_0$, where $c = 25$, with probability at least $3/5$. In addition, R may be assumed to be a power of 2. Our algorithm will implement the AMS algorithm on one track of the computation, and keep track of some extra quantities on another. In the sequel, we will add $2/5$ to the error probability, and assume that we have an estimate R that meets this bound.

Let $b_1, \ldots, b_{F_0}$ denote the F_0 distinct elements in the stream **a**, and let B denote the set $\{b_1, \ldots, b_{F_0}\}$. Consider a completely random map $h : [m] \to [R]$, and define $r = \mathrm{Pr}_h[h^{-1}(0) \cap B \neq \emptyset] = 1 - (1 - 1/R)^{F_0}$. We first show that if R and F_0 are within constant multiples of each other, then approximating r is a good way to approximate F_0.

Lemma 1. *Let $c = 25$ and let $\epsilon > 0$ be given. Let R and F_0 satisfy $1/(2c) \leq (F_0/R) \leq 1/2$. Then if $|r - \tilde{r}| \leq \gamma = \min\{1/e - 1/3, \epsilon/(6c)\}$, then $\tilde{F}_0$, defined by*

$$\tilde{F}_0 = \frac{\ln(1 - \tilde{r})}{\ln(1 - 1/R)}$$

satisfies $\left| F_0 - \tilde{F}_0 \right| \leq \epsilon F_0$.

Proof. Since $R \geq 2F_0$, we have $R \geq 2$, therefore $1/R \leq 1/2$, and so $1 - 1/R \geq e^{-2/R}$ (since $1 - x \geq e^{-2x}$ for $x \leq 1/2$). Hence $r = 1 - (1 - 1/R)^{F_0} \leq 1 - e^{-2F_0/R}$. Since $F_0/R \leq 1/2$, we have $r \leq 1 - 1/e$. By definition, $\gamma \leq 1/e - 1/3$, so we have $r + \gamma < 2/3$ and $1/(1 - (r + \gamma)) < 3$. Also for $R > 1$, we have $-1/\ln(1 - 1/R) \leq R$.

For a continuous function f, $|f(x) - f(x + \epsilon)| \leq \epsilon \left| \sup_{y \in (x, x+\epsilon)} f'(y) \right|$. Letting $f(x) = \ln(1 - x)$, we obtain $|f(x) - f(\tilde{x})| \leq |x - \tilde{x}| / (1 - \max\{x, \tilde{x}\})$. Therefore,

$$\left| F_0 - \tilde{F}_0 \right| = \frac{|\ln(1 - r) - \ln(1 - \tilde{r})|}{-\ln(1 - 1/R)} \leq \frac{R |r - \tilde{r}|}{1 - (r + \gamma)} \leq 3R\gamma \leq 3 \cdot (2cF_0) \cdot \frac{\epsilon}{6c} = \epsilon F_0.$$

Our idea is to approximate r by using hash functions h that are not totally random, but just random enough to yield the desired approximation. We will pick h from a family $\mathcal{H}$ of hash functions from $[m]$ into $[R]$, whose choice we will spell out shortly. Let $p = \mathrm{Pr}_{h \in \mathcal{H}}[h^{-1}(0) \cap B \neq \emptyset]$. For $i = 1, \ldots, F_0$, let $\mathcal{H}_i$ denote the set of hash functions in $\mathcal{H}$ that map the i-th distinct element b_i of B to 0. Note that $p = |\bigcup_i \mathcal{H}_i| / |\mathcal{H}|$, thus our goal is to estimate this union size. By inclusion–exclusion, we have

$$p = \left(\sum_i \Pr_{h \in \mathcal{H}} [h \in \mathcal{H}_i] \right) - \left(\sum_{i<j} \Pr_{h \in \mathcal{H}} [h \in \mathcal{H}_i \cap \mathcal{H}_j] \right)$$

$$+ \left(\sum_{i<j<k} \Pr_{h \in \mathcal{H}} [h \in \mathcal{H}_i \cap \mathcal{H}_j \cap \mathcal{H}_k] \right) - \cdots$$

Let P_ℓ denote the ℓ-th term in the above series. For any odd $t > 0$, we know that

$$\sum_{\ell=1}^{t-1} (-1)^{\ell+1} P_\ell \;\leq\; p \;\leq\; \sum_{\ell=1}^{t} (-1)^{\ell+1} P_\ell.$$

Our key observation is that if picking $h \in \mathcal{H}$ yields a t-wise independent hash function, then we know precisely what each P_ℓ is, and we have

$$\sum_{\ell=1}^{t-1} (-1)^{\ell+1} \binom{F_0}{\ell} R^{-\ell} \leq p \leq \sum_{\ell=1}^{t} (-1)^{\ell+1} \binom{F_0}{\ell} R^{-\ell}. \tag{1}$$

On the other hand, via binomial expansion we know that

$$r = 1 - \left(1 - \frac{1}{R} \right)^{F_0} = \sum_{i=1}^{F_0} (-1)^{i+1} \binom{F_0}{i} R^{-i},$$

whence we have for odd t that

$$\sum_{\ell=1}^{t-1} (-1)^{\ell+1} \binom{F_0}{\ell} R^{-\ell} \leq r \leq \sum_{\ell=1}^{t} (-1)^{\ell+1} \binom{F_0}{\ell} R^{-\ell}. \tag{2}$$

From Equations (1) and (2), it follows that both p and r are sandwiched inside an interval of width $\binom{F_0}{t} R^{-t} \leq (eF_0/(tR))^t \leq (1/5)^t$, which, with a choice of $t = \lceil \lg(2/\gamma)/\lg 5 \rceil$ implies that $|p - r| \leq \gamma/2$. Recall that $\gamma = \min\{1/e - 1/3, \epsilon/(6c)\}$, where $c = 25$.

Finally, we will show how to produce an estimate $\tilde{p}$ of p such that $|p - \tilde{p}| \leq \gamma/2$, so that $|\tilde{p} - r| \leq \gamma$, and we can apply Lemma 1.

The idea is to pick several hash functions $h_1, \ldots, h_k$ from a family $\mathcal{H}$ of t-wise independent hash functions. For a sequence $H_R = (h_1, \ldots, h_k)$ of hash functions from $[m]$ into $[R]$, define the estimator

$$X(H_R) = \frac{1}{k} \left| \{ j \mid h_j^{-1}(0) \cap B \neq \emptyset \} \right|.$$

Clearly, $\mathrm{E}[X(H_R)] = p$. If $k = O(1/\gamma^2) = O(1/\epsilon^2)$ is suitably large, then by Chebyshev's inequality, we can show that $\Pr[|X(H_R) - p| > \gamma/2] \leq 1/20$.

Each hash function can be described by $s = O(t \log m)$ bits. Instead of picking the k hash functions independently from $\mathcal{H}$, we will pick them pairwise independently; this requires only $2s$ bits (assuming $2^s \geq k$) that we dub the "master hash function." The idea is that we will keep only the master hash function;

as each element a_i of the stream is processed, we will construct each of the k hash functions in turn and compute whether it maps a_i to 0. Extracting the description of each hash function h_j from the master hash function can be done easily in space $O(s)$ and time $O(s \log s)$. Alternatively, we could use a master hash function of $O(s \log k)$ bits with the ability to extract each hash function in space and time $O(s \log k)$.

Lastly, we spell out how we handle the issue that we don't know R to begin with. Recall that R will be available to us through the AMS algorithm only after the stream has been processed. Thus, at the end of the stream, we need the ability to compute the estimator $X(H_R)$ for each $R = 1, \dots, \log m$. Here we use the fact that for standard hash functions where the size of the range $[m]$ is a power of 2, extracting the least significant z bits, $1 \leq z \leq \log m$, gives a hash function with range $[2^z]$. Thus, for each hash function h_j, we will maintain not just one bit indicating whether $h_j^{-1}(0) \cap B \neq \emptyset$, but we will keep track of the largest z such that for some element b in the stream, $h_j(b)$ had z least significant bits equal to zero.

To complete the correctness argument, note that the error probability is bounded by the error probability of the application of the AMS algorithm, which is $2/5$, plus the error probability of the estimation, which is $1/20$. Thus, with probability at least $11/20$, the algorithm produces an estimate $\tilde{F}_0$ of F_0 such that $\left| F_0 - \tilde{F}_0 \right| \leq \epsilon F_0$. Repeating this $O(\log 1/\delta)$ times and taking the median reduces the error probability to δ.

Let us summarize the space and time requirements:

1. Storing the master hash requires space either $O(s \log s) = O(\log(1/\epsilon) \cdot \log m \cdot (\log \log(1/\epsilon) + \log \log m))$ or $O(s \log k) = O(\log^2(1/\epsilon) \cdot \log m)$.
2. Storing the number of trailing zeros for each hash function needs $O(k \log \log m)$ space, which is $O(1/\epsilon^2 \cdot \log \log m)$ bits.
3. To process each item of the stream, the time required is dominated by accessing the master hash function k times, and is therefore $\tilde{O}(1/\epsilon^2 \cdot \log m)$, suppressing $\log \log m$ and $\log(1/\epsilon)$ factors.

4   The Third Algorithm

The algorithm in this section is a unified and improved version of two previous algorithms: one due to Bar-Yossef *et al.* [BKS02] and one due to Gibbons and Tirthapura [GT01].

Theorem 3. *There is an algorithm that for any $\epsilon, \delta > 0$, (ϵ, δ)-approximates F_0 using $S = \tilde{O}((1/\epsilon^2 + \log m) \log(1/\delta))$ bits of memory (suppressing $\log(1/\epsilon)$ and $\log \log m$ factors). The processing time per data item is $O(S)$ in the worst-case and $O((\log m + \log(1/\epsilon)) \log(1/\delta))$ amortized.*

Proof. For a bit string $s \in \{0,1\}^*$, we denote by TRAIL(s) the number of trailing 0's in s.

Let $b_1, \ldots, b_{F_0}$ the F_0 distinct elements in the input stream $a_1, \ldots, a_n$. Let $B = \{b_1, \ldots, b_{F_0}\}$ and for each $i \in [n]$, let $B_i = B \cap \{a_1, \ldots, a_i\}$.

The algorithm picks a random pairwise independent hash function $h : [m] \rightarrow [m]$; we will assume that m is a power of 2. For $t = 0, \ldots, \log m$, define $h_t : [m] \rightarrow [2^t]$ to be the projection of h on its last t bits. The algorithm finds the minimum t for which $r = \left| h_t^{-1}(0) \cap B \right| \leq c/\epsilon^2$, where $c = 576$. It then outputs $r \cdot 2^t$.

In order to find this t, the algorithm initially assumes $t = 0$, and while scanning the input stream, stores in a buffer all the elements b_j from the stream for which $h_t(b_j) = 0$. When the size of the buffer exceeds c/ϵ^2 (say, after reading a_i) the algorithm increases t by one. Note that $h_{t+1}(b_j) = 0$ implies $h_t(b_j) = 0$. Therefore, the algorithm does not have to rescan $a_1, \ldots, a_i$ in order to obtain $h_{t+1}^{-1}(0) \cap B_i$; rather, since $h_{t+1}^{-1}(0) \subseteq h_t^{-1}(0)$, the algorithm will simply extract it from the buffer, which contains $h_t^{-1}(0) \cap B_i$. At the end of the execution we are left with the minimum t for which the buffer size (i.e., $\left| h_t^{-1}(0) \cap B \right|$) does not exceed c/ϵ^2.

Up to this point, this is a simpler exposition of the Gibbons–Tirthapura algorithm. We further improve the algorithm by storing the elements in the buffer more efficiently. Instead of keeping the actual names of the elements, we keep their hash values, using a second hash function. Specifically, let $g : [m] \rightarrow [3 \cdot ((\log m + 1) \cdot c/\epsilon^2)^2]$ be a randomly chosen pairwise independent hash function. Note that since we apply g on at most $(\log m + 1) \cdot (c/\epsilon^2)$ distinct elements, g is injective on these elements with probability at least $5/6$. For each element b_j stored in the buffer, we need to store also the largest number t for which $h_t(b_j) = 0$ (which is basically TRAIL($h(b_j)$)); we use this value during the extraction of $h_{t+1}^{-1}(0) \cap B_i$ from $h_t^{-1}(0) \cap B_i$. In order to do this succinctly, we keep an array of balanced binary search trees T of size $\log m + 1$. The t-th entry in this array is to contain all the elements b_j in the buffer, for which TRAIL($h(b_j)$) $= t$.

We start by analyzing the space and time requirements of the algorithm. The space used by the algorithm breaks down as follows:

1. Hash function h: $O(\log m)$ bits.
2. Hash function g: $O(\log m + \log(1/\epsilon))$ bits.
3. Buffer T: $O(\log m)$ bits for the array itself, and $O(1/\epsilon^2 \cdot (\log(1/\epsilon) + \log \log m))$ for the elements stored in its binary search trees (because we always store at most $O(1/\epsilon^2)$ elements, and each one requires $O(\log(1/\epsilon) + \log \log m)$ bits).

The total space is, thus, $O(\log m + 1/\epsilon^2 \cdot (\log(1/\epsilon) + \log \log m))$.

The worst-case running time per item is $O(\log m + (1/\epsilon^2) \cdot (\log(1/\epsilon) + \log \log m))$, because this is the number of steps required to empty the buffer T. The amortized running time per item is, however, only $O(\log m + \log(1/\epsilon))$ because each element is inserted at most once and removed at most once from the buffer T.

We next prove the algorithm indeed produces a $1 \pm \epsilon$ relative approximation of F_0 with probability at least $2/3$.

One source of error in the algorithm is g having collisions. Note that for a given stream $a_1, \ldots, a_n$ and a given choice of h, the at most $(\log m + 1) \cdot (c/\epsilon^2)$ elements on which we apply g are totally fixed. Since the size of the range of g is thrice the square of the number of elements on which we apply it, and since g is pairwise independent, the probability that g has collisions on these elements is at most $1/6$. We thus assume from now on that g has no collisions, and will add $1/6$ to the final error.

For each $t = 0, 1, \ldots, \log m$, we define $X_t = \left| h_t^{-1}(0) \cap B \right|$; i.e., the number of distinct elements in the stream that h_t maps to 0. Further define for each such t and for each $j = 1, \ldots, F_0$, $X_{t,j}$ to be an indicator random variable, indicating whether $h_t(b_j) = 0$ or not. Note that $\mathrm{E}(X_{t,j}) = \Pr[h_t(b_j) = 0] = 1/2^t$ and $\mathrm{Var}[X_{t,j}] \leq \mathrm{E}[X_{t,j}]$. Therefore, $\mathrm{E}[X_t] = \sum_j \mathrm{E}[X_{j,t}] = F_0/2^t$ and $\mathrm{Var}[X_t] = \sum_j \mathrm{Var}[X_{j,t}] \leq \mathrm{E}[X_t]$ (the latter follows from the pairwise independence of $\{X_{t,j}\}_j$).

Let t^* be the final value of t produced by the algorithm; that is, t^* is the smallest t for which $X_t \leq c/\epsilon^2$. Note that the algorithm's output is $X_{t^*} \cdot 2^{t^*}$.

If $t^* = 0$, then it means that $F_0 \leq c/\epsilon^2$, in which case our algorithm computes F_0 exactly. Assume, then, that $t^* \geq 1$. We write the algorithm's error probability as follows: (in the derivation we define $\bar{t}$ to be the t for which $12/\epsilon^2 \leq F_0/2^{\bar{t}} < 24/\epsilon^2$; note that such a $\bar{t}$ always exists).

$$\Pr\left[\left| X_{t^*} \cdot 2^{t^*} - F_0 \right| > \epsilon F_0 \right] = \Pr\left[\left| X_{t^*} - \frac{F_0}{2^{t^*}} \right| > \epsilon \frac{F_0}{2^{t^*}} \right] =$$

$$= \sum_{t=1}^{\log m} \Pr\left[\left| X_t - \frac{F_0}{2^t} \right| > \epsilon \frac{F_0}{2^t} \mid t^* = t \right] \cdot \Pr[t^* = t]$$

$$= \sum_{t=1}^{\log m} \Pr\left[|X_t - \mathrm{E}[X_t]| > \epsilon \mathrm{E}[X_t] \mid X_t \leq \frac{c}{\epsilon^2}, X_{t-1} > \frac{c}{\epsilon^2} \right] \cdot \Pr\left[X_t \leq \frac{c}{\epsilon^2}, X_{t-1} > \frac{c}{\epsilon^2} \right]$$

$$\leq \sum_{t=1}^{\bar{t}-1} \Pr\left[|X_t - \mathrm{E}[X_t]| > \epsilon \mathrm{E}[X_t] \right] + \sum_{t=\bar{t}}^{\log m} \Pr\left[X_t \leq \frac{c}{\epsilon^2}, X_{t-1} > \frac{c}{\epsilon^2} \right]$$

$$\leq \sum_{t=1}^{\bar{t}-1} \frac{\mathrm{Var}[X_t]}{\epsilon^2 \mathrm{E}^2[X_t]} + \Pr\left[X_{\bar{t}-1} > \frac{c}{\epsilon^2} \right] \leq \sum_{t=1}^{\bar{t}-1} \frac{1}{\epsilon^2 \mathrm{E}[X_t]} + \mathrm{E}[X_{\bar{t}-1}] \cdot \frac{\epsilon^2}{c}$$

$$= \sum_{t=1}^{\bar{t}-1} \frac{2^t}{\epsilon^2 F_0} + \frac{F_0}{2^{\bar{t}-1}} \cdot \frac{\epsilon^2}{c} \leq \frac{2^{\bar{t}}}{\epsilon^2 F_0} + \frac{F_0}{2^{\bar{t}}} \cdot \frac{2\epsilon^2}{c} \leq \frac{1}{\epsilon^2} \cdot \frac{\epsilon^2}{12} + \frac{24}{\epsilon^2} \cdot \frac{2\epsilon^2}{576} = \frac{1}{6}.$$

Thus, the total error probability is $1/3$, as required. As before, this probability can be amplified to $1 - \delta$ by running in parallel $O(\log(1/\delta))$ copies of the algorithm, and outputting the median of the resulting approximations.

References

[AMS99]   N. Alon, Y. Matias, and M. Szegedy. The space complexity of approximating the frequency moments. *Journal of Computer and System Sciences*, 58(1):137–147, 1999.

[BKS02]   Z. Bar-Yossef, R. Kumar, and D. Sivakumar. Reductions in streaming algorithms, with an application to counting triangles in graphs. In *Proceedings
of the 13th Annual ACM-SIAM Symposium on Discrete Algorithms*, pages
623–632, 2002.

[Coh97]   E. Cohen. Size-estimation framework with applications to transitive closure
and reachability. *Journal of Computer and System Sciences*, 55(3):441–453,
1997.

[CW77]   L. Carter and M. Wegman. Universal classes of hash functions. In *Proceedings of the 9th ACM Annual Symposium on Theory of Computing*, pages
106–112, 1977. Journal version in *Journal of Computer and System Sciences*, 18(2) 143–154, 1979.

[FM85]   P. Flajolet and G. N. Martin. Probabilistic counting algorithms for data
base applications. *Journal of Computer and System Sciences*, 31:182–209,
1985.

[GT01]   P. Gibbons and S. Tirthapura. Estimating simple functions on the union
of data streams. In *Proceedings of the 13th ACM Symposium on Parallel
Algorithms and Architectures*, pages 281–291, 2001.

[HNSS96]   P.J. Haas, J.F. Naughton, S. Seshadri, and A.N. Swami. Selectivity and
cost estimation for joins based on random sampling. *Journal of Computer
and System Sciences*, 52(3), 1996.

[WC79]   M. Wegman and L. Carter. New classes and applications of hash functions.
In *Proceedings of the 20th IEEE Annual Symposium on Foundations of
Computer Science*, pages 175–182, 1979. Journal version titled "New Hash
Functions and Their Use in Authentication and Set Equality" in *Journal
of Computer and System Sciences*, 22(3): 265-279, 1981.

[WVT90]   K.-Y. Whang, B. T. Vander-Zanden, and H. M. Taylor. A linear-time probabilistic counting algorithm for database applications. *ACM Transactions
on Database Systems*, 15(2):208–229, 1990.

On Testing Convexity and Submodularity

Michal Parnas[1], Dana Ron[2*], and Ronitt Rubinfeld[3]

[1] The Academic College of Tel-Aviv-Yaffo. michalp@mta.ac.il
[2] Department of EE – Systems, Tel-Aviv University. danar@eng.tau.ac.il
[3] NEC Research Institute, Princeton, NJ. ronitt@research.nj.nec.com

Abstract. Submodular and convex functions play an important role in many applications, and in particular in combinatorial optimization. Here we study two special cases: convexity in one dimension and submodularity in two dimensions. The latter type of functions are equivalent to the well known *Monge matrices*. A matrix $V = \{v_{i,j}\}_{i,j=0}^{i=n_1, j=n_2}$ is called a Monge matrix if for every $0 \leq i < r \leq n_1$ and $0 \leq j < s \leq n_2$, we have $v_{i,j} + v_{r,s} \leq v_{i,s} + v_{r,j}$. If inequality holds in the opposite direction then V is an *inverse Monge* matrix (supermodular function). Many problems, such as the traveling salesperson problem and various transportation problems, can be solved more efficiently if the input is a Monge matrix.

In this work we present a testing algorithm for Monge and inverse Monge matrices, whose running time is $O\left((\log n_1 \cdot \log n_2)/\epsilon\right)$, where ϵ is the distance parameter for testing. In addition we have an algorithm that tests whether a function $f : [n] \rightarrow \mathbb{R}$ is convex (concave) with running time of $O\left((\log n)/\epsilon\right)$.

1 Introduction

Convex functions and their combinatorial analogs, submodular functions, play an important role in many disciplines and applications, including combinatorial optimization, game theory, probability theory, and electronic trade. Such functions exhibit a rich mathematical structure (see Lovász [14]), which often makes it possible to efficiently find their minimum [10,12,18], and thus leads to efficient algorithms for many important optimization problems.

Submodular functions are defined as follows: Let $\mathcal{I} = I_1 \times I_2 \times \ldots \times I_d$, $d \geq 2$, be a product space where $I_q \subseteq \mathbb{R}$. In particular, we are interested in discrete domains $I_q = \{0, \ldots, n_q\}$. The *join* and *meet* operations are defined for every $x, y \in \mathcal{I}$:

$$(x_1, \ldots, x_d) \vee (y_1, \ldots, y_d) \stackrel{\text{def}}{=} (\max\{x_1, y_1\}, \ldots, \max\{x_d, y_d\})$$

and

$$(x_1, \ldots, x_d) \wedge (y_1, \ldots, y_d) \stackrel{\text{def}}{=} (\min\{x_1, y_1\}, \ldots, \min\{x_d, y_d\}),$$

respectively.

Definition 1 (Submodularity and Supermodularity) *A function $f : \mathcal{I} \rightarrow \mathbb{R}$ is* submodular *if for every $x, y \in \mathcal{I}$, $f(x \vee y) + f(x \wedge y) \leq f(x) + f(y)$. The function f is* supermodular *if for every $x, y \in \mathcal{I}$, $f(x \vee y) + f(x \wedge y) \geq f(x) + f(y)$.*

---

* Supported by the Israel Science Foundation (grant number 32/00-1).

J.D.P. Rolim and S. Vadhan (Eds.): RANDOM 2002, LNCS 2483, pp. 11–25, 2002.
© Springer-Verlag Berlin Heidelberg 2002

Certain subclasses of submodular functions are of particular interest. One such subclass is that of *submodular set* functions, which are defined over binary domains. That is, $I_q = \{0, 1\}$ for every $1 \le q \le d$, and so each $x \in \mathcal{I}$ corresponds to a subset of $\{1, \dots, d\}$. Another important subclass is the class of *Monge* functions, which are obtained when the domain is large but the dimension is $d = 2$. Since such functions are 2-dimensional, it is convenient to represent them as 2-dimensional matrices, which are referred to as *Monge matrices*. When the function is a 2-dimensional supermodular function the corresponding matrix is called an *inverse Monge matrix*.

The first problem that was shown to be solvable more efficiently if the underlying cost matrix is a Monge matrix is the classical Hitchcock transportation problem (see Hoffman [11]). Since then it has been shown that many other combinatorial optimization problems can be solved more efficiently in this case (e.g. weighted bipartite matching, and NP-hard problems such as the traveling salesperson problem). See [2] for a comprehensive survey on Monge matrices and their applications.

Testing Submodularity and Convexity. In this paper we approach the question of submodularity and convexity from within the framework of property testing [17,9]. Let f be a fixed but unknown function, and let $\mathcal{P}$ be a fixed property of functions (such as the convexity or submodularity of a function). A testing algorithm for the property $\mathcal{P}$ should determine, by querying f, whether f has the property $\mathcal{P}$, or whether it is ϵ-*far* from having the property for a given distance parameter ϵ. By ϵ-*far* we mean that more than an ϵ–fraction of the values of f should be modified so that f obtains the desired property $\mathcal{P}$.

Our Results. We present efficient testing algorithms for Monge matrices and for discrete convexity in one dimension. Specifically:

- We describe and analyze a testing algorithm for Monge and inverse Monge matrices whose running time is $O\left((\log n_1 \cdot \log n_2)/\epsilon\right)$, when given an $n_1 \times n_2$ matrix.

  Furthermore, the testing algorithm for inverse Monge matrices can be used to derive a testing algorithm, with the same complexity, for an important subfamily of Monge matrices, named *distribution matrices*. A matrix $V = \{v_{i,j}\}$ is said to be a distribution matrix, if there exists a non-negative *density matrix* $D = \{d_{i,j}\}$, such that every entry $v_{i,j}$ in V is of the form $v_{i,j} = \sum_{k \le i} \sum_{\ell \le j} d_{k,\ell}$. In other words, the entry $v_{i,j}$ corresponds to the cumulative density of all entries $d_{k,\ell}$ such that $k \le i$ and $\ell \le j$.

- We provide an algorithm that tests whether a function $f : [n] \to \mathbb{R}$ is convex (concave). The running time of this algorithm is $O\left(\log n/\epsilon\right)$.

Techniques. As stated above, it is convenient to represent 2-dimensional submodular functions as 2-dimensional Monge matrices. Thus a function $f : \{0, \dots, n_1\} \times \{0, \dots, n_2\} \to \mathbb{R}$ can be represented as the matrix $V = \{v_{i,j}\}_{i,j=0}^{i=n_1, j=n_2}$ where $v_{i,j} = f(i, j)$. Observe that for every pair of indices

$(i, s), (r, j)$ such that $i < r$ and $j < s$ we have that $(i, s) \vee (r, j) = (r, s)$ and $(i, s) \wedge (r, j) = (i, j)$. It follows from Definition 1 that V is a Monge matrix (f is a 2-dimensional submodular function) if and only if:

$$\forall i, j, r, s \text{ s.t. } i < r, \ j < s : \quad v_{i,j} + v_{r,s} \leq v_{i,s} + v_{r,j}$$

and V is an inverse Monge matrix (f is a 2-dimensional supermodular function) if and only if: $\forall i, j, r, s$ s.t. $i < r, \ j < s : \quad v_{i,j} + v_{r,s} \geq v_{i,s} + v_{r,j}$.

That is, in both cases we have a constraint for every quadruple $v_{i,j}$, $v_{r,s}$, $v_{i,s}$, $v_{r,j}$ such that $i < r$ and $j < s$.[1] Our algorithm selects such quadruples according to a particular (non-uniform) distribution and verifies that the constraint is satisfied for every quadruple selected. Clearly the algorithm always accepts Monge matrices. The main thrust of the analysis is in showing that if the matrix V is far from being Monge then the probability of obtaining a "bad" quadruple is sufficiently large.

A central building block in proving the above, is the following combinatorial problem, which may be of independent interest. Let C be a given matrix, possibly containing negative values, and let R be a subset of positions in C. We are interested in refilling the entries of C that reside in R with *non-negative* values, such that the following constraint is satisfied: for every position (i, j) that does not belong to R, the sum of the modified values in C that are below[2] (i, j), is the same as in the original matrix C. That is, the sum of the modified values in entries (k, ℓ), such that $k \leq i$ and $j \leq \ell$, remains as it was.

We provide sufficient conditions on C and R under which the above is possible, and describe the corresponding procedure that refills the entries of C that reside in R. Our starting point is a simple special case in which R corresponds to a sub-matrix of C. In such a case it suffices that for each row and each column in R, the sum of the corresponding entries in the original matrix C is non-negative. Under these conditions a simple greedy algorithm can modify C as required. Our procedure for general subsets R is more involved but uses the sub-matrix case as a subroutine.

Previous Work. Property testing was first defined in the context of algebraic properties of functions [17]. It was extended in [9] and in particular applied to properties of graphs. In recent years it has been studied in a large variety of contexts. For surveys on property testing see [16,6].

One testing problem that is probably most related to ours is testing whether a function is monotone [8,4,5,7,1]. A function $f : [n] \rightarrow \mathbb{R}$ is monotone non-decreasing if for every $0 \leq i < n$, the differences $f(i+1) - f(i)$ are non-negative, whereas f is convex if and only if for every $1 \leq i \leq n - 1$, the differences of differences $[f(i + 1) - f(i)] - [f(i) - f(i - 1)]$ are non-negative. In other words, f is convex if and only if the function $f'(i) = f(i) - f(i - 1)$, is monotone non-decreasing. We note though that testing f for convexity cannot be done by

---

[1] It is easy to verify that for all other i, j, r, s (with the exception of the symmetric case where $r < i$ and $s < j$), the constraint holds trivially (with equality).

[2] We denote the lower left position of the matrix C by $(0, 0)$.

simply testing f' for monotonicity. Specifically, if f' is close to monotone then it does not necessarily follow that f is close to convex.

Further Research. We suggest the following open problems. First it remains open to determine the complexity of testing submodular functions when the dimension d of the input domain is greater than 2. It seems that our algorithm and its analysis can be extended to work for testing the special case of distribution matrices of dimension $d > 2$, where the complexity of the resulting algorithm is $O\left((\prod_{q=1}^{d} \log n_q)/\epsilon\right)$. However, as opposed to the $d = 2$ case, where Monge matrices are only slightly more general than distribution matrices, for $d > 2$ Monge matrices seem to be much more expressive. Hence it is not immediately clear how to adapt our algorithm to testing Monge matrices of higher dimensions.

It would also be interesting to find an efficient testing algorithm for the sub-class of submodular set functions, which are used for example in combinatorial auctions on the internet (e.g. [3],[13]).

Finally, in many optimization problems it is enough that the underlying cost matrix is a permutation of a Monge matrix. In such cases it may be useful to test whether a given matrix is a permutation of some Monge matrix or far from any permuted Monge matrix.

Organization. In Section 2 we describe several building blocks that will be used by our testing algorithm for Monge matrices. In Section 3 we describe a testing algorithm for Monge matrices whose complexity is $O(n/\epsilon)$, where we assume for simplicity that the matrix is $n \times n$. Building on this algorithm and its analysis, in Section 4 we present a significantly faster algorithm whose complexity is $O\left((\log^2 n)/\epsilon\right)$. We conclude this section with a short discussion concerning distribution matrices. The testing algorithm for convexity can be found in the full version of this paper [15]. All missing details of the analysis together with helpful illustrations appear in [15] as well.

2   Building Blocks for Our Algorithms for Testing Inverse Monge

From this point on we focus on inverse Monge matrices. Analogous claims hold for Monge matrices. We also assume for simplicity that the dimensions of the matrices are $n_1 = n_2 = n$. In what follows we provide a characterization of inverse Monge matrices that is exploited by our algorithm. Given any real valued matrix $V = \{v_{i,j}\}_{i,j=0}^{i,j=n}$ we define an $(n+1) \times (n+1)$ matrix $C_V' = \{c_{i,j}\}_{i,j=0}^{i,j=n}$ as follows:

- $c_{0,0} = v_{0,0}$; $\forall i > 0 : c_{i,0} = v_{i,0} - v_{i-1,0}$; $\forall j > 0 : c_{0,j} = v_{0,j} - v_{0,j-1}$;
- $\forall i,j > 0 : c_{i,j} = (v_{i,j} - v_{i-1,j}) - (v_{i,j-1} - v_{i-1,j-1}) = (v_{i,j} - v_{i,j-1}) - (v_{i-1,j} - v_{i-1,j-1})$.

Let $C_V = \{c_{i,j}\}_{i,j=1}^{i,j=n}$ be the sub-matrix of C_V' that includes all but the first (0'th) row and column of C_V'. The following two claims are well known and easy to verify.

Claim 1 *For every $0 \leq i, j \leq n$, $v_{i,j} = \sum_{i'=0}^{i} \sum_{j'=0}^{j} c_{i',j'}$.*

Claim 2 *A matrix V is an inverse Monge matrix if and only if C_V is a non-negative matrix.*

It follows from Claim 2 that if we find some entry of C_V that is negative, then we have evidence that V is not an inverse Monge matrix. However, it is not necessarily true that if V is far from being an inverse Monge matrix, then C_V contains many negative entries. For example, suppose C_V is 1 in all entries except the entry $c_{n/2,n/2}$ which is $-n^2$. Then it can be verified that V is very far from being an inverse Monge matrix (this can be proved by showing that there are $\Theta(n^2)$ disjoint quadruples $v_{i,j}, v_{r,s}, v_{i,s}, v_{r,j}$ in V such that from any such quadruple at least one value should be changed in order to transform V into a Monge matrix). However, as our analysis will show, in such a case there are many sub-matrices in C_V whose sum of elements is negative. Thus our testing algorithms will sample certain sub-matrices of C_V and check that the sum of elements in each sub-matrix sampled is non-negative. We first observe that it is possible to check this efficiently.

Claim 3 *Given access to V it is possible to check in time $O(1)$ if the sum of elements in a given sub-matrix A of C_V is non-negative. In particular, if the lower-left entry of A is (i, j) and its upper-right entry is (r, s) then the sum of elements of A is $v_{r,s} - v_{r,j} - v_{i,s} + v_{i,j}$.*

2.1 Filling Sub-matrices

An important building block for the analysis of our algorithms is a procedure for "filling in" a sub-matrix. That is, given constraints on the sum of elements in each row and column of a given sub-matrix, we are interested in assigning values to the entries of the sub-matrix so that these constraints are met.

Specifically, let $a_1, ..., a_s$ and $b_1, ..., b_t$ be non-negative real numbers such that $\sum_{i=1}^{s} a_i \geq \sum_{j=1}^{t} b_j$. Then it is possible to construct an $s \times t$ non-negative real matrix T, such that the sum of elements in column j is exactly b_j and the sum of elements in row i is at most a_i. In the special case that $\sum_{i=1}^{s} a_i = \sum_{j=1}^{t} b_j$, the sum of elements in row i will equal a_i. In particular, this can be done by applying the following procedure, which is the same as the one applied to obtain an initial feasible solution for the linear-programming formulation of the transportation problem.

Procedure 1 [Fill Matrix $T = (t_{i,j})_{i,j=1}^{i=s,j=t}$.]
Initialize $\bar{a}_i = a_i$ for $i = 1, ..., s$ and $\bar{b}_j = b_j$ for $j = 1, ..., t$.
(In each of the following iterations, $\bar{a}_i$ is an upper bound on what remains to be filled in row i, and $\bar{b}_j$ is what remains to be filled in column j.)
for $j = 1,...,t$:
  for $i = 1,...,s$:
    Assign to entry (i, j) the value $x = \min\{\bar{a}_i, \bar{b}_j\}$
    Update $\bar{a}_i = \bar{a}_i - x$, $\bar{b}_j = \bar{b}_j - x$.

3   A Testing Algorithm for Inverse Monge Matrices

We first present a simple algorithm for testing if a matrix V is an inverse Monge Matrix, whose running time is $O(n/\epsilon)$. In the next section we show a significantly faster algorithm that is based on the ideas presented here. We may assume without loss of generality that n is a power of 2. This is true since our algorithms probe the coefficients matrix C_V, and we may simply "pad" it by 0's to obtain rows and columns that have lengths which are powers of 2 and run the algorithm with $\epsilon \leftarrow \epsilon/4$. We shall need the following two definitions for both algorithms.

Definition 2 (Sub-Rows, Sub-Columns and Sub-Matrices.) *A sub-row in an $n \times n$ matrix is a consecutive sequence of entries that belong to the same row. The sub-row $((i,j),(i,j+1),\ldots,(i,j+t-1))$ is denoted by $[\ ]_{i,j}^{1,t}$. A sub-column is defined analogously, and is denoted by $[\ ]_{i,j}^{s,1} = ((i,j),(i+1,j),\ldots,(i+s-1,j))$. More generally, an $s \times t$ sub-matrix whose bottom-left entry is (i,j) is denoted $[\ ]_{i,j}^{s,t}$.*

Definition 3 (Legal Sub-Matrices.) *A sub-row in an $n \times n$ matrix is a legal sub-row if it can result from bisecting the row of length n that contains it in a recursive manner. That is, a complete (length n) row is legal, and if $[\ ]_{i,j}^{1,t}$ is legal, then so are $[\ ]_{i,j}^{1,t/2}$ and $[\ ]_{i,j+t/2}^{1,t/2}$. A legal sub-column is defined analogously. A sub-matrix is legal if both its rows and its columns are legal.*

Note that the legality of a sub-row $[\ ]_{i,j}^{1,t}$ is independent of the actual row i it belongs to, but rather it depends on its starting position j and ending position $j+t-1$ within its row. An analogous statement holds for legal sub-columns.

Although a sub-matrix is just a collection of positions (entries) in an $n \times n$ matrix, we talk throughout the paper about sums of elements in certain sub-matrices A of C_V. In this we mean the sum of elements of C_V determined by the set of positions in A.

Definition 4 (Good sub-matrix.) *We say that a sub-matrix A of C_V is good if the sum of elements in each row and each column of A is non-negative.*

Definition 5 (Good Point.) *We say that point (i,j) is good if all legal square sub-matrices A of C_V which contain (i,j) are good.*

Algorithm 1 [Test Monge I.]

1. *Choose $8/\epsilon$ points in the matrix C_V and check that they are good.*

2. *If all points are good then accept, otherwise reject.*

By Claim 3, it is possible to check in constant time that the sum of elements in a sub-row (sub-column) of C_V is non-negative. Therefore, it is possible to test that an $s \times s$ square sub-matrix A of C_V is good in time $\Theta(s)$. Notice that every point in an $n \times n$ matrix is contained in $\log n$ square sub-matrices. Hence the time required to check whether a point is good is $O(n) + O(n/2) + \ldots + O(n/2^i) + \ldots + O(1) = O(n)$, and the complexity of the algorithm is $O(n/\epsilon)$.

Theorem 1 *If V is an inverse Monge matrix then it is always accepted, and if V is ϵ-far from being an inverse Monge matrix, then the algorithm rejects with probability at least $2/3$.*

Proof. The first part of the theorem follows directly from Claim 2. In order to prove the second part of the theorem, we show that if V is ϵ-far from being inverse Monge, then C_V contains more than $(\epsilon/4)n^2$ bad points. The second part of the theorem directly follows because the probability in such a case that no bad point is selected by the algorithm, is at most $(1 - \epsilon/4)^{(8/\epsilon)} < e^{-2} < 1/3$.

Assume contrary to the claim that C_V contains at most $(\epsilon/4)n^2$ bad points. We shall show that by modifying at most ϵn^2 entries in V we obtain an inverse Monge matrix (in contradiction to our assumption concerning V). Let us look at the set of bad points in C_V, and for each such bad point look at the largest bad square sub-matrix in C_V which contains this bad point. By our assumption on the number of bad points, it must be the case that the area of all these maximal bad sub-matrices is at most $(\epsilon/4)n^2$, because all the points in a bad sub-matrix are bad.

For each maximal bad (legal square) sub-matrix B of C_V we will look at the smallest good (legal square) sub-matrix A which contains B. First observe that such a good sub-matrix must exist. Indeed, since B is maximal, if it is of size $s \times s$ where $s < n$, then the legal square sub-matrix of size $2s \times 2s$ that contains it must be good. But if $s = n$, then $B = C_V$ implying that all n^2 points in C_V are bad, contradicting our assumption on the number of bad points. Next observe that for every two maximal sub-matrices B and B', the corresponding good sub-matrices A and A' that contain them are either the same sub-matrix, or are totally disjoint. Finally, the sum of areas of all these good sub-matrices is at most $4 \cdot (\epsilon/4)n^2 = \epsilon n^2$.

We now correct each such good sub-matrix A so that it contains only non-negative elements, and the sum of elements in each row and column of A remains as it was. This can be done by applying Procedure 1 to A as described in Section 2.1.

Note that after correcting all these good sub-matrices of C_V, the new matrix C_V is non-negative, and thus the corresponding new matrix V must be an inverse Monge matrix. We must show however, that at most ϵn^2 values were changed in V following the changes to C_V. Notice that we made sure that the sum of elements in each row and column of each corrected sub-matrix A remains as it was. Therefore the values of all points $v_{k,\ell}$ in V that are outside A are not affected by the change to A, since by Claim 1 we have that $v_{k,\ell} = \sum_{i=0}^{k} \sum_{j=0}^{\ell} c_{i,j}$.

4   A Faster Algorithm for Inverse Monge Matrices

Though the above algorithm has running time sub-linear in the size of the matrix, which is n^2, we would further like to improve its dependence on n. We next suggest a variant of the algorithm whose running time is $O(\log^2 n/\epsilon)$ and explain what needs to be proved in order to argue its correctness. We first redefine the concepts of good sub-matrices and good points.

Definition 6 (Good sub-matrix.) *A (legal) sub-matrix T of C_V is* **good** *if the sum of all its elements is non-negative. Otherwise, T is* **bad**.

Definition 7 (Good Point.) *We say that a point is* **good** *if every legal sub-matrix of C_V that contains it is good. Otherwise the point is* **bad**.

For the sake of the presentation, we shall assume that every row and every column in C_V (that is, every sub-row and sub-column of length n) have a non-negative sum. In the full version of this paper [15] we explain how this assumption can be easily removed. Note that this assumption implies that every $s \times n$ sub-matrix is good, and similarly for every $n \times s$ sub-matrix (but of course it has no implications on smaller sub-matrices).

Algorithm 2 [Test Monge II.]

1. *Uniformly select $8/\epsilon$ points in the matrix C_V and check that they are good.*
2. *If all points are good then accept, otherwise reject.*

Note that by Definition 3, each point in an $n \times n$ matrix is contained in $O(\log^2 n)$ legal sub-matrices. Thus by Claim 3, checking that a point is good takes time $O(\log^2 n)$. Therefore the running time of the algorithm is $O((\log^2 n)/\epsilon)$.

Theorem 2 *If V is an inverse Monge matrix then it is always accepted, and if V is ϵ-far from being an inverse Monge matrix, then the algorithm rejects with probability at least $2/3$.*

4.1   Outline of the Proof of Theorem 2

If V is an inverse Monge matrix then by Claim 2 all elements in C_V are non-negative, and so the algorithm always accepts. Suppose V is ϵ-far from being inverse Monge. We claim that in such a case C_V must contain more than $(\epsilon/4)n^2$ bad points, causing the algorithm to reject with probability at least $1 - (1 - \epsilon/4)^{(8/\epsilon)} > 1 - e^{-2} > 2/3$. Assume contrary to the claim that C_V contains at most $(\epsilon/4)n^2$ bad points. Our goal from this point on is to show that in such a case V is ϵ-close to being an inverse Monge matrix.

Consider the union of all bad legal sub-matrices of C_V. Since within each bad legal sub-matrix, all points are bad, then the area occupied by this union is at most $(\epsilon/4)n^2$.

Definition 8 (Maximal bad legal sub-matrix.) *A bad legal sub-matrix T of C_V is a* **maximal bad legal sub-matrix** *of C_V if it is not contained in any larger bad legal sub-matrix of C_V.*

Now consider all such maximal bad legal sub-matrices of C_V. For each such sub-matrix B let us take the legal sub-matrix that contains it and has twice the number of rows and twice the number of columns. Then by the maximality of B

(and our assumption that all full rows and columns have a non-negative sum), the resulting sub-matrix is good. We now take the union of all these good legal sub-matrices, and get a total area of size at most ϵn^2. Denote the union of all these sub-matrices by R. See for example Figure 1.

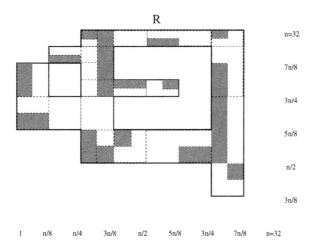

Fig. 1. An example of the structure of a subset R (outlined by a bold line). The bad legal sub-matrices determining R are the gray sub-matrices. Each is contained inside a good legal sub-matrix that has twice the number of rows and twice the number of columns (marked by dashed rectangles). Observe that maximal bad-legal sub-matrices may overlap.

Definition 9 (Maximal (legal) sub-row/column.) *Given a subset R of entries in an $n \times n$ matrix, a sub-row T is a* maximal (legal) sub-row with respect to R *if T is contained in R and there is no larger (legal) sub-row T' such that $T \subset T' \subseteq R$. A* maximal (legal) sub-column with respect to R *is defined analogously.*

For sake of succinctness, whenever it is clear what R is, we shall just say maximal (legal) sub-row and drop the suffix, "with respect to R". Note that a maximal sub-row is simply a maximal consecutive sequence of entries in R that belong to the same row, while a maximal legal sub-row is a more constrained notion. In particular, a maximal sub-row may be a concatenation of several maximal legal sub-rows.

We would like to show that it is possible to change the at most ϵn^2 entries of C_V within R to non-negative values so that the following property holds:

Property 1 (Sum Property for R.) *For every point (i, j) outside of R, the sum of the elements in the modified entries (i', j') within R such that $i' \leq i$ and $j' \leq j$ is as it was in the original matrix C_V.*

Let $\tilde{C}_V$ be the matrix obtained from C_V by modifying R so that Property 1 holds, and let $\tilde{V}$ be the matrix which corresponds to $\tilde{C}_V$. Then it follows from Claim 1 that $\tilde{V}$ is at most ϵ-far from the original matrix V.

4.2   Fixing R

Let R be the subset of entries in the matrix C_V that consists of a union of good legal sub-matrices. In the following discussion, when we talk about elements in sub-matrices of R we mean the elements in C_V determined by the corresponding set of positions in R.

Lemma 4 *The sum of elements in every maximal legal sub-row and every maximal legal sub-column in R is non-negative.*

Proof. Assume, contrary to the claim, that R contains some maximal legal sub-row $L = [\;]_{i,j}^{1,t}$ whose sum of elements is negative. Let T be the maximal bad legal sub-matrix in C_V that contains L. By the maximality of L, necessarily $T = [\;]_{i',j}^{s,t}$ for some $i' \leq i$ and $s \geq 1$. That is, the rows of T (one of which is L) are of length t. By the construction of R, R must contain a good legal sub-matrix T' that contains T and is twice as large in each dimension. But this contradicts the maximality of L.

Maximal Blocks. We will partition R into disjoint blocks (sub-matrices) and fill each block separately with non-negative values, so that the sum property for R is maintained (see Property 1). We define blocks as follows.

Definition 10 (Maximal Block.) *A maximal block $B = [\;]_{i,j}^{s,t}$ in R is a sub-matrix contained in R which has the following property: It consists of a maximal consecutive sequence of maximal legal sub-columns of the same height. The maximality of each sub-column is as in Definition 9. That is, for every $j \leq r \leq j+t-1$, the column $[\;]_{i,r}^{s,1}$ is a maximal legal sub-column (with respect to R).*

The maximality of the sequence of sub-columns in a block means that we cannot extend the sequence of columns neither to the left nor to the right. That is, neither $[\;]_{i,j-1}^{s,1}$ nor $[\;]_{i,j+t}^{s,1}$ are maximal legal sub-columns in R. (Specifically, each is either not fully contained in R or R contains a larger legal sub-column that contains it.)

We shall sometimes refer to maximal blocks simply as blocks. Observe that by this definition, R is indeed partitioned in a unique way into maximal disjoint blocks. See Figure 2 for an example of R and its partition into maximal blocks.

Definition 11 (Size of a Maximal Block.) *Let B be a maximal block. The size of B is the height of the columns in B (equivalently, the number of rows in B).*

R

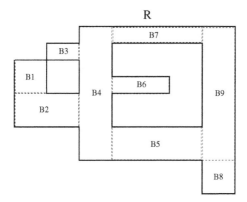

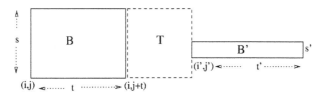

Fig. 2. An example of the partition of R into maximal blocks (numbered B_1–B_9). Note that the ratio between the sizes of any two blocks is always a power of 2. Furthermore, the blocks are "oriented" in the following sense. Suppose a block B has size s and a block B' has size $s' \leq s$ and some of their sub-rows belong to the same row of the matrix (e.g., B_4 and B_2, or B_9 and B_6). Then the smaller block B' must be aligned either with the first or second half of B, or with one of the quarters of B, or with one of its eighth's, etc.

Fig. 3. An illustrations for Lemma 5.

Bounded Sub-matrices. As we have shown in Lemma 4, the sum of elements in every maximal legal sub-column in R is non-negative. It directly follows that every maximal block has a non-negative sum. We would like to characterize other sub-matrices of R whose sum is necessarily non-negative.

Definition 12 *For a given sub-matrix T, we denote the sum of the elements in T by $sum(T)$.*

Lemma 5 *Consider any two maximal blocks $B = [\ ]_{i,j}^{s,t}$ and $B' = [\ ]_{i,j'}^{s,t'}$ where $j' > j + t$. That is, both blocks have the same size s and both start at row i and end at row $i + s - 1$. Consider the sub-matrix $T = [\ ]_{i,j+t}^{s,j'-(j+t)}$ "between them". Suppose that $T \subset R$. Then $sum(T) \geq 0$.*

See Figure 3 for an illustration of Lemma 5.

Definition 13 (Covers.) *We say that a collection A of sub-rows* covers *a given block B with respect to R, if $B \subset A \subset R$ and the number of rows in A equals the* *size of* B. *We say that* A *is a* maximal row-cover *with respect to* R *if* A *consists of maximal sub-rows with respect to* R.

Definition 14 (Borders.) *We say that a sub-matrix* $A = [\ ]_{i,j}^{s,t}$ *in* R, borders *a maximal block* $B = [\ ]_{i',j'}^{s',t'}$ *if* $i \leq i' \leq i+s-1$, $i'+s' \leq i+s$, *and either* $j' = j+t$ *(so that* A *borders* B *from the left), or* $j'+t' = j$ *(so that* A *borders* B *from then right).*

By Lemma 5 and using the above terminology, we get the following corollary whose proof is illustrated in Figure 4.

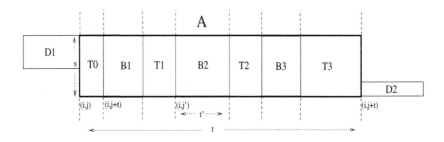

Fig. 4. An illustration for Corollary 6. Here A covers the blocks B_1, B_2 and B_3, and borders the blocks D_1 and D_2. The sub-matrices T_0–T_4 are parts of larger blocks (that extend above and/or below A).

Corollary 6 *Let* A *be a sub-matrix of* R *which covers a given block* B. *If on each of its sides* A *either borders a block smaller than* B *or its border coincides with the border of* R, *then* $sum(A) \geq sum(B)$.

The Procedure for Refilling R. We now describe the procedure that refills the entries of R with non-negative values. Recall that R is a disjoint union of maximal blocks. Hence if we remove a maximal block from R, then the maximal blocks of the remaining structure, are simply the remaining maximal blocks of R. The procedure described below will remove the blocks of R one by one, in order of increasing size, and refill each block separately using Procedure 1. After removing each block, the sum of the elements in each remaining column in R remains the same, however the row sums must be updated. Procedure 1 is used here as well.

Procedure 2 [Refill R.]

1. We assign with each maximal sub-row L in R a designated sum of elements for that row, which is denoted by $\overline{sum}(L)$, and initially set to be $\overline{sum}(L) = sum(L)$.

2. Let m be the number of maximal blocks in R, and let $R_1 = R$.
3. for $p = 1, ..., m$:

   a) Let B_p be a maximal block in R_p whose size is minimum among all maximal blocks of R_p, and assume that B_p is an $s \times t$ sub-matrix. Let A_p be a maximal row-cover of B_p with respect to R_p. For $1 \leq \ell \leq s$, let L_ℓ denote the sub-row of A_p that covers the ℓ'th sub-row of B_p.

   b) Refill B_p by applying Procedure 1 (see Section 2.1), where the sum filled in the k'th sub-column of B_p, $1 \leq k \leq t$, should be the original sum of this sub-column in C_V, and the sum filled in the ℓ'th sub-row of B_p, $1 \leq \ell \leq s$, is at most $\overline{sum}(L_\ell)$.

   For each $1 \leq \ell \leq s$, let x_ℓ denote the sum of elements filled by Procedure 1 in the ℓ'th sub-row of B_p.

   c) Let $R_{p+1} = R_p \setminus B_p$ and assign designated sums to the rows of R_{p+1} that have been either shortened or broken into two parts by the removal of B_p from R_p. This is done as follows:

   The set $A_p \setminus B_p$ is the union of two non-consecutive sub-matrices, A' and A'', so that A' borders B_p from the left and A'' borders B_p from the right. Let L'_ℓ and L''_ℓ be the sub-rows in A' and A'' respectively that are contained in sub-row L_ℓ of A_p. We assign to L'_ℓ and L''_ℓ non-negative designated sums, $\overline{sum}(L'_\ell)$ and $\overline{sum}(L''_\ell)$, that satisfy the following:

   $$\overline{sum}(L'_\ell) + \overline{sum}(L''_\ell) = \overline{sum}(L_\ell) - x_\ell,$$

   and furthermore,

   $$\sum_{\text{row } L \in A'} \overline{sum}(L) = sum(A'), \qquad \sum_{\text{row } L \in A''} \overline{sum}(L) = sum(A'').$$

   This is done by applying Procedure 1 to a $2 \times s$ matrix whose sums of columns are $sum(A')$ and $sum(A'')$ and sums of rows are $\overline{sum}(L_\ell) - x_\ell$, where $1 \leq \ell \leq s$.

   (Note that one or both of A' and A'' may not exist. This can happen if B_p bordered $A_p \setminus B_p$ on one side and its boundary coincided with R_p, or if $A_p = B_p$. In this case, if, for example, A' does not exist then we view it as a sub-matrix of size 0 where $sum(A') = 0$.)

4.3   Proving That Procedure 2 Works

Recall that for each $1 \leq p \leq m$, R_p is what remains of R at the start of the p'th iteration of Procedure 2. In particular, $R_1 = R$. We would first like to show that the procedure does not "get stuck". That is, for each iteration p, Procedure 1 can be applied to the block B_p selected in this iteration, and the updating of the designated sum for the rows that have been shortened by the removal of B_p can be performed. Note that since the blocks are selected according to increasing size, then in each iteration the maximal row cover A_p of B_p must actually be a sub-matrix.

Proving that Procedure 2 Does not Get Stuck. For every $1 \leq p \leq m$, let s_p be the minimum size of the maximal blocks of R_p, where $s_0 = 1$. Observe that whenever s_p increases, it does so by a factor of 2^k for some k. This is true because the columns of maximal blocks are legal sub-columns.

Lemma 7 *For every $1 \leq p \leq m$, Procedure 1 can be applied to the block B_p selected in R_p, and the updating process of the designated sum of rows can be applied. Moreover, if A is a sub-matrix of R_p with height of at least s_{p-1}, whose columns are legal sub-columns and whose rows are maximal rows with respect to R_p, then $\sum_{row\ L \in A} \overline{sum}(L) = sum(A)$.*

Proving that Procedure 2 is Correct. Let $\tilde{C}_V = \{\tilde{c}_{i,j}\}$ be the matrix resulting from the application of Procedure 2 to the matrix $C_V = \{c_{i,j}\}$. For any sub-matrix T of C_V (and in particular of R), we let $\widetilde{sum}(T)$ denote the sum of elements of T in $\tilde{C}_V$. By definition of the procedure, $\widetilde{sum}(K) = sum(K)$ for every maximal legal sub-column K of R. Hence this holds also for every maximal sub-column of R. We next state a related claim concerning rows.

Lemma 8 *For every sub-row L in R, such that L is assigned $\overline{sum}(L)$ as a designated sum at some iteration of Procedure 2, we have that $\widetilde{sum}(L) = \overline{sum}(L)$.*

Observe that in particular we get that for every maximal sub-row L of R, $\widetilde{sum}(L) = \overline{sum}(L) = sum(L)$.

Definition 15 (Boundary.) *We say that a point (i,j) is on the boundary of R if $(i,j) \in R$, but either $(i+1,j) \notin R$, or $(i,j+1) \notin R$, or $(i+1,j+1) \notin R$. We denote the set of boundary points by $\mathcal{B}$.*

Definition 16 *For a point (i,j), $1 \leq i,j \leq n$ let $R^{\leq}(i,j)$ denote the subset of points $(i',j') \in R$, $i' \leq i, j' \leq j$, and let $sum^R(i,j) = \sum_{(i',j') \in R^{\leq}(i,j)} c_{i',j'}$ and $\widetilde{sum}^R(i,j) = \sum_{(i',j') \in R^{\leq}(i,j)} \tilde{c}_{i',j'}$.*

Property 1 and therefore Theorem 2 will follow directly from the next two lemmas.

Lemma 9 *For every point $(i,j) \in \mathcal{B}$, $\widetilde{sum}^R(i,j) = sum^R(i,j)$.*

Lemma 10 *Let (i,j) be any point such that $(i,j) \notin R$. Then $\widetilde{sum}^R(i,j) = sum^R(i,j)$.*

4.4   Distribution Matrices

As noted in the introduction, a sub-family of inverse Monge matrices that is of particular interest is the class of *distribution matrices*. A matrix $V = \{v_{i,j}\}$ is said to be a distribution matrix, if there exists a non-negative *density matrix* $D = \{d_{i,j}\}$, such that every entry $v_{i,j}$ in V is of the form $v_{i,j} = \sum_{k \leq i} \sum_{\ell \leq j} d_{k,\ell}$. In particular, if V is a distribution matrix then the corresponding density matrix D is simply the matrix C'_V (as defined in Section 2). Hence, in order to test that V is a distribution matrix, we simply run our algorithm for inverse Monge matrix on C'_V instead of C_V.

Acknowledgments. We would like to thank Noam Nisan for suggesting to examine combinatorial auctions in the context of property testing.

References

1. T. Batu, R. Rubinfeld, and P. White. Fast approximate pcps for multidimensional bin-packing problems. In *Proceedings of RANDOM*, pages 245–256, 1999.
2. R.E. Burkard, B. Klinz, and R. Rudolf. Perspectives of monge properties in optimization. *Discrete Applied Mathematics and Combinatorial Operations Research and Computer Science*, 70, 1996.
3. S. de Vries and R. Vohra. Combinatorial auctions: a survey. available from: http://www.kellogg.nwu.edu/faculty/vohra/htm/res.htm, 2000.
4. Y. Dodis, O. Goldreich, E. Lehman, S. Raskhodnikova, D. Ron, and A. Samorodnitsky. Improved testing algorithms for monotonocity. In *Proceedings of RANDOM*, pages 97–108, 1999.
5. F. Ergun, S. Kannan, S. R. Kumar, R. Rubinfeld, and M. Viswanathan. Spot-checkers. In *Proceedings of the Thirty-Second Annual ACM Symposium on the Theory of Computing*, pages 259–268, 1998.
6. E. Fischer. The art of uninformed decisions: A primer to property testing. *Bulletin of the European Association for Theoretical Computer Science*, 75:97–126, 2001.
7. E. Fischer, E. Lehman, I. Newman, S. Raskhodnikova, R. Rubinfeld, and A. Samrodnitsky. Monotonicity testing over general poset domains. In *Proceedings of the Thirty-Sixth Annual ACM Symposium on the Theory of Computing*, 2002.
8. O. Goldreich, S. Goldwasser, E. Lehman, D. Ron, and A. Samordinsky. Testing monotonicity. *Combinatorica*, 20(3):301–337, 2000.
9. O. Goldreich, S. Goldwasser, and D. Ron. Property testing and its connection to learning and approximation. *JACM*, 45(4):653–750, 1998.
10. M. Grotschel, L. Lovasz, and A. Schrijver. The ellipsoid method and its consequences in combinatorial optimization. *Combinatorica*, 1, 1981.
11. A.J. Hoffman. On simple linear programming problems. In *In Proceedings of Symposia in Pure Mathematics, Convexity*, volume 7, pages 317–327, 1963. American Mathematical Society.
12. S. Iwata, L. Fleischer, and S. Fujishige. A combinatorial strongly polynomial algorithm for minimizing submodular functions. In *Proceedings of the Thirty-Fourth Annual ACM Symposium on the Theory of Computing*, pages 96–107, 2000. To appear in JACM.
13. B. Lehmann, D. Lehmann, and N. Nisan. Combinatorial auctions with decreasing marginal utilities. In *ACM Conference on Electronic Commerce*, pages –, 2001.
14. L. Lovász. Submodular functions and convexity. *Mathematical Programming: The State of the Art*, pages 235–257, 1983.
15. M. Parnas, D. Ron, and R. Rubinfeld. On testing convexity and submodularity. Available from: http://www.eng.tau.ac.il/~danar/papers.html, 2002.
16. D. Ron. Property testing. In *Handbook on Randomization, Volume II*, pages 597–649, 2001.
17. R. Rubinfeld and M. Sudan. Robust characterization of polynomials with applications to program testing. *SIAM Journal on Computing*, 25(2):252–271, 1996.
18. A. Schrijver. A combinatorial algorithm minimizing submodular functions in strongly polynomial time. *Journal of Combinatorial Theory B*, 80:346–355, 2000.

ω-Regular Languages Are Testable with a Constant Number of Queries

Hana Chockler and Orna Kupferman

Hebrew University, School of Engineering and Computer Science, Jerusalem 91904, Israel
{hanac,orna}@cs.huji.ac.il, http://www.cs.huji.ac.il/{~hanac,~orna}

Abstract. We continue the study of combinatorial property testing. For a property ψ, an ε-test for ψ, for $0 < \varepsilon \le 1$, is a randomized algorithm that given an input x, returns "yes" if x satisfies ψ, and returns "no" with high probability if x is ε-far from satisfying ψ, where ε-far essentially means that an ε-fraction of x needs to be changed in order for it to satisfy ψ. In [AKNS99], Alon et al. show that regular languages are ε-testable with a constant (depends on ψ and ε and independent of x) number of queries. We extend the result in [AKNS99] to ω-regular languages: given a nondeterministic Büchi automaton $\mathcal{A}$ on infinite words and a small $\varepsilon > 0$, we describe an algorithm that gets as input an infinite lasso-shape word of the form $x \cdot y^{\omega}$, for finite words x and y, samples only a constant number of letters in x and y, returns "yes" if $w \in L(\mathcal{A})$, and returns "no" with probability 2/3 if w is ε-far from $L(\mathcal{A})$. We also discuss the applicability of property testing to formal verification, where ω-regular languages are used for the specification of the behavior of nonterminating reactive systems, and computations correspond to lasso-shape words.

1 Introduction

Property testing was first introduced in [RS96], where Rubinfeld and Sudan checked whether a given function computes a low-degree polynomial or is far from computing it. The work in [RS96] have led to the study of combinatorial property testing, defined by Goldreich et al. in [GGR98]. Generally speaking, given a property ψ, an input x, and $0 < \varepsilon \le 1$, we say that x is ε-far from satisfying ψ if we need to change an ε-fraction of x in order for it to satisfy ψ. Then, an ε-test for ψ is a randomized algorithm that given ψ, x, and ε, accepts x if x satisfies ψ, and rejects x with probability at least 2/3 if x is ε-far from satisfying ψ. An ε-test has no obligation for x that neither satisfies ψ nor is ε-far from ψ. We say that a property ψ is ε-*testable* if there exists an ε-test for ψ that uses only $f(\varepsilon)$ queries on the input, where f is independent of the size of the input.[1] It turned out that several properties are ε-testable. For example, it is possible to check bipartiteness by randomly testing $poly(1/\varepsilon)$ edges of the graph [GGR98], and similar results hold for k-connectivity, acyclicity, k-colorability, and more [GR97,AK,GR99, BR00,PRR01,PRS01].

---

[1] Alternative definitions of ε-test allow two-sided error, a number of queries that depend on the input (usually in some sub-linear way), and other bounds on the error. The definition above is common in many cases, and is the one we use in this paper.

J.D.P. Rolim and S. Vadhan (Eds.): RANDOM 2002, LNCS 2483, pp. 26–38, 2002.
© Springer-Verlag Berlin Heidelberg 2002

Most of today's knowledge on ε-testable properties is based on a collection of examples, and general results are quite limited [AFKS99]. One of the few general results is described in [AKNS99], which studies the testability of formal languages. For a word $w \in \{0,1\}^n$ and a regular language L, we say that w is ε-far from a language $L \subseteq \{0,1\}^*$, if no word of length n that differs from w in at most εn positions is a member of L. Alon et al. proved that regular languages are ε-testable with query complexity $\tilde{O}(1/\varepsilon)$. More precisely, for every deterministic automaton $\mathcal{A}$ on finite words, integer n, and small enough $\varepsilon > 0$, there is an algorithm that gets as input a word $w \in \{0,1\}^n$, samples only $c \log^3(1/\varepsilon)/\varepsilon$ letters in w, where c depends only on $\mathcal{A}$, returns "yes" if $w \in L(\mathcal{A})$, and returns "no" with probability 2/3 if w is ε-far from $L(\mathcal{A})$ [AKNS99].

In this paper we extend the result of [AKNS99] to ω-*regular languages*. An ω-regular language over an alphabet Σ is a set $L \subseteq \Sigma^\omega$ of infinite words over Σ. ω-regular languages are described by automata on infinite words, first introduced in the 1960's. Motivated by decision problems in mathematical logic, Büchi, McNaughton, and Rabin developed a framework of automata on infinite words and infinite trees [Büc62,McN66, Rab69]. The framework has proven to be very powerful. Automata, and their tight relation to second-order monadic logics were the key to the solution of several fundamental decision problems in mathematical logic [Tho90]. Today, automata on infinite objects are used for specification and verification of nonterminating programs. Like automata on finite words, automata on infinite words either accept or reject an input word. Since a run on an infinite word does not have a final state, acceptance is determined with respect to the set of states visited infinitely often during the run. There are various ways to refer to this set. In *Büchi* automata, some of the states are designated as accepting states, and a run is accepting iff it visits states from the accepting set infinitely often [Büc62].

Nondeterministic Büchi automata recognize all the ω-regular languages [Lan69] and our algorithm assumes that L is given by such an automaton. The input to our algorithm are infinite words. We consider infinite words that have a finite representation. A general such representation maps each letter $\sigma \in \Sigma$ to a predicate $P_\sigma \subseteq \mathbb{N}$ that describes the positions of the word that are labeled σ. A special case we consider here is of *lasso-shape* (also known as *ultimately periodic*) infinite words, which are of the form $x \cdot y^\omega$, for $x \in \Sigma^*$ and $y \in \Sigma^*$. Thus, every lasso-shaped word has a position from which it is cyclic. As we discuss in Section 4, this special case is of particular interest in the context of specification and verification. In particular, it is easy to see that the language of a Büchi automaton is not empty iff the automaton accepts some lasso-shape word. Following similar considerations, if a system violates an ω-regular property, it has a lasso-shape computation violating the property [CD88,VW94]. Given a lasso-shaped word w, our algorithm tests the membership of w in the language of a nondeterministic Büchi automaton[2].

For some problems on automata, the transition from finite to infinite words is complicated. For example, one cannot determinize Büchi automata [Lan69], making the com-

---

[2] So, we actually extend [AKNS99] by three aspects: we consider a general (rather than binary) alphabet, we consider languages given by nondeterministic (rather than deterministic) automata, and we consider ω-regular, rather than regular, languages. It is not hard to see that the algorithm in [AKNS99] can be applied also to general alphabets and nondeterministic automata, thus the only real contribution is the extension to infinite words.

plementation problem for nondeterministic Büchi automata very challenging [Saf88]. For other problems, the transition is simple. For example, while the nonemptiness problem for automata on finite words can be reduced to one reachability test (from an initial state to the accepting set α), the nonemptiness problem for Büchi automata can be reduced to $2|\alpha|$ reachability tests (from an initial to an accepting state and from the accepting state to itself). It is easy to see then, that given an oracle for the nonemptiness problem for automata on finite words, the nonemptiness problem for Büchi automata can be solved by $2|\alpha|$ calls to the oracle.

Consider a nondeterministic Büchi automaton $\mathcal{A}$ and a lasso-shape word $w = x \cdot y^\omega$. The membership of w in $\mathcal{A}$ can be reduced to the nonemptiness of the product of w and $\mathcal{A}$. As we explain in Section 3, the latter can be reduced to a sequence of membership tests for automata on finite words, where the ε-test of [AKNS99] can be used as an oracle. The problem with this simple reduction is that the number of calls to the oracle depends on the length, $|x| + |y|$, of w. Finding an algorithm with a query complexity that does not depend on w turns out to be much more difficult, and is the main technical contribution of this paper. Essentially, we show that for every word w there is a set D of positions such that the size of D depends only on $\mathcal{A}$ and the following holds: if $w \in \mathcal{L}(\mathcal{A})$ then there is a word $v \in \mathcal{L}(\mathcal{A})$ such that v differs from w only in positions in D and the membership of v in $\mathcal{L}(\mathcal{A})$ can be verified by a constant number of applications of (some variant of) the algorithm of [AKNS99]. Moreover, if $w \notin L(\mathcal{A})$, then the above check would fail for all the words v that differ from w only in positions in D. The full details are described in Section 3. The query complexity of our algorithm is $\tilde{O}(1/\varepsilon)$, as the one of [AKNS99]. In addition, we study the special case where the language of the Büchi automaton is a *safety language*; that is, every word not in L has a finite "bad" prefix that cannot be extended to a word in a *safety language*[3]. In the full version, we also prove an $\Omega(1/\varepsilon)$ lower bound for the problem.

We hope that the ε-test for ω-regular languages would stimulate further efforts to apply the study of combinatorial property testing to *formal verification*. In formal verification, we verify that a system meets a desired behavior by executing an algorithm that checks whether a mathematical model of the system satisfies a formal specification that describes the behavior [CGP99]. Almost all current efforts and heuristics to cope with the large state spaces that commercial formal-verification tools handle do not deviate from the strict definition of formal verification, where the algorithm is not allowed to err. We believe that a major improvement of currently used heuristics should involve a deviation from the strict definition of formal verification. The setting of property testing seems very appealing for this task: the specifications are small, the systems are exceedingly large, and it is the complexity in terms of the system that we wish to bound, which is exactly what property testing does. In Section 4, we discuss this direction in detail. Due to lack of space, many details are omitted from this version. The full version can be found in authors' URL.

---

[3] We note that the result of [AKNS99] does not immediately apply ε-testability for *safety properties*, even though such properties can be characterized by a regular language of bad prefixes. The reason is that there is no *a-priori* bound on the length of the prefix that needs to be checked.

2 Definitions

Automata. A finite word over an alphabet Σ is a finite sequence $w \in \Sigma^*$ of letters from Σ. We can view a finite word as a function $w : \{1, \ldots, n\} \to \Sigma$, where n is the length of w. An infinite word over Σ is an infinite sequence $w \in \Sigma^\omega$ of letters from Σ, and it can be viewed as a function $w : \mathbb{N} \setminus \{0\} \to \Sigma$. For a word $w \in \Sigma^\omega$ and positions $0 \le x \le y$ we denote by $w[x, y]$ the sub-word of w that starts at position x and ends at position y. A *nondeterministic automaton* $\mathcal{A}$ is $\mathcal{A} = \langle \Sigma, Q, \delta, q_0, \alpha \rangle$, where Σ is an alphabet, Q is a set of states, $\delta : Q \times \Sigma \to 2^Q$ is a transition relation, $q_0 \in Q$ is the initial states, and $\alpha \subseteq Q$ is a set of accepting states. Given a finite word $w \in \Sigma^*$, a run r of $\mathcal{A}$ on w is a function $r : \{0, \ldots, n\} \to Q$ such that $r(0) = q_0$ and for all $0 \le i \le n$, we have $r(i + 1) \in \delta(r(i), w(i))$. The run r is *accepting* iff $r(n) \in \alpha$. If for all $q \in Q$ and $\sigma \in \Sigma$ we have that $|\delta(q, \sigma)| = 1$, then $\mathcal{A}$ is *deterministic*.

The automaton $\mathcal{A}$ can also get as input infinite words over Σ. Given such a word $w \in \Sigma^\omega$, a run r of $\mathcal{A}$ on w is a function $r : \mathbb{N} \to Q$ such that $r(0) = q_0$ and for all $i \ge 0$, we have $r(i + 1) \in \delta(r(i), w(i))$. Since the run has no final states, acceptance is determined with respect to the set $inf(r)$, of states that appear in r infinitely often. Formally, $q \in inf(r)$ iff $r(i) = q$ for infinitely many i's. When $\mathcal{A}$ is a *Büchi* automaton, the run r is *accepting* iff $inf(r) \cap \alpha \ne \emptyset$ [Büc62]. That is, a run is accepting iff it visits some accepting state infinitely often. Otherwise, r is rejecting. A path in $\mathcal{A}$ that corresponds to an accepting run is called an *accepting path*. The language of $\mathcal{A}$, denoted $\mathcal{L}(\mathcal{A})$, is the set of words w such that there is an accepting run of $\mathcal{A}$ on w. Note that $\mathcal{L}(\mathcal{A}) \subseteq \Sigma^*$ for automata on finite words and $\mathcal{L}(\mathcal{A}) \subseteq \Sigma^\omega$ for automata on infinite words. We assume that $\mathcal{L}(\mathcal{A}) \ne \emptyset$.

An automaton $\mathcal{A}$ induces a directed graph $G_\mathcal{A} = \langle V, E \rangle$ in the following way. The set of vertices of $G(\mathcal{A})$ is $V = Q$, and for each q and q' in V, we have $\langle q, q' \rangle \in E$ iff there exists $\sigma \in \Sigma$ such that $q' \in \delta(q, \sigma)$. For a graph G, the *period* of G is the greatest common divisor of cycle lengths in G. Note that if $\mathcal{A}$ is a Büchi automaton with $\mathcal{L}(\mathcal{A}) \ne \emptyset$, then there is at least one cycle in $G_\mathcal{A}$, thus the period of $G_\mathcal{A}$ is a finite $g \ge 1$.

Infinite words. We say that an infinite word $w \in \Sigma^\omega$ is *lasso-shaped* if there are $w_1 \in \Sigma^*$ and $w_2 \in \Sigma^*$ such that $w = w_1 \cdot (w_2)^\omega$. That is, there exists a position from which w is cyclic. The word w_1 is called the *prefix* of w, and the word w_2 is called the *lasso* of w. When $|w_1| = n_1$ and $|w_2| = n_2$, we say that w is (n_1, n_2)-*lasso-shaped*. As we discuss in Section 4, lasso-shaped words are of special interest in the context of formal verification.

For two finite words w and v of the same length, the *distance* between w and v, denoted $dist(w, v)$, is the number of letters that have to be changed in w in order to obtain v [AKNS99]. We say that two finite words w and v of the same length n are ε-*far*, for $0 > \varepsilon \ge 1$, if $\frac{dist(w,v)}{n} \ge \varepsilon$.

For infinite words, the number of letters that have to be changed in one word in order to obtain the other can be infinite, thus we cannot extend the definition of [AKNS99] to infinite words in a straightforward way. Instead, the definition of distance should refer to the finite representation of an infinite word, thus to the prefix and lasso of lasso-shaped words. When we define the distance between lasso-shaped words, we want our definition of distance to be insensitive to a particular representation: the distance between different representations of the same word should be 0. Let w and v be two lasso-shaped words with prefixes of length n_1 and n_1' and lassos of length n_2 and n_2', respectively. Without loss of generality, assume that $n_1' \geq n_1$. Let $i = n_1' - n_1$ and $n = lcm(n_2, n_2')$ (least common multiplier). We define

$$dist(w, v) = dist(w[n_1 + i + 1, n_1 + i + n], v[n_1' + 1, n_1' + n]).$$

We say that a lasso-shaped word w is ε-*far* from a lasso-shaped word v if $dist(w, v) \geq \varepsilon n$, where n is the least common multiplier of the lengths of lassos of w and v. For a lasso-shaped word w and a language $L \subseteq \Sigma^\omega$, we say that w is ε-far from L with respect to (n_1, n_2) if w is ε-far from all lasso-shaped words $v \in L$. Our definition of distance does not compare the prefixes, as their weight in the infinite words is negligible, but it does take the lengths of the prefixes into an account. The following lemma shows that the distance between two different representations of the same lasso-shaped word is 0, as required.

Lemma 1. *Let $w = w_1 \cdot (w_2)^\omega$ be a lasso-shaped word. Let n_1 be the length of the prefix w_1, and n_2 be the length of the lasso w_2. Then, for all numbers $k \geq 0$, $l \geq 1$, and all partitions of w_2 to w_3 and w_4, we have $dist(w, w_1 \cdot (w_2)^k \cdot w_3 \cdot ((w_4 \cdot w_3)^l)^\omega) = 0$.*

3   The Algorithm

In this section we describe an ε-test for lasso-shaped words with respect to properties given by a nondeterministic Büchi automaton. The query complexity of the test is $\tilde{O}(1/\varepsilon)$. We first analyze the structure of accepting runs of nondeterministic Büchi automata on infinite words. We argue that for lasso-shaped infinite words it suffices to examine a finite prefix of a word in order to decide its membership in $\mathcal{L}(\mathcal{A})$.

Let $\mathcal{A} = \langle \Sigma, Q, \delta, q_0, \alpha \rangle$ be a Büchi automaton. For simplicity, we describe our algorithm for the case $\alpha = \{q_{acc}\}$ is a singleton. Later we show how to extend our results to the case of multiple accepting states. We denote by $\mathcal{A}_{fin}$ the automaton $\mathcal{A}$ viewed as an automaton on finite words. Let C_{acc} be the maximal strongly connected component of $G_{\mathcal{A}}$ that contains q_{acc}, and let S be the set of immediate successors of q_{acc} in C_{acc}. Thus, $q \in S$ iff there is $\sigma \in \Sigma$ such that $q \in \delta(q_{acc}, \sigma) \cap C_{acc}$. We define the automaton $\mathcal{A}_{fin}^S = \langle \Sigma, C_{acc}, \delta_{fin}^S, S, \{q_{acc}\} \rangle$ as the automaton on finite words that is derived from the graph C_{acc}, with initial state S and accepting state q_{acc}. Formally, $\mathcal{A}_{fin}^S = \langle \Sigma, Q \cap C_{acc}, \delta_{fin}^S, S, \{q_{acc}\} \rangle$, where $\delta_{fin}^q : C_{acc} \times \Sigma \to 2^{C_{acc}}$ is such that $q' \in \delta_{fin}^S(q, \sigma)$ iff $q' \in \delta(q, \sigma) \cap C_{acc}$. In Lemma 2 we show that for lasso-shaped words, the membership problem of a word in the language can be reduced to the membership problem of two prefixes of a bounded length in languages on finite words.

Lemma 2. *An (n_1, n_2)-lasso-shaped word w belongs to $\mathcal{L}(\mathcal{A})$ iff there exist $n_1 \leq p \leq n_1 + |Q|n_2$ and $1 \leq i \leq |Q|$ such that $w[1, p] \in \mathcal{L}(\mathcal{A}_{fin})$, and $w[p + 1, p + i \cdot n_2] \in \mathcal{L}(\mathcal{A}_{fin}^S)$.*

So, by iterating over all the possible values of p and i, we can reduce the membership problem for Büchi automata and lasso-shaped infinite words to a sequence of membership tests for finite words. However, since the number of possible values for p depends on

n_2, so is the query complexity of an algorithm that is based on such a reduction. We now describe the long journey required in order to avoid this dependency in n_2. Fortunately, some of the techniques developed in [AKNS99] in order to ε-test finite words turned out to be useful also for bounding the number of calls to the algorithm in [AKNS99]. In particular, we need the following lemma about strongly connected graphs.

Lemma 3. [AKNS99] *Let $G = \langle V, E \rangle$ be a nonempty, strongly connected graph with a finite period g. Then there exists a partition of V to pairwise disjoint sets $V(G) = V_0, \ldots, V_{g-1}$ and a constant $m \le 3|V|^2$ such that:*

1. *For every $0 \le i, j \le g - 1$, and for every $u \in V_i$, $v \in V_j$, the length of every directed path from u to v in G is $(j - i) \bmod g$.*
2. *For every $0 \le i, j \le g - 1$, and for every $u \in V_i$, $v \in V_j$, and for every $l \ge m$ such that $l = (j - i) \bmod g$, there exists a directed path from u to v in G of length l.*

The constant m from Lemma 3 is called the *reachability constant* of G.

We use Lemma 3 in order to change the input word slightly in a way that enables us to restrict attention to runs of $\mathcal{A}$ that visit q_{acc} at specific positions whose number depends on the period of C_{acc} rather than on n_2. This involves two arguments. First, in Lemma 4 we show that when w is in $\mathcal{L}(\mathcal{A})$, we can change w slightly so that $\mathcal{A}$ accepts the resulted word v by visiting q_{acc} at specific positions. Then, in Lemma 5, we prove that if w is ε-far from $\mathcal{L}(\mathcal{A})$, then all words v that are slightly different from w cannot be accepted by runs that visit q_{acc} in these specific positions. In what follows, we fix g and m to denote the period and the reachability constant of C_{acc}, respectively. Also, let

$$D = \bigcup_{k \ge 0} \{n_1 + (|Q| + k)n_2, \ldots, n_1 + (|Q| + k)n_2 + 2m + g - 1\}.$$

When we formalize in Lemmas 4 and 5 the notion of "change w slightly", we mean that w can be changed only in positions in D.

Lemma 4. *Let $w \in \mathcal{L}(\mathcal{A})$ be an (n_1, n_2)-lasso-shaped word with $n_2 > 2m + g$. There exists a lasso-shaped word $v \in \mathcal{L}(\mathcal{A})$ that satisfies the following.*

1. *For all $j \notin D$, we have $w(j) = v(j)$, and*
2. *The length of the prefix of v is p^v, where $n_1 + |Q|n_2 + m \le p^v \le n_1 + |Q|n_2 + m + g - 1$, and the length of the lasso of v is $i^v n_2$, where $1 \le i^v \le |Q|$, and we have $v[1, p^v] \in \mathcal{L}(\mathcal{A}_{fin})$ and $v[p^v + 1, p^v + i^v \cdot n_2] \in \mathcal{L}(\mathcal{A}_{fin}^S)$.*

In particular, Lemma 4 implies that $dist(w, v) \le m + g$.

Lemma 5. *For each $0 < c < 1$, there exists $N_c \in \mathbb{N}$ such that for every (n_1, n_2)-lasso-shaped word w that is ε-far from $\mathcal{L}(\mathcal{A})$, with $n_2 \ge N_c$, all the words v satisfy one of the following.*

1. *There is $j \notin D$ such that $w(j) \ne v(j)$, or*
2. *For all $n_1 + |Q|n_2 + m \le p \le n_1 + |Q|n_2 + m + g - 1$ and $1 \le i \le |Q|$, either $\mathcal{L}(\mathcal{A}_{fin})$ does not contain words of length p, or $v[p + 1, p + i \cdot n_2]$ is $c \cdot \varepsilon$-far from $\mathcal{L}(\mathcal{A}_{fin}^S)$.*

In fact, in the proof of Lemma 5, we show that $N_c = \frac{2m+g-1}{\varepsilon(1-c)}$.

We are now ready to prove the following theorem, which gives a reduction from infinite lasso-shaped words to finite words, where the complexity of the reduction is independent of the size of the input.

Theorem 1. *Consider a Büchi automaton* $\mathcal{A} = \langle \Sigma, Q, \delta, q_0, \{q_{acc}\} \rangle$. *For every* (n_1, n_2)-*lasso-shaped word* $w \in \Sigma^\omega$, *the following hold.*

1. *If* $w \in \mathcal{L}(\mathcal{A})$, *then there exists a lasso-shaped word* $v \in \mathcal{L}(\mathcal{A})$ *such that* $dist(w, v) \leq 2m+g-1$, *for all* $j \notin D$ *we have* $w(j) = v(j)$, *and there exist integers* $n_1 + |Q|n_2 + m \leq p \leq n_1 + |Q|n_2 + m + g - 1$ *and* $1 \leq i \leq |Q|$, *such that* $v[1, p] \in \mathcal{L}(\mathcal{A}_{fin})$ *and* $v[p+1, p+i \cdot n_2] \in \mathcal{L}(\mathcal{A}_{fin}^S)$.
2. *If* w *is* ε-*far from* $\mathcal{L}(\mathcal{A})$, *then for all lasso-shaped words* $v \in \Sigma^\omega$ *such that* $w(j) = v(j)$ *for all* $j \notin D$, *all integers* p *and* i *such that* $n_1 + |Q|n_2 + m \leq p \leq n_1 + |Q|n_2 + m + g - 1$ *and* $1 \leq i \leq |Q|$, *and all* $0 < c \leq \frac{\varepsilon n_2 - (2m+g-1)}{\varepsilon n_2}\}$, *either* $\mathcal{L}(\mathcal{A}_{fin})$ *does not contain words of length* p, *or* $dist(v[p+1, p+i \cdot n_2], \mathcal{L}(\mathcal{A}_{fin}^S)) \geq c \cdot \varepsilon \cdot i \cdot n_2$.

We note that the constant c is greater than 0 and is smaller than 1.

Theorem 1 leads to our ε-test, which is described in Figure 1. The algorithm gets five parameters: a lasso-shaped word w, the length of the prefix of w, the length of the lasso of w, a Büchi automaton $\mathcal{A}$, and $0 < \varepsilon \leq 1$. It invokes two algorithms: an algorithm Check_Length, which gets as input an automaton $\mathcal{A}_{fin}$ and an integer p and checks whether the language $\mathcal{L}(\mathcal{A}_{fin})$ contains words of length p, and an algorithm Fin_Test, which, from reasons we explain below, differs slightly from the algorithm of [AKNS99]. The algorithm Check_Length is deterministic, and it checks whether $\mathcal{A}_{fin}$ contains words of length p by computing several *modulo* operations on p. The algorithm Fin_Test gets four parameters: a finite word w', the length of w', an automaton on finite words $\mathcal{A}_{fin}^S$, and $0 < \varepsilon' \leq 1$. Like the algorithm in [AKNS99], when $w' \in \mathcal{L}(\mathcal{A}_{fin}^S)$, it outputs "yes", and when w' is ε'-far from $\mathcal{L}(\mathcal{A}_{fin}^S)$, it outputs "no" with probability at least $2/3$. We will describe the algorithm Fin_Test in more detail later. Essentially, Fin_Test is a modification of the algorithm in [AKNS99] that discards from the random sample the letters of the input that are located in D. Thus, the algorithm Fin_Test gives the same answer for all words that differ one from another only in letters that are located in these areas. We note that Fin_Test can handle general alphabet and nondeterministic automata, nevertheless it is not hard to see that the algorithm in [AKNS99] can be applied also to general alphabets and nondeterministic automata.

---

procedure LS_Test $(w, n_1, n_2, \mathcal{A}, \varepsilon)$
    for $p = n_1 + |Q|n_2 + m$ **to** $n_1 + |Q|n_2 + m + g - 1$ **do**
        if Check_Length $(\mathcal{A}_{fin}, p)$ **then**
            for $i = 1$ **to** $|Q|$ **do**
                if Fin_Test $(w[p+1, p+i \cdot n_2], p, \mathcal{A}_{fin}^S, \varepsilon - \frac{2m+g-1}{n_2})$ **then return** "yes";
    return "no".

Fig. 1. Testing of Lasso-Shaped Words

The algorithm LS_Test first calls the algorithm Check_Length with the automaton $\mathcal{A}_{fin}$ and the length p, for all $n_1 + |Q|n_2 + m \leq p \leq n_1 + |Q|n_2 + m + g - 1$. Thus, the first part of LS_Test invokes Check_Length at most g times. If LS_Test gets a positive answer for some p, it calls Fin_Test with the word $w[p+1, p+i \cdot n_2]$ of length $i \cdot n_2$, the automaton $\mathcal{A}^S_{fin}$, and $\varepsilon' = \varepsilon - \frac{2m+g-1}{n_2}$, for all $1 \leq i \leq |Q|$. Now, since the the query complexity of Check_Length is 0, and the query complexity of Fin_Test is $\tilde{O}(1/\varepsilon')$, and since m, g, and $|Q|$ do not depend on w, the query complexity of LS_Test is $\tilde{O}(1/\varepsilon)$.

Readers not familiar with [AKNS99] may be happy at this point and wonder about the need to modify the algorithm in [AKNS99] and the need for the complicated expressions that wait in the full version. The need for them arises from two technical difficulties. The first difficulty has to do with the fact that LS_Test calls Fin_Test several times, and it returns "yes" if for some p and i the language $\mathcal{L}(\mathcal{A}_{fin})$ contains words of length p and the corresponding call to Fin_Test returned "yes". Consequently, if we want to bound the probability of error of LS_Test by $1/3$, the probability of error of Fin_Test has to be much lower. The second difficulty has to do with the *transition intervals* of [AKNS99] and the way they interfere with the intervals in the set D of positions in which we allowed a modification of w.

To understand the second difficulty, let us outline briefly the algorithm of [AKNS99]. The main idea of the algorithm is to consider the possible traversal paths (of an automaton $\mathcal{A}$ on the input word w) that finish in an accepting state, and then to check whether the input word can be accepted by following one of these paths. The graph of $\mathcal{A}$ is partitioned into strongly connected components. A possible traversal path of $\mathcal{A}$ is described by a triplet $\langle A, P, \Pi \rangle$, where A is the list of traversed components, P is the sequence of entrance and exit states for each component, and Π is the sequence of integers that describe the positions in which the path exits each component. A triplet is called *admissible* if the last component in A contains an accepting state, which is the exit state of the last pair in P.

The algorithm then chooses a number of random subwords of w, of length that depends on A and ε and is independent of n. For each admissible triplet and for each subword, the algorithm checks whether the subword can be completed to a word that belongs to the language of the corresponding component. A subword that can be completed in this way is called *feasible*. If for all admissible triplets at least one of the subwords is infeasible, the algorithm outputs "no". Otherwise (that is, there exists an admissible triplet that matches all chosen subwords), the algorithm outputs "yes". The correctness of the algorithm follows from the fact that if the input word w is ε-far from the language of $\mathcal{A}$, then for each admissible triplet, w has many short infeasible subwords, one of which is be detected by the algorithm with high probability.

The number of admissible triplets for a word of length n depends on n. The algorithm in [AKNS99] gets rid of this dependency by placing evenly in the interval $\{1, \ldots, n\}$ a bounded number (depending only on ε and the parameters of $\mathcal{A}$) of *transition intervals*, and considering only admissible triplets for which the transitions between the components of G occur inside these transition intervals. Then, [AKNS99] show that if $w \in \mathcal{L}(\mathcal{A})$, it can be modified slightly to be accepted by a traversal path whose admissible triplet meets these restrictions, whereas if w is ε-far from $\mathcal{L}(\mathcal{A})$, the probability of detecting a short infeasible subword is still high.

Consider an automaton $\mathcal{A}$ on finite words. Let m be the maximal reachability constant for all strongly connected components of $G(\mathcal{A})$, let l be the least common multiplier of the periods of strongly connected components of $G(\mathcal{A})$, and let k is the number of strongly connected components of $G(\mathcal{A})$. When the algorithm of [AKNS99] gets as an input a word w of length n, it places S transition intervals $T_s = \{[a_s, b_s]\}_{s=1}^{S}$ evenly in $[n]$, where $S = 129km\log(1/\varepsilon)/\varepsilon$, and the length of each interval is $(k-1)(l+m)$. For $1 \le i \le \log(8km/\varepsilon)$ we define $r_i = \frac{2^{8-i}k^2m\log^2(1/\varepsilon)}{\varepsilon}$. Then the algorithm proceeds as follows.

1. For each $1 \le i \le \log(8km/\varepsilon)$ choose r_i random subwords in w of length 2^{i+1} each.
2. For each admissible triplet $\langle A, P, \Pi \rangle$ with $A = (C_{i_1}, \ldots, C_{i_t})$, $P = (p_j^1, p_j^2)_{j=1}^{t}$, and $\Pi = (n_j)_{j=1}^{t+1}$ such that each n_j is inside an interval T_s for some $1 \le s \le S$, do the following.
   2.1. Discard chosen subwords that have one of their ends closer than $\varepsilon n/(128km\log(1/\varepsilon))$ from some $n_j \in \Pi$.
   2.2. For each remaining subword R, if R falls between n_j and n_{j+1}, check whether R is feasible for the automaton $\mathcal{A}_j$ obtained by restricting $\mathcal{A}$ to the strongly connected component C_j.
3. If for some admissible triplet all checked subwords are feasible, output "yes". Otherwise, (i.e. in the case where for all admissible triplets at least one infeasible subword was found), output "no".

Our Fin_Test algorithm differs from the algorithm in [AKNS99] in steps **1** and **2.1**. In step **1**, we increase the number of random subwords we choose from r_i to xr_i, for $x = \lfloor 1-\log(1-(\frac{2}{3})^{\frac{1}{|S|\cdot|Q|\cdot g}})/2 \rfloor$. The reason for that is that an increase in the number of chosen subwords reduces the probability of missing an infeasible subword, and thus also the probability of error in the algorithm of [AKNS99]. Formally, the probability of error for Fin_Test is bounded by $1/4^x$, where x is the factor by which we increase the size of the number of chosen subwords. Taking x as above then bounds probability of error for LS_Test by $1/3$. In step **2.1**, in addition to discarding the subwords that have one of their ends closer than $\varepsilon n/(128km\log(1/\varepsilon))$ from some $n_j \in \Pi$, we also discard subwords that intersect with the intervals $[n_1 + (x+|Q|)n_2, n_1 + (x+|Q|)n_2 + 2m + g - 1]$, for $1 \le x \le |Q|$. The reason for discarding those subwords is that the algorithm should give the same answer for the word w and all words v that differ from w only in letters that are located in these intervals. Discarding these subwords does not affect the probability of the correct answer.

We are now ready to prove the correctness of the algorithm LS_Test.

Theorem 2. *For a Büchi automaton $\mathcal{A} = \langle \Sigma, Q, \delta, q_0, \{q_{acc}\} \rangle$, and an (n_1, n_2)-lasso-shaped word $w \in \Sigma^\omega$, if $w \in \mathcal{L}(\mathcal{A})$, then the algorithm LS_Test always outputs "yes", and if w is ε-far from $\mathcal{L}(\mathcal{A})$, then the algorithm LS_Test outputs "no" with probability at least $2/3$.*

Remark 1. The algorithm LS_Test can be extended to handle multiple accepting states by running it for each accepting state separately. This increases the running time by at most the size of α. The algorithm LS_Test, as well as the algorithm of [AKNS99], are described for the case of a single initial state. They can be extended to handle multiple initial states by running it for each initial state separately. This increases the running time by at most the number of initial states.

Remark 2. Of special interest in formal verification are *safety properties*, asserting that the observed behavior of the system always stay within some allowed region, in which nothing "bad" happens. Intuitively, a property ψ is a safety property if every violation of ψ occurs after a finite execution of the system. Consider a language L of infinite words over Σ. A finite word x over Σ is a *bad prefix* for L iff for all infinite words y over Σ, the concatenation $x \cdot y$ of x and y is not in L. Thus, a bad prefix for L is a finite word that cannot be extended to an infinite word in L. A language L is a *safety language* if every word not in L has a finite bad prefix [AS85].

Given a nondeterministic Büchi automaton $\mathcal{A}$ that recognizes a safety language (the latter can be checked in PSPACE [Sis94]), it is possible to construct a deterministic automaton $\mathcal{A}_{bad}$ on finite words that accepts exactly all the bad prefixes of $\mathcal{L}(\mathcal{A})$ [KV01]. Clearly, an infinite word w belongs to L iff there is no finite prefix of w that belongs to $\mathcal{L}(\mathcal{A}_{bad})$. While this still does not imply ε-testability of safety properties, it can be shown that for lasso-shaped words, it is enough to check a single prefix. Indeed, as we prove in the full version, if w is an (n_1, n_2)-lasso-shaped word, then $w \in \mathcal{L}(\mathcal{A})$ iff $w[1, n_1 + |Q|n_2] \notin \mathcal{L}(\mathcal{A}_{bad})$, where $|Q|$ is the number of states of the automaton $\mathcal{A}_{bad}$. Thus, an ε-test for the membership of w in the language of $\mathcal{A}$ can invokes the ε-test of regular languages of [AKNS99] with the word $w[1, n_1 + |Q|n_2]$ and the automaton $\mathcal{A}_{bad}$.

4   Discussion

The main technical contribution of this paper is an ε-test for ω-regular languages and lasso-shaped words. This result is an extension of the ε-test for regular languages presented in [AKNS99]. The extension is not trivial. In fact, already the definition of distance, which is straightforward for finite words, involves subtle considerations, and discussion has to be restricted to infinite words with a finite representation. We describe a reduction from ε-test of infinite words to a constant number of ε-tests for finite words. The main difficulty that is posed by the fact the word is infinite is that, unlike the case of finite words, we do not know the position in the word in which an accepting run of $\mathcal{A}$ on the word visits an accepting state. For general (not lasso-shaped) words, this difficulty cannot be circumvented. We show that for lasso-shaped words, we can bound the number of positions where an accepting run can visit an accepting state. Moreover, we show that a word in the language of $\mathcal{A}$ can be modified slightly to ensure that an accepting run would visit an accepting state inside a specific interval of a constant length.

Today's rapid development of complex and safety-critical systems requires *formal verification* methods. In *model checking*, we verify that a system meets a desired behavior by executing an algorithm that checks whether a mathematical model of the system satisfies a formal specification that describes the behavior. The systems we verify are non-terminating, and their specifications describe an on-going behavior of the system (c.f., "every request is eventually granted"). The algorithmic nature of model checking makes it fully automatic, and thus attractive to practitioners. At the same time, model checking is very sensitive to the size of the mathematical model of the system. Commercial model-checking tools need to cope with the exceedingly large state spaces that are present in real-life designs, making the so-called *state-explosion problem* perhaps the most challenging issue in computer-aided verification [CGP99].

Almost all previous efforts and heuristics for coping with the state-explosion problem, such as symbolic methods [McM93] and modular verification [dRLP98], do not deviate from the strict definition of model checking, where the algorithm is not allowed to err. Consequently, complexity lower bounds for the model-checking problem apply also for these heuristics. We believe that a major improvement of currently used heuristics should involve a deviation from the strict definition of model checking. The setting of property testing seems very appealing for this task: the specifications are small, the systems are exceedingly large, and it is the complexity in terms of the system that we wish to bound, which is exactly what property testing does. Indeed, the complexity of ε-testing algorithms depends only on ε and on the size of the property, and is independent of the size of the input.

The one-sided error allowed in property testing means that if the system is correct, the testing algorithm always say it is correct, yet when the system is incorrect, the testing algorithm may say it is correct. This at first seems like the unfortunate side, as a model-checking algorithm that reports the correctness of an incorrect system may be more harmful than an algorithm that reports the incorrectness of a correct system. It is now agreed, however, that the anticipated goal of formal verification, of providing a "correctness stamp" on the system, has turned out to involve too many obstacles (e.g., exact modeling of the system, comprehensive specification, etc.), and the primary use of model checking nowadays is *falsification*. There, as in debugging, the goal is to detect errors, rather than to serve as a correctness stamp [BCCZ99,Sip99]. The one-sided error of property testing fits well in this approach: whenever the testing algorithm reports a dissatisfaction of the property, an error indeed exists. In addition, as with all other randomized algorithms, repeated runs can be used in order to reduce the error to any desirable constant.

Naturally, the ultimate goal of applying property testing to model checking is an ε-test that gets as input a Kripke structure that models an entire system (rather than a single computation), and distinguishes between the case where the system satisfies a specification and the case where it is far from satisfying it. Technically, this can be achieved by extending the ε-test here to ω-*regular tree languages*. Our efforts in this direction are still unsuccessful. The difficulty is already in an extension of [AKNS99] to regular tree languages, and it lies in the fact that local changes in the tree influence several paths in it. While the ultimate goal seems hard to achieve, some helpful applications can arise already from our result here.

Our ε-test can replace model checking of lasso-shaped words. While the complexity of the ε-test is independent of n_1 and n_2, the best time complexity known for the model-checking problem of (n_1, n_2)-lasso-shaped words is $O((n_1 + n_2) * m)$, where m is the size of the specification, represented as a nondeterministic Büchi automaton. The restriction to lasso-shape words is not really restrictive in the context of model checking: as proven in [CD88], if a system violates an LTL property ψ, there is a lasso-shaped computation of the system that violates ψ. In addition, counter examples for LTL properties (that is, descriptions of computations that violate the specification) are given by model-checking tools in terms of lasso-shaped computations [VW94], and several *random simulation* algorithms that are based on sampling and checking individual computations of the system consider, or can be easily modify to consider, lasso-shaped words [Wes89,Hol91].

Our ε-test, however, has an additional crucial requirement on the input, namely the ability to perform local queries of the input in a constant time (also known as *random access* to the input). On the other hand, current random-simulation methods construct the sampled computations "on-the-fly," and a naive application of our algorithm on top of them involves storing of the sampled computation and would cause the time and space complexity to depend on the size of the input word. Still, our ε-test is useful in combination with a random simulator as running the model checker on the output of the random simulator causes a significant slow-down in the performance of the simulator [Fix02]. Thus, storing the sampled computations and then running our ε-test on them seems a promising way to reduce the complexity of model checking of sampled computations.

Acknowledgment. We thank David Dill, Limor Fix, Ilan Newman, and Dana Ron for helpful discussions.

References

[AFKS99] N. Alon, E. Fischer, M. Krivelevich, and M. Szegedy. Efficient testing of large graphs. In *40th IEEE FOCS*, pp. 645–655, 1999.

[AK] N. Alon and M. Krivelevich. Testing k-colorability. *SIAM Journal on Discrete Mathematics*. to appear.

[AKNS99] N. Alon, M. Krivelevich, I. Newman, and M. Szegedy. Regular languages are testable with a constant number of queries. In *40th IEEE FOCS*, pp. 645–655, 1999.

[AS85] B. Alpern and F.B. Schneider. Defining liveness. *IPL*, 21:181–185, 1985.

[BCCZ99] A. Biere, A. Cimatti, E.M. Clarke, and Y. Zhu. Symbolic model checking without BDDs. In *TACAS, LNCS* 1579, 1999.

[BR00] M.A. Bender and D. Ron. Testing acyclicity of directed graphs in sublinear time. In *27th ICALP*, 2000.

[Büc62] J.R. Büchi. On a decision method in restricted second order arithmetic. In *ICLMPS 1960*, pp. 1–12, Stanford, 1962. Stanford University Press.

[CD88] E.M. Clarke and I.A. Draghicescu. Expressibility results for linear-time and branching-time logics. In J.W. de Bakker, W.P. de Roever, and G. Rozenberg, editors, *REX Workshop, LNCS* 354, pp. 428–437. 1988.

[CGP99] E.M. Clarke, O. Grumberg, and D. Peled. *Model Checking*. MIT Press, 1999.

[dRLP98] W-P. de Roever, H. Langmaack, and A. Pnueli, editors. *Compositionality: The Significant Difference. COMP, LNCS* 1536, 1998.

[Fix02] L. Fix. Application of ε-test to random simulation. Private communication, 2002.

[GGR98] O. Goldreich, S. Goldwasser, and D. Ron. Property testing and its connection to learning and approximation. *Journal of ACM*, 45(4):653–750, 1998.

[GR97] O. Goldreich and D. Ron. Property testing in bounded degree graphs. In *25th STOC*, pp. 406–415, 1997.

[GR99]   O. Goldreich and D. Ron. A sublinear bipartite tester for bounded degree graphs. *Combinatorica*, 19(3):335–373, 1999.

[Hol91]   G. Holzmann. *Design and Validation of Computer Protocols*. Prentice-Hall International Editions, 1991.

[KV01]   O. Kupferman and M.Y. Vardi. Model checking of safety properties. *Formal methods in System Design*, 19(3):291–314, November 2001.

[Lan69]   L.H. Landweber. Decision problems for ω–automata. *Mathematical Systems Theory*, 3:376–384, 1969.

[McM93]   K.L. McMillan. *Symbolic Model Checking*. Kluwer Academic Publishers, 1993.

[McN66]   R. McNaughton. Testing and generating infinite sequences by a finite automaton. *Information and Control*, 9:521–530, 1966.

[PR99]   M. Parnas and D. Ron. Testing the diameter of graphs. In *RANDOM*, pp. 85–96, 1999.

[PRR01]   M. Parnas, D. Ron, and R. Rubinfeld. Testing parenthesis languages. In *5th RANDOM*, pp. 261–272, 2001.

[PRS01]   M. Parnas, D. Ron, and A. Samorodnitsky. Proclaiming dictators and juntas or testing boolean formulae. In *5th RANDOM*, pp. 273–284, 2001.

[Rab69]   M.O. Rabin. Decidability of second order theories and automata on infinite trees. *Transaction of the AMS*, 141:1–35, 1969.

[RS96]   R. Rubinfeld and M. Sudan. Robust characterization of polynomials with applications to program testing. *SIAM Journal on Computing*, 25(2):252–271, 1996.

[Saf88]   S. Safra. On the complexity of ω-automata. In *29th FOCS*, pp. 319–327, 1988.

[Sip99]   H.B. Sipma. *Diagram-based Verification of Discrete, Real-time and Hybrid Systems*. PhD thesis, Stanford University, Stanford, California, 1999.

[Sis94]   A.P. Sistla. Safety, liveness and fairness in temporal logic. *Formal Aspects of Computing*, 6:495–511, 1994.

[Tho90]   W. Thomas. Automata on infinite objects. *Handbook of Theoretical Computer Science*, pp. 165–191, 1990.

[VW94]   M.Y. Vardi and P. Wolper. Reasoning about infinite computations. *Information and Computation*, 115(1):1–37, November 1994.

[Wes89]   C.H. West. Protocol validation in complex systems. In *8th PODC*, pp. 303–312, 1989.

Optimal Lower Bounds for 2-Query Locally Decodable Linear Codes

Kenji Obata[*]

Optimal Lower Bounds for 2-Query Locally Decodable Linear Codes

Kenji Obata[*]

Computer Science Division
University of California, Berkeley
kenjioba@eecs.berkeley.edu.

Abstract. This paper presents essentially optimal lower bounds on the size of linear codes

$$\mathbf{C} : \{0,1\}^n \to \{0,1\}^m$$

which have the property that, for constants $\delta, \epsilon > 0$, any bit of the message can be recovered with probability $\frac{1}{2} + \epsilon$ by an algorithm reading only 2 bits of a codeword corrupted in up to δm positions. Such codes are known to be applicable to, among other things, the construction and analysis of information-theoretically secure private information retrieval schemes. In this work, we show that m must be at least $2^{\Omega(\frac{\delta}{1-2\epsilon}n)}$. Our results extend work by Goldreich, Karloff, Schulman, and Trevisan [GKST02], which is based heavily on methods developed by Katz and Trevisan [KT00]. The key to our improved bounds is an analysis which bypasses an intermediate reduction used in both prior works. The resulting improvement in the efficiency of the overall analysis is sufficient to achieve a lower bound optimal within a constant factor in the exponent. A construction of a locally decodable linear code matching this bound is presented.

1 Introduction

This paper presents essentially optimal lower bounds on the size of linear codes

$$\mathbf{C} : \{0,1\}^n \to \{0,1\}^m$$

which have the property that, for constants $\delta, \epsilon > 0$, any bit of the message can be recovered with probability $\frac{1}{2} + \epsilon$ by an algorithm reading only 2 bits of a codeword corrupted in up to δm positions. Such codes are known to be applicable to, among other things, the construction and analysis of information-theoretically secure private information retrieval schemes.

In this work, we show that m must be at least $2^{\Omega(\frac{\delta}{1-2\epsilon}n)}$. Our results extend work by Goldreich, Karloff, Schulman, and Trevisan [GKST02], who show that m must be at least $2^{\Omega(\epsilon\delta n)}$. Note that the prior bound does not grow arbitrarily

---

[*] Work supported by an NSF graduate fellowship.

J.D.P. Rolim and S. Vadhan (Eds.): RANDOM 2002, LNCS 2483, pp. 39–50, 2002.
© Springer-Verlag Berlin Heidelberg 2002

large as the error probability of the decoder goes to zero ($\epsilon \to \frac{1}{2}$), as intuitively it should; our new results have the correct qualitative behavior.

The key to our improved bounds is an analysis which bypasses an intermediate reduction used in both prior works. The resulting improvement in the efficiency of the overall analysis is sufficient to achieve a lower bound optimal within a constant factor in the exponent. A construction of a locally decodable linear code matching this bound is presented.

Our work is structured as follows: In the remainder of this section, we briefly review the definitions and techniques employed in [KT00] and [GKST02]. In Section 2, we prove the main technical result of this paper, which establishes a relationship between the probability that an edge of a graph sampled from any distribution intersects any vertex-subset of a given size, and the size of a maximum matching in the graph. The analysis in this result seems independently interesting, and may be applicable in other contexts. In Section 3, we show how the combination of this result with the techniques of [GKST02] establishes lower bounds for this class of locally decodable codes, and present a construction of a family of these codes with size matching our bounds within a constant factor in the exponent.

1.1   Locally Decodable and Smooth Codes

Let Σ_1, Σ_2 be arbitrary finite alphabets, Σ_i^n the set of strings of elements from Σ_i of length n, and for $x, y \in \Sigma_i^n$, $d(x, y)$ the number of positions i such that $x_i \neq y_i$.

Definition 1 (Locally Decodable Code). *For fixed constants δ, ϵ, q, a mapping*

$$\mathbf{C} : \Sigma_1^n \to \Sigma_2^m$$

is a (q, δ, ϵ)-locally decodable code if there exists a probabilistic oracle machine A such that:

- *A makes at most q queries (without loss of generality, A makes exactly q queries).*
- *For every $x \in \Sigma_1^n$, $y \in \Sigma_2^m$ with $d(y, \mathbf{C}(x)) \leq \delta m$, and $i \in \{1, \dots, n\}$,*

$$\Pr\left[A^y(i) = x_i\right] \geq \frac{1}{|\Sigma_1|} + \epsilon$$

where the probability is over the randomness of A.

In this paper, we consider codes $\mathbf{C}$ satisfying the above properties where, in addition, Σ_1, Σ_2 are fields and $\mathbf{C}$ is a *linear* mapping from $\Sigma_1^n \to \Sigma_2^m$. While all of our results are applicable to finite fields in general, and some to non-linear codes, we will for simplicity narrow our current discussion to linear codes on $\mathbf{Z}_2$. Also, while we have observed that our results are equally applicable to reconstruction algorithms making queries adaptively, we limit our comments in this abstract to algorithms making non-adaptive queries, and summarize some details of our proofs.

We begin by reviewing the definitions and techniques of [KT00] and [GKST02], which our results build upon.

It was observed in [KT00] that a locally decodable code should have the property that a decoding algorithm A reads from each location in the code word with roughly uniform probability. This motivated the following definition:

Definition 2 (Smooth Code). *For fixed constants c, ϵ, q, a mapping*

$$\mathbf{C} : \Sigma_1^n \rightarrow \Sigma_2^m$$

is a (q, c, ϵ)-smooth code if there exists a probabilistic oracle machine A such that:

- *A makes at most q queries (without loss of generality, A makes exactly q queries).*
- *For every $x \in \Sigma_1^n$ and $i \in \{1, \dots, n\}$,*

$$\Pr\left[A^{\mathbf{C}(x)}(i) = x_i\right] \geq \frac{1}{|\Sigma_1|} + \epsilon.$$

- *For every $i \in \{1, \dots, n\}, j \in \{1, \dots, m\}$, the probability that on input i machine A queries index j is at most $\frac{c}{m}$.*

Intuitively, if a code is insufficiently smooth, so that a particular small subset of indices is queried with too high a probability, then corrupting that subset causes the decoding algorithm to fail with too high a probability. Thus, a locally decodable code must have a certain smoothness. Specifically, [KT00] proved:

Theorem 1. *If $\mathbf{C} : \Sigma_1^n \rightarrow \Sigma_2^m$ is a (q, δ, ϵ)-locally decodable code, then $\mathbf{C}$ is also a $(q, \frac{q}{\delta}, \epsilon)$-smooth code.*

The lower bounds for linear locally decodable codes in [GKST02] are proved by establishing lower bounds for smooth codes. The result for locally decodable codes follows by application of Theorem 1.

Smooth codes are closely related to the concept of information-theoretically secure private information retrieval schemes introduced in [CGKS98]. Briefly, the idea in these constructions is to allow a user to retrieve a value stored in a database in such a way that the database server does not learn significant information about what value was queried. It is easy to see that, in the information-theoretic setting, achieving privacy in this sense with a single database server requires essentially that the entire database be transferred to the user on any query. [CGKS98] showed, however, that by using 2 (non-colluding) servers, one can achieve privacy in this sense with a single round of queries and communication complexity $O(n^{1/3})$. [KT00] observed that if one interprets the query bits sent to the databases as indexes into a 2-query decodable code, then the smoothness parameter of a code can be interpreted as a statistical indistinguishability condition in the corresponding retrieval scheme. In this way, one can construct and analyze smooth codes, and therefore locally decodable codes, from private information retrieval schemes and vice versa. We refer the reader to [GKST02] for a detailed discussion.

The basic technique for proving lower bounds for smooth codes introduced in [KT00] and extended in [GKST02] is to study, for each $i \in \{1, \ldots, n\}$, the *recovery graph* G_i defined on vertex set $\{1, \ldots, m\}$ where (q_1, q_2) is an edge of G_i iff for all $x \in \{0, 1\}^n$,

$$\Pr\left[A^{\mathbf{C}(x)}(i) = x_i | A \text{ queries } (q_1, q_2)\right] > \frac{1}{2}.$$

Such edges are called *good* edges. Then, one shows a lower bound on the size of a maximum matching in the recovery graphs G_i which is a function of the smoothness parameter of $\mathbf{C}$:

Lemma 1 ([KT00], [GKST02]). *If* $\mathbf{C}$ *is a* $(2, c, \epsilon)$-*smooth code with recovery graphs* $\{G_i\}_i$ *then, for every* i, G_i *has a matching of size at least* $\frac{\epsilon m}{c}$.

For *linear* smooth codes, it is easy to see that an edge (q_1, q_2) can be good for x_i iff x_i is a linear combination of q_1, q_2. To simplify matters, one narrows the analysis to codes in which these linear combinations are non-trivial:

Definition 3 (Non-Degenerate Code). *A linear code* $\mathbf{C}$ *is non-degenerate if none of the entries in the range of* $\mathbf{C}$ *is a scalar multiple of an input entry.*

We can assume non-degeneracy in smooth codes with only a constant factor modification in length and recovery parameters:

Theorem 2 ([GKST02]). *For* $n > \frac{4c}{\epsilon}$, *let* $\mathbf{C} : \{0, 1\}^n \to \{0, 1\}^m$ *be a* (q, c, ϵ)-*smooth code. Then there exists a* $(q, c, \frac{\epsilon}{2})$-*smooth code* $\mathbf{C}' : \{0, 1\}^{n'} \to \{0, 1\}^{m'}$ *with* $n' \geq \frac{n}{2}, m' \leq m$ *in which for all* $i \in \{1, \ldots, n'\}, j \in \{1, \ldots, m'\}$, *the* j-*th bit of* $\mathbf{C}'(x)$ *is not a scalar multiple of* x_i.

Putting the pieces together, we have that a non-degenerate $(2, c, \epsilon)$-smooth code has for every $i \in \{1, \ldots, n\}$ a recovery graph G_i containing a matching of size at least $\frac{\epsilon m}{c}$, and for each of the edges (q_1, q_2) in this matching, x_i is in the span of q_1, q_2, but is not a scalar multiple of q_1 or q_2. Thus, the preconditions for the following key result of [GKST02] are satisfied:

Lemma 2. *Let* $q_1, \ldots, q_m$ *be linear functions on* $x_1, \ldots, x_n \in \{0, 1\}^n$ *such that for every* $i \in \{1, \ldots, n\}$ *there is a set* M_i *of at least* γm *disjoint pairs of indices* j_1, j_2 *such that* $x_i = q_{j_1} + q_{j_2}$. *Then* $m \geq 2^{\gamma n}$.

Composing this with the degenerate to non-degenerate reduction of Theorem 2, we have:

Theorem 3 ([GKST02]). *Let* $\mathbf{C} : \{0,1\}^n \to \{0,1\}^m$ *be a* $(2, c, \epsilon)$-*smooth linear code. Then* $m \geq 2^{\frac{\epsilon n}{4c}}$.

Finally, composing this with the locally decodable to smooth reduction, this says:

Theorem 4 ([GKST02]). *Let* $\mathbf{C} : \{0,1\}^n \to \{0,1\}^m$ *be a* $(2, \delta, \epsilon)$-*locally decodable linear code. Then* $m \geq 2^{\frac{\epsilon \delta n}{8}}$.

Note that Lemma 2 yields a lower bound which is exponential in the fraction of vertices in $\{0,1\}^m$ covered by a matching in every recovery graph of the code. Thus, if we can prove a tighter lower bound on the size of these matchings, then we get a corresponding improvement in the exponent in the final lower bound. This is exactly the method used in this paper. In particular, we achieve an optimized bound on the size of the matchings in the recovery graphs by bypassing the reduction to smooth codes and instead arguing directly about locally decodable codes. The resulting direct reduction is strong enough to yield a tight final lower bound.

2   Blocking Distributions and Matchings

In this section, we prove a combinatorial theorem regarding the relationship between the probability that an edge of a graph sampled from any distribution intersects any vertex-subset of a given size, and the size of a maximum matching in the graph. This is the primary technical tool which allows us to optimize the lower bounds of [GKST02]. Further, the analysis seems independently interesting and may be applicable in other contexts.

Let $G(V, E)$ be an undirected graph on n vertices, $w : E \to \mathbf{R}^+$ a probability distribution on the edges of G, $\mathcal{W}$ the set of all such distributions, and S a subset of V. Our concern in this section is to establish a bound on the following parameter of G based on the size of a maximum matching in G:

Definition 4 (Blocking Probability). *Let* X^w *denote a random edge of G sampled according to distribution w. Define the* blocking probability $\beta_\delta(G)$ *as*

$$\beta_\delta(G) = \min_{w \in \mathcal{W}} \left(\max_{S \subseteq V, |S| \leq \delta n} \Pr\left[X^w \cap S \neq \emptyset \right] \right).$$

One can think of $\beta_\delta(G)$ as the value of a game in which the goal of the first player (the decoding algorithm) is to sample an edge from G which avoids a vertex in a δn-set selected by the second player (the channel adversary), whose goal is to maximize the probability of blocking the edge selected by the first player.

For a graph G, the *independence number* $\alpha(G)$ of G is the size of a maximum independent set of vertices in G, and the *defect* $d(G)$ of G is the number of vertices left uncovered in a maximum matching in G. We begin our analysis by observing that $d(G)$ is a lower bound on $\alpha(G)$. We then define a relaxation of the optimization problem for the blocking probability on graphs with a given independence number. For this relaxed problem, we define a special family of distributions and show that some distribution in this family optimizes the blocking probability. Finally, we exhibit a lower bound on the blocking probability of a particular set of δn vertices with respect to any distribution in this family of distributions.

Lemma 3. *For a graph G, $\alpha(G) \geq d(G)$.*

Proof. Choose any maximum matching in G. The vertices left uncovered in this matching must be an independent set, for an edge between any of these vertices would allow us to increase the size of the matching by at least one.

Define the graph $K(n, \alpha)$ with vertex set

$$K_1(n, \alpha) \cup K_2(n, \alpha), |K_1(n, \alpha)| = \alpha n, |K_2(n, \alpha)| = (1 - \alpha)n,$$

such that the edge set of $K(n, \alpha)$ is the union of the edge set of the complete bipartite graph with bipartition $(K_1(n, \alpha), K_2(n, \alpha))$ and the $(1 - \alpha)n$-clique on $K_2(n, \alpha)$.

Lemma 4. *Let G be a graph with defect αn. Then*

$$\beta_\delta(G) \geq \beta_\delta(K(n, \alpha)).$$

Proof. By Lemma 3, G has an independent set S of size at least αn. With a labeling of vertices of $K(n, \alpha)$ which sets $K_1(n, \alpha)$ to an arbitrary αn-subset of S, it is easy to see that the edge set of $K(n, \alpha)$ contains the edge set of G. Therefore, the optimization of w on $K(n, \alpha)$ is a relaxation of the optimization of w on G (a distribution w on G can be expressed as a distribution w' on $K(n, \alpha)$ in which any edge of $K(n, \alpha)$ not in G has probability 0). The claim follows.

We will focus on the following special class of distributions on $K(n, \alpha)$ and show that the blocking probability of $K(n, \alpha)$ is always optimized by some distribution in this class:

Definition 5 ((λ_1, λ_2)-Symmetric Distribution). *An edge distribution w on the graph $K(n, \alpha)$ is (λ_1, λ_2)-symmetric if for every edge $e \in (K_1(n, \alpha), K_2(n, \alpha))$, $w(e) = \lambda_1$, and for every edge $e \in (K_2(n, \alpha), K_2(n, \alpha))$, $w(e) = \lambda_2$.*

Lemma 5. *Let $w_1, \ldots, w_k$ be edge distributions on G such that*

$$\max_{S \subseteq V, |S| \leq \delta n} \Pr\left[X^{w_i} \cap S \neq \emptyset\right] = \beta_\delta(G).$$

Then for any convex combination of the distributions $w = \sum_i \gamma_i w_i$,

$$\max_{S \subseteq V, |S| \leq \delta n} \Pr\left[X^w \cap S \neq \emptyset\right] = \beta_\delta(G).$$

Proof. For every $S \subseteq V$,

$$\Pr[X^w \cap S \neq \emptyset] = \sum_i \gamma_i \Pr[X^{w_i} \cap S \neq \emptyset]$$

since this is simply the sum over edge weights of edges of G incident to S. By the condition on the w_i, for any subset S with $|S| \leq \delta n$,

$$\Pr[X^w \cap S \neq \emptyset] \leq \sum_i \gamma_i \beta_\delta(G)$$

$$= \beta_\delta(G) \sum_i \gamma_i$$

$$= \beta_\delta(G).$$

Therefore,

$$\max_{S \subseteq V, |S| \leq \delta n} \Pr[X^w \cap S \neq \emptyset] \leq \beta_\delta(G).$$

However, by definition of $\beta_\delta(G)$, this must be at least $\beta_\delta(G)$. Therefore,

$$\max_{S \subseteq V, |S| \leq \delta n} \Pr[X^w \cap S \neq \emptyset] = \beta_\delta(G).$$

The *automorphism group* of a graph G is the set of permutations π on the vertices of G such that $(\pi(i), \pi(j)) \in E \iff (i, j) \in E$. Let Γ be the automorphism group of $K(n, \alpha)$.

Lemma 6. *There exists a (λ_1, λ_2)-symmetric distribution w such that*

$$\max_{S \subseteq V, |S| \leq \delta n} \Pr[X^w \cap S \neq \emptyset] = \beta_\delta(K(n, \alpha)).$$

Proof. Let w' be any distribution which optimizes the blocking probability of $K(n, \alpha)$. It is obvious that if w' is such a distribution, then so is $\pi(w')$ for $\pi \in \Gamma$ (where we extend the action of Γ to the edges of G in the natural way). By Lemma 5, the distribution

$$w = \frac{1}{|\Gamma|} \sum_{\pi \in \Gamma} \pi(w')$$

optimizes the blocking probability of $K(n, \alpha)$. We claim that w is a (λ_1, λ_2)-symmetric distribution: For any edge $e \in E$ and $\sigma \in \Gamma$,

$$w(e) = \frac{1}{|\Gamma|} \sum_{\pi \in \Gamma} w'(\pi(e))$$

$$= \frac{1}{|\Gamma|} \sum_{\pi \in \Gamma} w'(\pi\sigma(e))$$

$$= w(\sigma(e))$$

where the second step is the usual group-theoretic trick of permuting terms in summations over Γ. Therefore, if $e, e' \in E$ are in the same orbit under the action of Γ, $w(e) = w(e')$. It is easy to verify that Γ is the direct product of the group of permutations of the vertices of $K_1(n, \alpha)$ and $K_2(n, \alpha)$, and so there are exactly two edge-orbits of $K(n, \alpha)$ under Γ, one consisting of the edges $(K_1(n, \alpha), K_2(n, \alpha))$ and the other $(K_2(n, \alpha), K_2(n, \alpha))$. This is exactly the condition for a (λ_1, λ_2)-symmetric distribution.

Finally, we need to compute a lower bound on the blocking probability for a (λ_1, λ_2)-symmetric distribution:

Lemma 7. *Let w be a (λ_1, λ_2)-symmetric distribution on $K(n, \alpha)$. Then there exists a subset $S \subseteq V$ with $|S| \leq \delta n$ such that*

$$\Pr\left[X^w \cap S \neq \emptyset\right] \geq \min\left(\frac{\delta}{1 - \alpha}, 1\right).$$

Proof. We will study a blocking set which selects any δn vertices of $K_2(n, \alpha)$. Note that, by (λ_1, λ_2)-symmetry, it does not matter which δn vertices we select. Further, we can assume that $\delta < 1 - \alpha$, for if $\delta \geq 1 - \alpha$ we can cover all of $K_2(n, \alpha)$ and thereby achieve blocking probability 1.

Placing a blocking set in this manner and summing up over edges and weights, we achieve blocking probability

$$(\delta n)(\alpha n)\lambda_1 + \frac{1}{2}(\delta n)(\delta n - 1)\lambda_2 + (\delta n)(1 - \alpha - \delta)n\lambda_2.$$

Since w is a probability distribution, we must have

$$(\alpha n)(1 - \alpha)n\lambda_1 + \frac{1}{2}(1 - \alpha)n\left((1 - \alpha)n - 1\right)\lambda_2 = 1.$$

Using this to eliminate λ_1 from the first expression, we obtain blocking probability

$$\delta\left(\frac{1}{1 - \alpha} + \frac{1}{2}n^2(1 - \alpha - \delta)\lambda_2\right).$$

Since $\delta < 1 - \alpha$, the second term in the sum is positive (and, obviously, optimized when $\lambda_2 = 0$), so the blocking probability must be at least

$$\frac{\delta}{1 - \alpha}.$$

It is now easy to prove our main result:

Theorem 5. *Let G be a graph with defect αn. Then*

$$\beta_\delta(G) \geq \min\left(\frac{\delta}{1 - \alpha}, 1\right).$$

Proof. By Lemma 4, $\beta_\delta(G) \geq \beta_\delta(K(n,\alpha))$. By Lemma 6, the blocking probability of $K(n,\alpha)$ is optimized by some (λ_1, λ_2)-symmetric distribution. By Lemma 7, there exists a subset of δn vertices which blocks any such distribution with probability at least $\min\left(\frac{\delta}{1-\alpha}, 1\right)$. Therefore, $\beta_\delta(G) \geq \beta_\delta(K(n,\alpha)) \geq \min\left(\frac{\delta}{1-\alpha}, 1\right)$.

2.1 Probabilistic Proof of Theorem 5

An anonymous referee observed the following probabilistic proof of Theorem 5. This argument does not characterize the optimal strategies for the blocking game, as in our original analysis, but is sufficient to prove our ultimate result.

Fix an arbitrary edge-distribution w on $K(n,\alpha)$ and, for $\delta < 1 - \alpha$ as before, select a subset S of δn vertices of $K_2(n,\alpha)$ uniformly at random. The resulting blocking probability β can be written as a sum $\beta = \sum_e \beta_e$ over edges e, where β_e is a random variable with value w_e if S intersects e, or 0 otherwise. By linearity of expectation,

$$E(\beta) = \sum_e E(\beta_e) = \sum_e w_e \Pr\left[S \cap e \neq \emptyset\right]$$

where the randomness is over the selection of the subset S. Clearly, S intersects each edge e with probability at least $\frac{\delta n}{(1-\alpha)n} = \frac{\delta}{1-\alpha}$, so this expectation is at least

$$\sum_e w_e \frac{\delta}{1-\alpha} = \frac{\delta}{1-\alpha} \sum_e w_e = \frac{\delta}{1-\alpha}.$$

In particular, there must exist some subset S achieving this expectation, proving the theorem. In fact, our original analysis shows that, for the natural family of symmetric optimal strategies w, *every* δn-subset S of $K_2(n,\alpha)$ achieves this blocking probability.

3 Lower Bounds

In this section, we apply Theorem 5 to our original problem of finding lower bounds for locally decodable linear codes. We also present a construction of a family of 2-query decodable linear codes with size matching our bounds within a constant factor in the exponent.

3.1 Degenerate to Non-degenerate Reduction

We require a degenerate to non-degenerate reduction analogous to Theorem 2. Note that we cannot use Theorem 2 directly as this argues about smooth codes, whereas the point in our analysis is to bypass the use of smooth codes.

Theorem 6. *Let* $\mathbf{C} : \{0,1\}^n \rightarrow \{0,1\}^m$ *be a* $(2, \delta, \epsilon)$*-locally decodable linear code where n is large enough so that* $\frac{2(n+1)}{2^n} \leq \delta/201$. *Then there exists a non-degenerate* $(2, \frac{\delta}{2.01}, \epsilon)$*-locally decodable code* $\mathbf{C}' : \{0,1\}^n \rightarrow \{0,1\}^{2m}$.

Proof. Write the ith entry y_i of the codeword as $y_i = a_i \cdot x$ where $a_i \in \{0,1\}^n$. A straightforward probabilistic argument shows that there exists a vector $r \in \{0,1\}^n$ such that the Hamming weights of r and $a_i + r$ are at least 2 for a fraction at least $\left(1 - \frac{2(n+1)}{2^n}\right)$ of the y_i. Let S be the set of y_i satisfying this property. Note that for $y_i \in S$, $(a_i + r) \cdot x$ is a not a scalar multiple of an input entry. We form a non-degenerate code $\mathbf{C}' : \{0,1\}^n \rightarrow \{0,1\}^{2m}$ from $\mathbf{C}$ by setting $y_i' = (a_i + r) \cdot x$ for $y_i \in S$, $y_i' = (1, \ldots, 1) \cdot x$ for all other indices, and adding a set of m codeword bits $y_i'' = r \cdot x$ for all $i \in \{1, \ldots, m\}$.

   We claim that $\mathbf{C}'$ is a $(2, \frac{\delta}{2.01}, \epsilon)$-locally decodable code. Let A be a recovery algorithm for $\mathbf{C}$, and recall that an edge for A can be good only if the answer of A is a linear combination of the entries it queries. Without loss of generality, we can assume that A only queries good edges (otherwise, we can ignore the answers to the queries and output a random coin flip). We implement a recovery algorithm A' for $\mathbf{C}'$ as follows: If A takes a non-trivial linear combination of queries y_i, y_j, then A' simulates A but executes queries y_i', y_j'; if A is the identity on a query y_i, then A' makes queries y_i', y_i'' and takes the (non-trivial) linear combination $y_i' + y_i''$, which for $i \in S$ equals $(a_i + r) \cdot x + r \cdot x = a_i \cdot x = y_i$. Finally, we note that if at most $\frac{\delta}{2.01}$ entries of a codeword of $\mathbf{C}'$ are corrupted, then A' exactly simulates the behavior of A when interacting with some code word with at most

$$\frac{\delta}{2.01}2m + |\overline{S}| \leq \frac{200}{201}\delta m + \frac{2(n+1)}{2^n}m \leq \frac{200}{201}\delta m + \frac{1}{201}\delta m = \delta m$$

corrupt entries. By the decoding condition on A, A' succeeds with probability at least $\frac{1}{2} + \epsilon$.

   So, we can essentially assume that we have a non-degenerate locally decodable code.

3.2   Lower Bound for $(2, \delta, \epsilon)$-Locally Decodable Linear Codes on $\mathbf{Z}_2$

Theorem 7. *Let* $\mathbf{C} : \{0,1\}^n \rightarrow \{0,1\}^m$ *be a* (q, δ, ϵ)*-locally decodable linear code for* $0 < \delta, \epsilon < \frac{1}{2}$. *Then for sufficiently large n,* $m \geq 2^{\frac{1}{4.03}\frac{\delta}{1-2\epsilon}n}$.

Proof. By Theorem 6, for sufficiently large n, the existence of $\mathbf{C}$ implies the existence of a non-degenerate $(2, \frac{\delta}{2.01}, \epsilon)$-locally decodable code $\mathbf{C}' : \{0,1\}^n \rightarrow \{0,1\}^{2m}$. As before, we can assume that the recovery algorithm A' for $\mathbf{C}'$ only queries good edges. On one hand, for all $i \in \{1, \ldots, n\}$ and $y \in \{0,1\}^{2m}$ such that $d(y, \mathbf{C}'(x)) \leq \frac{\delta}{2.01}(2m)$,

$$\Pr\left[A'^y(i) \neq x_i\right] \leq \frac{1}{2} - \epsilon.$$

On the other, if $\alpha(2m)$ is the minimum over all $i \in \{1, \ldots, n\}$ of the defect in the recovery graph G_i of A', then

$$\Pr\left[A'^y(i) \neq x_i\right] \geq \frac{1}{2} \frac{\delta/2.01}{1-\alpha}$$

for, by Theorem 5, there exists a fraction $\frac{\delta}{2.01}$ of vertices S such that an adversary which sets the values of S to random coin flips causes A' to read a blocked edge, and therefore have probability $\frac{1}{2}$ of outputting an incorrect response, with probability at least $\frac{\delta/2.01}{1-\alpha}$. Therefore,

$$\frac{1}{2} \frac{\delta/2.01}{1-\alpha} \leq \frac{1}{2} - \epsilon \implies \alpha \leq 1 - \frac{\delta/2.01}{1-2\epsilon}$$

which is equivalent to saying that there exists for all $i \in \{1, \ldots, n\}$ a set of at least $\frac{1}{2} \frac{\delta/2.01}{1-2\epsilon}(2m)$ disjoint pairs of indices j_1, j_2 such that $x_i = q_{j_1} + q_{j_2}$. Then by Lemma 2, $2m \geq 2^{\frac{1}{2} \frac{\delta/2.01}{1-2\epsilon} n}$ or $m \geq 2^{\frac{1}{4.02} \frac{\delta}{1-2\epsilon} n - 1}$ which is at least, say, $2^{\frac{1}{4.03} \frac{\delta}{1-2\epsilon} n}$ for sufficiently large n.

3.3 Matching Upper Bound

Finally, we show that the lower bound of Theorem 7 is optimal within a constant factor in the exponent. The following construction was observed earlier by Luca Trevisan.

The *Hadamard code* on $x \in \{0,1\}^n$ is given by

$$y_i = a_i \cdot x$$

where a_i runs through all 2^n vectors in $\{0,1\}^n$. Hadamard codes are locally decodable with 2 queries as, for any $i \in \{1, \ldots, n\}$ and $r \in \{0,1\}^n$,

$$x_i = r \cdot x + (r + e_i) \cdot x = e_i \cdot x$$

where e_i is the ith unit vector in $\{0,1\}^n$. It is easy to see that the recovery graphs of this code are perfect matchings on the n-dimensional hypercube, and the code has recovery parameter $\epsilon = \frac{1}{2} - 2\delta$.

For given δ, ϵ, let $c = \frac{1-2\epsilon}{4\delta}$. It can be shown that for feasible values of δ, ϵ, $1 - 2\epsilon \geq 4\delta$ so that $c \geq 1$. We divide the input bits into c blocks of $\frac{n}{c}$ bits, and encode each block with the Hadamard code on $\{0,1\}^{\frac{n}{c}}$. The resulting code has length $\frac{1-2\epsilon}{4\delta} 2^{\frac{4\delta}{1-2\epsilon} n}$ which is, say, less than $2^{4.01 \frac{\delta}{1-2\epsilon} n}$ for sufficiently large n. Finally, since each code block has at most a fraction $c\delta$ of corrupt entries, the code achieves recovery parameter

$$\frac{1}{2} - 2c\delta = \frac{1}{2} - 2\left(\frac{1-2\epsilon}{4\delta}\right)\delta = \epsilon$$

as required.

Acknowledgements. The author thanks Luca Trevisan for many helpful discussions and the anonymous referees for several corrections and observations, including the probabilistic argument in Section 2.1.

References

[CGKS98]   B. Chor, O. Goldreich, E. Kushilevitz, and M. Sudan. Private information retrieval. *Journal of the ACM*, 45(6):965–982, 1998.

[GKST02]   O. Goldreich, H. Karloff, L. Schulman, and Luca Trevisan. Lower bounds for linear locally decodable codes and private information retrieval. In *Proc. of the 17th IEEE CCC*, 2002.

[KT00]   J. Katz and Luca Trevisan. On the efficiency of local decoding procedures for error-correcting codes. In *Proc. of the 32nd ACM STOC*, 2000.

Counting and Sampling H-Colourings[*]

Martin Dyer[1], Leslie A. Goldberg[2], and Mark Jerrum[3]

[1] School of Computing, University of Leeds, Leeds LS2 9JT, United Kingdom,
dyer@comp.leeds.ac.uk,
http://www.comp.leeds.ac.uk/~dyer/
[2] Department of Computer Science, University of Warwick, Coventry, CV4 7AL,
United Kingdom,
leslie@dcs.warwick.ac.uk,
http://www.dcs.warwick.ac.uk/~leslie/
[3] Division of Informatics, University of Edinburgh, JCMB, The King's Buildings,
Edinburgh EH9 3JZ, United Kingdom,
mrj@dcs.ed.ac.uk,
http://www.dcs.ed.ac.uk/~mrj/

Abstract. For counting problems in #P which are "essentially self-reducible", it is known that sampling and approximate counting are equivalent. However, many problems of interest do not have such a structure and there is already some evidence that this equivalence does not hold for the whole of #P. An intriguing example is the class of H-colouring problems, which have recently been the subject of much study, and their natural generalisation to vertex- and edge-weighted versions. Particular cases of the counting-to-sampling reduction have been observed, but it has been an open question as to how far these reductions might extend to any H and a general graph G. Here we give the first completely general counting-to-sampling reduction. For every fixed H, we show that the problem of approximately determining the partition function of weighted H-colourings can be reduced to the problem of sampling these colourings from an approximately correct distribution. In particular, any rapidly-mixing Markov chain for sampling H-colourings can be turned into an FPRAS for counting H-colourings.

1 Introduction

Jerrum, Valiant and Vazirani [13] showed that for *self-reducible* problems in #P, approximate counting and approximate sampling are of similar computational complexity. In particular, a problem has a *fully polynomial randomised approximation scheme* (FPRAS) if and only if it has a *fully polynomial approximate sampler* (FPAS). The techniques of [13] have been applied even to problems that

[*] This work was partially supported by the EPSRC grant "Sharper Analysis of Randomised Algorithms: a Computational Approach", the EPSRC grant GR/R44560/01 "Analysing Markov-chain based random sampling algorithms" and the IST Programme of the EU under contract numbers IST-1999-14186 (ALCOM-FT) and IST-1999-14036 (RAND-APX).

do not seem to be self-reducible, and a generalization of [13] was given by Dyer and Greenhill [4]. In general, however, the situation seems more complicated, as exemplified by the following observation of Brightwell and Goldberg [1].

Observation 1. *There exists a problem in #P which has an FPRAS but no FPAS, unless there is a polynomial time algorithm for computing the discrete logarithm.*

Proof. Consider the problem with instances $(p, r, C(p, r), y)$, where $C(p, r)$ is a certificate that p is a prime with primitive root r, and $y \in \{1, \ldots, p - 1\}$. The input can be verified in polynomial time. (See Section 10.2 and Example 12.2 of [14].) The solution set is defined to be $\{x \mid 0 \leq x \leq p - 2, r^x = y \pmod{p}\}$. This problem trivially has an FPRAS, since the solution set is always of size 1 exactly. Furthermore, the problem is in #P, since $r^x \bmod p$ can be computed in polynomial time. However, an FPAS would clearly give a polynomial-time solution to the discrete logarithm problem.

In fact, the proof of Observation 1 does not rely on the details of the discrete logarithm problem. Any *one-way permutation*[1] could be used to construct a #P-problem with an FPRAS but no FPAS. Thus it seems likely that there exist problems in #P which have an FPRAS but no FPAS. On the other hand, it is an open question as to whether, under any reasonable complexity assumption, there exist problems in #P which possess an FPAS but no FPRAS. A candidate problem might be the *orbit-counting problem* [7]. If a sampling algorithm were discovered which did not essentially implement Burnside's lemma, it would be unclear how to use it for approximate counting.

Despite these issues, it is widely believed that approximate counting and approximate sampling are inter-reducible for most, or even all, "reasonable" problems in #P. H-colouring[2] provides a convenient setting for investigating this issue. On the one hand, reductions between approximate counting and sampling are known for several of the best-known instances of H-colouring. These include the (usual) vertex-colouring problem [11] (see also section 3 below) and the independent set problem or, more generally, its vertex-weighted version the *hard core lattice gas* model. (See, for instance, Examples 3.3 and 3.4 in [4].) On the other hand, straightforward attempts to apply the method of [13] to H-colouring seem to fail.

Dyer, Jerrum, and Vigoda [6] have shown how to extend the counting-to-sampling reduction from the vertex-colouring setting to the H-colouring setting, but their proof works only if H is dismantleable (which is quite a strong restriction, see [2]) and the input graph, G, has bounded degree. This paper extends their result to any H and to general graphs G. We show that, for every fixed H,

---

[1] The definition of a "one-way permutation" is beyond our scope — think of a one-way function which, for each n, is a permutation on inputs of size n. Details can be found in [10].

[2] See Section 2 for a precise definition of this and other basic notions mentioned in this introduction.

the problem of approximately-counting H-colourings can be reduced to the problem of sampling H-colourings from an approximately-correct distribution. Thus, the MCMC method is applicable to H-colouring. In particular, any rapidly-mixing Markov chain for sampling H-colourings can be turned into an FPRAS for counting H-colourings. In fact, we express our results in the more general setting from Section 1.1 of [5] in which vertices and edges of H may have weights. Thus, we show that an algorithm for sampling from the Gibbs distribution leads to an FPRAS for the partition function.

The other direction is still open. The natural reduction from sampling H-colourings to counting H-colourings suffers from the defect that the resultant counting sub-problems correspond to *list colouring* problems rather than to unrestricted colouring problems. Thus, sampling may be reduced to the problem of (approximately) counting *list H-colourings*, but possibly not to the (presumably easier) problem of counting H-colourings. Thus, it is not clear for which graphs H negative sampling results such as [3,9] yield negative results for approximability. Approximate counting could be easier than approximate sampling for H-colouring. Note that for almost every H, it is #P-hard to *exactly* count H-colourings (see [5]).

2   Definitions and Statement of Theorem 2

Our definitions are from Section 1.1 of [5]. Let $H = (V(H), E(H))$ be a fixed graph. We will allow H to have self-loops, but not multiple edges between a pair of vertices. Let $V(H) = \{c_1, \ldots, c_h\}$. We refer to the vertices of $V(H)$ as "colours". Every colour c_j has a weight $\lambda_{c_j} > 0$. If an unordered pair of colours (c_i, c_j) is in $E(H)$ then it has a weight $\lambda_{c_i, c_j} > 0$. Otherwise, it has zero weight, i.e. $\lambda_{c_i, c_j} = 0$. Let $\lambda_{\max}$ be the maximum of all vertex and edge weights in H.

Suppose that σ is a function from $V(G)$ to $V(H)$, where G is a simple graph, without multiple edges or self-loops. We assign the weight $w_\sigma(G)$ to σ, where $w_\sigma(G)$ is given by

$$w_\sigma(G) = \prod_{v \in V(G)} \lambda_{\sigma(v)} \prod_{(u,v) \in E(G)} \lambda_{\sigma(u), \sigma(v)}.$$

Note that $w_\sigma(G) > 0$ if and only if σ is a *homomorphism* from G to H. (A homomorphism from G to H is just a function σ from $V(G)$ to $V(H)$ which has the property that for every edge (u, v) of G, $(\sigma(u), \sigma(v))$ is an edge of H. A homomorphism from G to H is also known as an "H-colouring of G".) Let $\Omega_H(G)$ be the set of H-colourings of G. That is,

$$\Omega_H(G) = \{\sigma : V(G) \to V(H) \mid w_\sigma(G) > 0\}.$$

The partition function $Z_H(G)$ is given by

$$Z_H(G) = \sum_{\sigma \in \Omega_H(G)} w_\sigma(G). \tag{1}$$

The Gibbs distribution on H-colourings of G is the distribution in which each colouring σ has probability

$$\pi_{H,G}(\sigma) = \frac{w_\sigma(G)}{Z_H(G)}.$$

If u is a vertex of G and c_i is a colour in $V(H)$, we use the notation $Z_H(G)\{u \to c_i\}$ to denote $\sum_{\sigma \in \Omega_H(G), \sigma(u)=c_i} w_\sigma(G)$. We will use similar notation when we want to restrict more vertices of G to have particular colours.

As a technical matter, we can assume without loss of generality that there are not distinct colours $c_\alpha \in V(H)$ and $c_\beta \in V(H)$ with identical edge weights. That is, we do not have c_α and c_β such that, for all i, $\lambda_{c_\alpha,c_i} = \lambda_{c_\beta,c_i}$. It is straightforward to see that any two such colours can be treated as a single colour with effective vertex weight $\lambda_{c_\alpha} + \lambda_{c_\beta}$.

Since we are interested in computation (which is inherently discrete), we will assume that all of the weights λ_{c_j} and λ_{c_i,c_j} are rational. Now suppose that K is the least common multiple of the denominators of all of the positive weights. Consider what happens when replace the weights with $\hat{\lambda}_{c_j} = K\lambda_{c_j}$ and $\hat{\lambda}_{c_i,c_j} = K\lambda_{c_i,c_j}$. The weight of a colouring is then $\hat{w}_\sigma(G) = K^{n+m}w_\sigma(G)$, where $n = |V(G)|$ and $m = |E(G)|$. Similarly, $\hat{Z}_H(G) = K^{n+m}Z_H(G)$ and $\hat{\pi}_{H,G}(\sigma) = \pi_{H,G}(\sigma)$. Thus, we can assume without loss of generality that all weights λ_{c_j} and λ_{c_i,c_j} are natural numbers. We will make this assumption in the rest of this paper. See [4] and [8] for a further discussion of this issue.

We will consider the complexity of the following problems.

Name. H-PARTITION.
Instance. A graph G.
Output. The value of the partition function $Z_H(G)$.

Name. H-GIBBSSAMPLE.
Instance. A graph G.
Output. An H-colouring σ of G chosen from distribution $\pi_{H,G}$.

Note that if all vertex and edge weights of H are set to 1 then H-PARTITION is simply the problem of counting H-colourings of G and H-GIBBSSAMPLE is the problem of sampling an H-colouring of G uniformly at random.

Figure 1 gives an example. The triangle xyz forces x to be coloured with a or one of its neighbours. It follows that, for large N, there are $\Theta(10^N)$ colourings where x is coloured a and v is coloured b, and $\Theta(9^N)$ other colourings. Thus "almost all" colourings are of the former type.

A *randomised approximation scheme* for H-PARTITION is a randomised algorithm that takes as input a graph G and an error tolerance $\varepsilon > 0$, and outputs a number $\hat{Z} \in \mathbb{N}$ (a random variable of the "coin tosses" made by the algorithm) such that

$$\Pr\left[e^{-\varepsilon}Z_H(G) \leq \hat{Z} \leq e^\varepsilon Z_H(G)\right] \geq \frac{3}{4}. \tag{2}$$

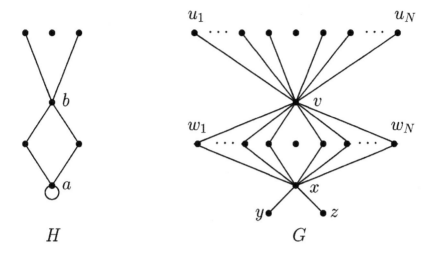

Fig. 1. An H-colouring problem.

The algorithm is a *fully polynomial randomised approximation scheme*, or *FPRAS*, if it runs in time bounded by a polynomial in $|V(G)|$ and ε^{-1}.

In this paper we will simplify the presentation of our FPRAS by presenting it in a slightly different form. Our randomised algorithm will take as input an n-vertex graph G and an error tolerance $\varepsilon > 0$. With probability at least $1 - 2^{-n^5}$, it will succeed. In this case, the running time will be bounded from above by a polynomial in n and ε^{-1}. Also, it will output a number $\widehat{Z} \in \mathbb{N}$ such that

$$\Pr\left[e^{-\varepsilon} Z_H(G) \le \widehat{Z} \le e^{\varepsilon} Z_H(G)\right] \ge \frac{7}{8}. \tag{3}$$

If the algorithm fails, the running time might be as large as $\mathrm{poly}(n, \varepsilon^{-1})\, 2^{\binom{n'}{2}} \times |V(H)|^{n'}$, where $n' \in O(n^2)$. Note that the expected running time of our algorithm is at most a polynomial in n and ε^{-1}. Furthermore, our algorithm can be converted into a standard FPRAS by truncating long runs after polynomially many steps (and outputting an arbitrary answer after truncation).

The total variation distance between two distributions π and π' on a countable set Ω is given by

$$\mathrm{d_{TV}}(\pi, \pi') = \frac{1}{2} \sum_{\omega \in \Omega} |\pi(\omega) - \pi'(\omega)| = \max_{A \subseteq \Omega} |\pi(A) - \pi'(A)|.$$

An *approximate sampler* [4,12,13] for H-GIBBSSAMPLE is a randomised algorithm that takes as input a graph G and an accuracy parameter $\varepsilon \in (0, 1]$ and gives an output (a random variable) such that the variation distance between the output distribution of the algorithm and the Gibbs distribution $\pi_{H,G}$ is at most ε. The algorithm is a *fully polynomial approximate sampler (FPAS)* if its running time is bounded from above by a polynomial in $|V(G)|$ and $\log(\varepsilon^{-1})$.

Theorem 2. *If there is an FPAS for H-GIBBSSAMPLE then there is an FPRAS for H-PARTITION.*

3   An Easy Reduction

Our general strategy will be to reduce G to a tree by removing edges one by one, but unfortunately the reduction is not straightforward. We will need to attach "gadgets" to the vertices of G in order to exclude some undesirable colourings. These are discussed in section 4 below. But first, to illustrate some of the difficulties, we will sketch a simpler reduction, which suffices for two special cases of counting *unweighted* H-colourings. These are problems in which either *every* or *no* vertex of H has a loop. The usual vertex colouring problem provides an example.

Recall that $h = |V(H)|$. If G has h or fewer vertices, we will count its H-colourings by exhaustive enumeration. Otherwise, by applying the pigeonhole principle to any subset of $V(G)$ of size $(h + 1)$, there must exist two vertices $u, v \in V(G)$ such that

$$\Pr\left(\sigma(u) = \sigma(v)\right) = \sum_{\sigma:\sigma(u)=\sigma(v)} \pi_{H,G}(\sigma) \geq \binom{h+1}{2}^{-1}.$$

Take sufficiently many samples to locate any pair u, v with $\Pr(\sigma(u) = \sigma(v)) \geq 1/h^2$. Now let G_{uv} be the graph obtained from G by identifying u and v as a single vertex uv. Parallel edges may be removed from G_{uv} since all edge weights of H are 1. However, there may be a loop on the vertex uv, which means it must be coloured with a looped vertex of H. In the case where H has no looped vertices, the situation does not arise (u and v will not be adjacent in G). In the case where H has all looped vertices, the uv loop is no restriction and we may remove the loop. By sampling colourings of G, we can estimate the ratio $\tau_{uv} = |\Omega_H(G_{uv})|/|\Omega_H(G)| \geq 1/h^2$. Now we estimate $|\Omega_H(G_{uv})|$ recursively, and hence estimate $|\Omega_H(G)|$ as $|\Omega_H(G_{uv})|/\tau_{uv}$.

This reduction is clearly invalid if H has vertex weights, since the vertex uv must receive a squared weighting in G_{uv}. Thus G itself becomes vertex-weighted. To proceed further, we must assume that we can sample H-colourings when G is a vertex-weighted graph. Similarly, if H has edge weights, the parallel edges in G_{uv} are significant, and we are soon obliged to deal with edge-weighted G. Even in the case where all weights are 1, but H has both looped and unlooped vertices, the reduction may be invalid, as illustrated by Figure 1. Here $y, z \in V(G)$ are both coloured a with frequency almost one-fifth. But, if we identify y and z, the vertex yz has a loop, signifying that it can only be coloured with the looped vertex $a \in V(H)$. If we ignore the loop on yz (in order to make G_{yz} a simple graph), the number of H-colourings of G_{yz} explodes to $\Theta(25^N)$, by colouring both x and v with b. The ratio τ_{yz} is now an exponentially large quantity rather than a fraction.

In general, this reduction is valid if we assume that the class of graphs from which G can be chosen includes the class from which H can be chosen. But this assumption is not true of the H-colouring problem as usually stated, particularly in its weighted variants. Therefore we need to proceed more carefully to obtain our reduction.

4   Gadgets

Let $t = 2|V(H)|$. Let P be a path of $2t$ edges from some vertex A to some vertex B. For any colour $c_i \in V(H)$ and any colour $c_j \in V(H)$, recall that $Z_H(P)\{A \to c_i, B \to c_j\}$ denotes $\sum_{\sigma \in \Omega_H(P), \sigma(A)=c_i, \sigma(B)=c_j} w_\sigma(P)$. Let $\delta(c_i, c_j)$ be the quantity

$$\delta(c_i, c_j) = \frac{Z_H(P)\{A \to c_i, B \to c_j\}}{\lambda_{c_i} \lambda_{c_j}}.$$

The quantity $\delta(c_i, c_j)$ is the total weight of all H-colourings of P which start at colour c_i and end at colour c_j except that we exclude the weight of the colours at the two endpoints. For any colour c_i, let $\delta(c_i) = \delta(c_i, c_i)$.

We will be assuming that H is connected and that it has more than one vertex. Thus, every colour $c_i \in V(H)$ has at least one neighbour so $\delta(c_i) > 0$. If all vertices $c_i \in V(H)$ and $c_j \in V(H)$ have $\delta(c_i) = \delta(c_j)$ we will define $\delta^*(H) = 1$. Otherwise, we define $\delta^*(H)$ to be the following positive quantity.

$$\delta^*(H) = \min\left\{\lg\left(\frac{\delta(c_i)}{\delta(c_j)}\right) \;\middle|\; c_i \in V(H), c_j \in V(H), \delta(c_i) > \delta(c_j)\right\}.$$

We will use the following technical lemma (cf. [5]).

Lemma 1. *If $\delta(c_i) \geq \delta(c_j)$ and $j \neq i$ then $\delta(c_i, c_j) < \delta(c_i)$.*

Proof. If c_i dominates c_j in the sense that $\lambda_{c_i c_\alpha} > \lambda_{c_j c_\alpha}$ for all α then the lemma follows from the definition of δ. Suppose that c_i does not dominate c_j. Let W be a symmetric $h \times h$ matrix in which the entry in row i and column j is λ_{c_i, c_j}. Let Λ be the diagonal matrix in which the entry in row i and column i is λ_{c_i}. Let Ψ be the positive diagonal matrix such that $\Psi^2 = \Lambda$. Let $[\,\cdot\,]_i$ denote the ith column of a matrix and $[\,\cdot\,]_{ij}$ its (i,j)th element. Note that

$$\delta(c_i, c_j) = [(W\Lambda)^{2t-1}W]_{ij} = [\Psi^{-1}(\Psi W \Psi)^{2t}\Psi^{-1}]_{ij}.$$

Since $\Psi W \Psi$ is symmetric, it can be written as $U^T L U$ where L is diagonal and U is orthonormal (i.e., $U^T U = I$).

Now

$$\delta(c_i, c_j) = [\Psi^{-1}(\Psi W \Psi)^{2t}\Psi^{-1}]_{ij} = [\Psi^{-1}U^T L^{2t}U\Psi^{-1}]_{ij} = [L^t U \Psi^{-1}]_i^T[L^t U \Psi^{-1}]_j$$
$$\leq \sqrt{[\Psi^{-1}U^T L^{2t}U\Psi^{-1}]_{ii}[\Psi^{-1}U^T L^{2t}U\Psi^{-1}]_{jj}} = \sqrt{\delta(c_i)\delta(c_j)} \leq \delta(c_i)$$

using Cauchy-Schwartz, with strict inequality unless $[L^tU\Psi^{-1}]_i$ is a multiple of $[L^tU\Psi^{-1}]_j$. But this condition is true if and only if $[LU\Psi^{-1}]_i$ is a multiple of $[LU\Psi^{-1}]_j$, which is true if and only if $[\Psi^{-1}U^TLU\Psi^{-1}]_i$ is a multiple of $[\Psi^{-1}U^TLU\Psi^{-1}]_j$, i.e. $[W]_i$ is a multiple of $[W]_j$. This is impossible since c_i does not dominate c_j and there are not distinct colours with identical edge weights (see Section 2). □

We will let $\delta'(H)$ be the following positive quantity.

$$\delta'(H) = \min\left\{ \lg\left(\frac{\delta(c_i)}{\delta(c_i,c_j)}\right) \,\Big|\, i \neq j, \delta(c_i) \geq \delta(c_j) \right\}.$$

Finally, we let $\delta^\dagger(H) = \min(\delta^*(H), \delta'(H))$.

Let S be a subset of $V(H)$ such that for every colour $c_i \in S$ and every colour $c_j \in S$, $\delta(c_i) = \delta(c_j)$. Let $\delta(S)$ denote $\delta(c_i)$ for $c_i \in S$.

A graph H' with a designated vertex u' is said to be "good" for S if it satisfies the following properties.

   i. For every $c \in S$, $Z_H(H')\{u' \to c\} > 0$, and
   ii. for every colour $c \in V(H)$ with $\delta(c) > \delta(S)$, $Z_H(H')\{u' \to c\} = 0$.

Informally, (H', u') is good for S if every colour $c \in S$ can be applied to u' in a valid colouring but no vertex of higher δ-value can be applied to u'.

The set S is said to be "good" if there exists an (H', u') which is good for S and κ_S is then defined to be the minimum number of vertices in a graph H' such that some pair (H', u') is good for S. κ is defined to be the maximum of κ_S over all good S. We will not assume that κ is known in our algorithm, but we will refer to it in our analysis.

Suppose we have a (fixed-size) graph H' with a designated vertex u' and we want to check whether the pair (H', u') is good for S. We do this by examining each of the (at most $|V(H)|^{|V(H')|}$) colourings in $\Omega_H(H')$. Thus, we can check every graph H' of size at most n' and every possible designated vertex u' by examining at most $(n')^2 2^{\binom{n'}{2}} |V(H)|^{n'}$ colourings.

As we observed in section 3, the function of these gadgets (H', u') is to exclude unwanted colourings. The triangle $H' = xyz$ in Figure 1, with distinguished vertex $u' = x$, illustrates the phenomenon. It has colourings in which x can be coloured a, but none in which it can receive the colour b of larger H-degree. Its attachment to G at x therefore excludes the (otherwise more numerous) colourings of G in which x would be coloured b.

5   Proof of Theorem 2

Suppose that G has multiple connected components, say G_A, G_B and G_C. It is immediate from Equation (1) that $Z_H(G) = Z_H(G_A)Z_H(G_B)Z_H(G_C)$. Thus, we may assume without loss of generality that G is connected. We will do so for the rest of the paper. For connected G, suppose that H has multiple connected components, say H_A, H_B and H_C. Inspection of Equation (1) reveals that $Z_H(G) = Z_{H_A}(G) + Z_{H_B}(G) + Z_{H_C}(G)$. Thus, we may assume without loss of generality that H is also connected. We will do so for the rest of the paper.

Let n denote $|V(G)|$. In the reduction, we will construct a sequence G_0, G_1, $\ldots$, G_p of connected graphs. As long as there is no failure in the (randomised) reduction, the following properties will hold.

(i) $G_0 = G$,
(ii) $Z_H(G_p)$ can be calculated in polynomial time (polynomial in n), and
(iii) the construction of $G_0, \ldots, G_p$ will take polynomial time.

Let

$$\varrho_i = \frac{Z_H(G_i)}{Z_H(G_{i+1})}.$$

Then

$$Z_H(G) = \varrho_0 \varrho_1 \cdots \varrho_{p-1} Z_H(G_p).$$

We will estimate $Z_H(G)$ using the method of Jerrum, Valiant and Vazirani. In particular, we will define a quantity s_i for each i such that

(iv) either ϱ_i or ϱ_i^{-1} is an easily-computable multiple of s_i (so an approximation to s_i gives an approximation to ϱ_i), and
(v) there is an experiment which can be performed using a perfect sampler for H-GIBBSSAMPLE with input G_i or G_{i+1} for which the output is a $0/1$ random variable with mean s_i, and
(vi) there is a polynomial q in n and ε^{-1} such that $s_i{}^{-1} \leq q(n, \varepsilon^{-1})$.

It follows (see the proof of Proposition 3.4 of [12]) that $O(q(n, \varepsilon^{-1}) p \varepsilon^{-2})$ samples taken from an approximate sampler for H-GIBBSSAMPLE with accuracy parameter

$$O\left(\frac{\varepsilon}{q(n, \varepsilon^{-1})p}\right)$$

give a sufficiently accurate approximation to s_i, and hence to ϱ_i. That is, if we use the sampler with this many samples, and multiply our estimates of the ϱ_is, the resulting estimate of $|\Omega_H(G)|$ is within the required accuracy with probability at least $7/8$.

As mentioned earlier, the overall probability of failure in our reduction (which could cause one or more of (i)–(vi) to fail) will be at most $m 2^{-n^6} \leq 2^{-n^5}$, where $m = |E(G)|$. Suppose that $E(G) = \{e_1, \ldots, e_m\}$, and that the edges are ordered in such a way that the graph $(V(G), \{e_1, \ldots, e_{n-1}\})$ is a tree. We will use the notation $u(e_i)$ and $v(e_i)$ to denote the endpoints of the edge e_i. We will now describe the construction of the sequence $G_0, \ldots, G_p$.

In order to shorten our description of the reduction, we will describe the construction of a sequence of graphs

$$\Gamma_0 = G_y, \Gamma_1 = G_{y+1}, \ldots, \Gamma_{2r+2} = G_{y+2r+2},$$

where y is a multiple of $2r + 2$ for some number r to be chosen later. We will assume that for some $j \in \{n, \ldots, m\}$, the graph $\Gamma_0 = G_y$ is identical to $(V(G), \{e_1, \ldots, e_j\})$ except that every vertex $u \in V(G)$ may have some gadgets attached to u in Γ_0. Each gadget is simply a graph H' of size at most $\kappa + 1$. One of the vertices of H' is identified with u. There are $2(m - j)$ gadgets in all. Thus, Γ_0 has $O(n^2)$ vertices. Also, it is connected. As an invariant in the construction, we will guarantee that each graph Γ_i has at least one H-colouring. That is $Z_H(\Gamma_i) > 0$. The first sequence of graphs that we will construct will start with $\Gamma_0 = G_0$ and $j = m$, so all of the invariants will be true initially.

Recall that Γ_0 looks like the graph $(V(G), \{e_1, \ldots, e_j\})$ except for possibly some small gadgets. Our goal will be to remove the edge e_j.

Part 1: Learning about the graph Γ_0 and constructing $H(S)$

Before we can remove e_j it will help us to know which colours in $V(H)$ can be used to colour $u(e_j)$ in Γ_0. More particularly, we would like to know which colours in $V(H)$ are good choices for $u(e_j)$ when we modify Γ_0 by attaching a certain structure to $u(e_j)$.

Thus, we will first define a graph Γ_0' which is the same as Γ_0 except that a certain structure (which we will call F) will be attached to $u(e_j)$. We will then use our sampling oracle to study the colourings of Γ_0', paying particular attention to which colours are applied to vertex $u(e_j)$. Once we know the colours, we will use this information in the construction of $\Gamma_1, \Gamma_2, \ldots$ We start with some definitions. Let M be a straightforward upper bound for $Z_H(\Gamma_0)$. In particular,

$$ M = |V(H)|^{|V(\Gamma_0)|} \lambda_{\max}^{|V(\Gamma_0)| + |E(\Gamma_0)|}. \tag{4} $$

Recall that $t = 2|V(H)|$ and that $\delta^\dagger(H)$ is the minimum of the quantities $\delta^*(H)$ and $\delta'(H)$ from Section 4. Let r be defined by the following equation.

$$ r = \left\lceil \frac{n^7 + \lg(M)}{\delta^\dagger(H)} \right\rceil. $$

For now, the reader should just think of r as being a sufficiently-large polynomial in n. Let F be the graph with the vertex set

$$ V(F) = \{f_0\} \cup \bigcup_{p \in [1, \ldots, r], q \in [1, \ldots, 2t-1]} \{f_{p,q}\} $$

and the edge set $E(F)$ which is defined to be

$$ \bigcup_{p \in [1, \ldots, r]} \{(f_0, f_{p,1})\} \cup \bigcup_{p \in [1, \ldots, r], q \in [1, \ldots, 2t-2]} \{(f_{p,q}, f_{p,q+1})\} \cup \bigcup_{p \in [1, \ldots, r]} \{(f_{p,2t-1}, f_0)\}. $$

F looks like a "flower" with vertex f_0 at the centre and r petals. Each petal is a cycle of length $2t$ which starts and ends at f_0.

Let Γ_0' be a graph constructed from Γ_0 by attaching F. Vertex f_0 of F should be identified with vertex $u(e_j)$.

We now need some notation to describe the colourings of Γ_0' and of Γ_0. For any $d \in \mathbb{R}$, let

$$Z_H(\Gamma_0)\{u(e_j) \to [d]\} = \sum_{c \in V(H), \delta(c)=d} Z_H(\Gamma_0)\{u(e_j) \to c\}.$$

Informally, $Z_H(\Gamma_0)\{u(e_j) \to [d]\}$ is the collective weight of all colourings in which $u(e_j)$ is coloured with a colour with δ-value d.

Define δ to be the quantity such that $Z_H(\Gamma_0)\{u(e_j) \to [\delta]\} > 0$ but, for all $d > \delta$, $Z_H(\Gamma_0)\{u(e_j) \to [d]\} = 0$. Informally, δ is the largest δ-value which can be applied to $u(e_j)$.

Let S^+ be the set of all colours c with $\delta(c) = \delta$ and $Z_H(\Gamma_0)\{u(e_j) \to c\} > 0$. Let S^- be $\{c \in S^+ \mid Z_H(\Gamma_0)\{u(e_j) \to c\} \geq (1/n)Z_H(\Gamma_0)\{u(e_j) \to [\delta]\}\}$. Thus, S^+ is the set of all value-δ colours which may be applied to $u(e_j)$ and S^- is the set of "frequently used" ones. Note that S^- is non-empty since there are fewer than n colours.

We will now describe an experiment which can be performed on Γ_0' to determine the "likely" colours that colour vertex $u(e_j)$. In the reduction, we will perform the experiment to learn about these colours. This knowledge will be used in the construction of Γ_1. Suppose that we run H-GIBBSSAMPLE with input Γ_0' and accuracy parameter $\gamma = 2^{-n^7}$ to collect $s = 2n^8$ samples from $\Omega_H(\Gamma_0')$. Let S be the collection of colours that are assigned to $u(e_j)$ in these samples.

We claim that, except with failure probability at most 2^{-n^6}, we have $S^- \subseteq S \subseteq S^+$. To see that the failure probability is this small first observe that the probability that a colour c with $\delta(c) < \delta$ is in S is at most

$$s\left(\frac{Z_H(\Gamma_0')\{u(e_j) \to c\}}{Z_H(\Gamma_0')} + \gamma\right). \tag{5}$$

Since $\delta(c)$ is the total weight of all colourings of a "petal" of F in which $u(e_j)$ is coloured c, the quantity (5) is at most

$$s\left(\frac{Z_H(\Gamma_0)\{u(e_j) \to c\}\delta(c)^r}{\delta^r} + \gamma\right).$$

Since $\delta^\dagger(H) \leq \delta^*(H)$ (see the definition of $\delta^*(H)$ in Section 4), the definition of r guarantees that the term $\frac{Z_H(\Gamma_0)\{u(e_j) \to c\}\delta(c)^r}{\delta^r} \leq \gamma$. Thus the probability that there exists a colour c with $\delta(c) < \delta$ in S is at most $s\,|V(H)|\,2\gamma$.

Also, the probability that a colour $c \in S^-$ is left out of S is at most

$$\left(1 - \frac{1}{n} + \sum_{c:\delta(c)<\delta} \frac{Z_H(\Gamma_0')\{u(e_j) \to c\}}{Z_H(\Gamma_0')} + \gamma\right)^s \leq \left(1 - \frac{1}{2n}\right)^s \leq \exp(-n^7),$$

so the probability that such a colour exists is at most 2^{-n^6}.

We have shown that, except with failure probability at most 2^{-n^6}, we have $S^- \subseteq S \subseteq S^+$. The reduction now begins searching for a graph $H(S)$ with a designated vertex $u'(S)$ which is good for S. (See Section 4.) If we do not have failure, then the pair $(H(S), u'(S))$ exists and $|V(H(S))| \leq \kappa$. Recall that κ is a constant depending only on H, and not on our input Γ_0. If there is no failure, then our input Γ_0 does provide an upper bound for $|V(H(S))|$ since $|V(H(S))| \leq |V(\Gamma_0)|$. The latter follows from the fact that $(\Gamma_0, u(e_j))$ is good for S. Thus we restrict the search to graphs with at most $|V(\Gamma_0)|$ vertices and the expected time of the search is at most a polynomial in n.

Part 2: Constructing the sequence $\Gamma_1, \dots, \Gamma_{2r+2}$ from Γ_0

In this part we will show how to construct $\Gamma_1, \dots, \Gamma_{2r+2}$ assuming that we did not have failure in Part 1.

First, the graphs $\Gamma_1, \dots, \Gamma_r$ are constructed. For $i \in [0, \dots, r-1]$, Γ_{i+1} is constructed from Γ_i by adding a length-$2t$ cycle $\{u(e_j), f_{i+1,1}, \dots, f_{i+1,2t-1}, u(e_j)\}$ where $f_{i+1,1}, \dots, f_{i+1,2t-1}$ are new vertices.

For every $\sigma \in \Omega_H(\Gamma_i)$, let $\text{ext}(\sigma)$ be the non-empty set

$$\text{ext}(\sigma) = \{\sigma' \in \Omega_H(\Gamma_{i+1}) \mid \forall\, v \in V(\Gamma_i), \sigma(v) = \sigma'(v)\}.$$

($\text{ext}(\sigma)$ is non-empty because every colour in H has at least one neighbour.) For every $\sigma' \in \text{ext}(\sigma)$, let $\hat{w}(\sigma, \sigma') = w_{\sigma'}(\Gamma_{i+1})/w_\sigma(\Gamma_i)$. Note that $\hat{w}(\sigma, \sigma') \geq 1$ since all vertex and edge weights are positive integers. Let $s_i = \varrho_i$. Note that (iv) is satisfied for the graph Γ_i — ϱ_i is s_i, so it is clearly "an easily-computable multiple of s_i". We now wish to establish (v) for the graph Γ_i. We wish to exhibit an experiment which can be performed using a perfect sampler for H-GIBBSSAMPLE with input Γ_i or Γ_{i+1} for which the output is a $0/1$ random variable with mean s_i. Here is the experiment: Choose σ' from $\pi_{H,\Gamma_{i+1}}$. Let σ be the restriction of σ' to $V(\Gamma_i)$. Output 1 with probability $(\hat{w}(\sigma, \sigma') |\text{ext}(\sigma)|)^{-1}$ and 0 otherwise. The probability that a 1 is output is

$$\frac{1}{Z_H(\Gamma_{i+1})} \sum_{\sigma' \in \Omega_H(\Gamma_{i+1})} w_{\sigma'}(\Gamma_{i+1}) \frac{1}{\hat{w}(\sigma, \sigma') |\text{ext}(\sigma)|}$$

$$= \frac{1}{Z_H(\Gamma_{i+1})} \sum_{\sigma \in \Omega_H(\Gamma_i)} \sum_{\sigma' \in \text{ext}(\sigma)} w_\sigma(\Gamma_i) \frac{1}{|\text{ext}(\sigma)|}$$

$$= \frac{Z_H(\Gamma_i)}{Z_H(\Gamma_{i+1})} = s_i.$$

Thus, (v) is satisfied. Finally, we must satisfy (vi). That is, we must show that there is a polynomial q in n and ε^{-1} such that $s_i^{-1} \leq q(n, \varepsilon^{-1})$. Since $Z_H(\Gamma_{i+1}) \leq Z_H(P)Z_H(\Gamma_i)$ where P is a length-$2t$ path, we have $s_i^{-1} \leq Z_H(P)$, so (vi) holds. We have now completed the construction of the graphs $\Gamma_1, \dots, \Gamma_r$ and the argument that these graphs satisfy our requirements. Note that the graph Γ_r is the same as the graph Γ_0' which we considered in Part 1.

Next, the graph Γ_{r+1} is constructed from Γ_r by attaching $H(S)$ to $u(e_j)$, identifying the vertex $u(e_j)$ of Γ_r with the vertex $u'(S)$ in the gadget $H(S)$. That is $V(\Gamma_{r+1}) = V(\Gamma_r) \cup V(H(S))$, but $|V(\Gamma_{r+1})| = |V(\Gamma_r)| + |V(H(S))| - 1$ since the vertex $u(e_j)$ of Γ_r is identified with the vertex $u'(S)$ of $H(S)$. Also,

$E(\Gamma_{r+1}) = E(\Gamma_r) \cup E(H(S))$. Since $|V(H(S))| \le \kappa$, the construction of Γ_{r+1} is fast.

We will now show that (iv), (v) and (vi) hold for Γ_{r+1} (i.e., for $i = r$). Let $s_r = \varrho_r^{-1} Z_H(H(S))^{-1}$. Consider the following experiment. Choose σ from $\pi_H(\Gamma_r)$. Output 1 with probability

$$\sum_{\sigma' \in \text{ext}(\sigma)} \frac{\hat{w}(\sigma, \sigma')}{Z_H(H(S))},$$

where

$$\text{ext}(\sigma) = \{\sigma' \in \Omega_H(\Gamma_{r+1}) \mid \forall\, v \in V(\Gamma_r), \sigma(v) = \sigma'(v)\}$$

as above and $\hat{w}(\sigma, \sigma') = w_{\sigma'}(\Gamma_{r+1})/w_\sigma(\Gamma_r)$. Output 0 otherwise. The probability that 1 is output is

$$\frac{1}{Z_H(\Gamma_r)} \sum_{\sigma \in \Omega_H(\Gamma_r)} w_\sigma(\Gamma_r) \sum_{\sigma' \in \text{ext}(\sigma)} \frac{\hat{w}(\sigma, \sigma')}{Z_H(H(S))} = \frac{1}{Z_H(H(S))} \frac{1}{Z_H(\Gamma_r)} Z_H(\Gamma_{r+1}) = s_r.$$

We must now establish (vi).

$$s_r^{-1} = \frac{Z_H(H(S)) Z_H(\Gamma_r)}{Z_H(\Gamma_{r+1})} \le \frac{Z_H(H(S)) Z_H(\Gamma_r)}{\sum_{c \in S} Z_H(\Gamma_r)\{u(e_j) \to c\}}, \tag{6}$$

where the inequality follows from the fact that $(H(S), u(e_j))$ is good for S. Now from our analysis in Part 1 we have $Z_H(\Gamma_r) = Z_H(\Gamma_0') \le 2\delta^r Z_H(\Gamma_0)\{u(e_j) \to [\delta]\}$. Also, since $S^- \subseteq S$,

$$\sum_{c \in S} Z_H(\Gamma_r)\{u(e_j) \to c\} \ge \delta^r \sum_{c \in S^-} Z_H(\Gamma_0)\{u(e_j) \to c\}$$

$$\ge \delta^r \sum_{c \in S^-} (1/n) Z_H(\Gamma_0)\{u(e_j) \to [\delta]\}, \tag{7}$$

where the final inequality follows from the definition of S^-. Thus,

$$s_r^{-1} \le \frac{Z_H(H(S)) Z_H(\Gamma_r)}{\sum_{c \in S} Z_H(\Gamma_r)\{u(e_j) \to c\}} \le Z_H(H(S)) 2n,$$

which gives us (vi).

The graph Γ_{r+2} is constructed from Γ_{r+1} as follows. Let H' be a new copy of the gadget $H(S)$. Let w be the designated vertex of H' so that (H', w) is good for S. To form Γ_{r+2}, we join together Γ_{r+1} and H'. Thus, $V(\Gamma_{r+2}) = V(\Gamma_{r+1}) \cup V(H')$. We do the "joining" by deleting the edges $(f_{i,2t-1}, u(e_j))$ (for $i \in [1, \ldots, r]$) and adding in edges $(f_{i,2t-1}, w)$ for each such i. Also, we delete the edge $(u(e_j), v(e_j))$ and add in edge $(w, v(e_j))$. See Figure 2.

Now let $s_{r+1} = \varrho_{r+1}$. Consider the following experiment. Choose σ' from the distribution $\pi_H(\Gamma_{r+2})$. If $\sigma'(w) = \sigma'(u(e_j))$ then output 1 with probability

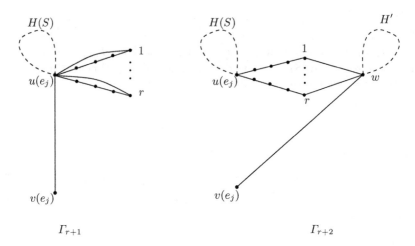

Fig. 2. The construction of Γ_{r+2}

$(Z_H(H')\{w \to \sigma'(w)\})^{-1}$. Otherwise output a 0. The probability that 1 is output is

$$\frac{1}{Z_H(\Gamma_{r+2})} \sum_{\sigma' \in \Omega_H(\Gamma_{r+2}), \sigma'(w)=\sigma'(u(e_j))} w_{\sigma'}(\Gamma_{r+2}) \frac{1}{Z_H(H')\{w \to \sigma'(w)\}} = \varrho_{r+1}$$
$$= s_{r+1}.$$

We must now establish (vi).
Now

$$Z_H(\Gamma_{r+2}) = \sum_{c_\alpha, c_\beta} Z_H(\Gamma_{r+2})\{u(e_j) \to c_\alpha, w \to c_\beta\}. \qquad (8)$$

Also,

$$Z_H(\Gamma_{r+2})\{u(e_j) \to c_\alpha, w \to c_\beta\} \le M Z_H(H(S)) Z_H(H')\delta(c_\alpha, c_\beta)^r,$$

where M is our upper bound for $Z_H(\Gamma_0)$ from Equation (4). On the other hand, we have just shown in our proof of (6) and (7) that

$$Z_H(\Gamma_{r+1}) \ge Z_H(\Gamma_{r+1})\{u(e_j) \to [\delta]\} \ge \delta^r(1/n).$$

Thus if $c_\alpha \neq c_\beta$

$$\frac{Z_H(\Gamma_{r+2})\{u(e_j) \to c_\alpha, w \to c_\beta\}}{Z_H(\Gamma_{r+1})} \le \frac{nM Z_H(H(S)) Z_H(H')\delta(c_\alpha, c_\beta)^r}{\delta^r}$$
$$\le n Z_H(H(S)) Z_H(H')\gamma, \qquad (9)$$

by the definition of r since $\delta^\dagger(H) \le \delta'(H)$ (see the definition of $\delta'(H)$ in Section 4).

Finally,

$$Z_H(\Gamma_{r+2})\{u(e_j) \to c_\alpha, w \to c_\alpha\} \le Z_H(\Gamma_{r+1})\{u(e_j) \to c_\alpha\}Z_H(H')$$
$$\le Z_H(\Gamma_{r+1})Z_H(H'). \tag{10}$$

Putting together (8) and (9) and (10) we get

$$s_{r+1}^{-1} = \varrho_{r+1}^{-1}$$
$$= \frac{Z_H(\Gamma_{r+2})}{Z_H(\Gamma_{r+1})}$$
$$\le \sum_{c_\alpha \neq c_\beta} (n Z_H(H(S))Z_H(H')\gamma) + \sum_{c_\alpha} Z_H(H'),$$

which gives us (vi).

For $i \in \{0, \ldots, r-2\}$, graph $\Gamma_{r+2+i+1}$ is constructed from graph Γ_{r+2+i} by deleting vertices $f_{i+1,1}, \ldots, f_{i+1,2t-1}$ (and the edges adjacent to theses vertices).

To establish Property (v) we define a notion which is analogous to ext(σ). In particular, for every $\sigma \in \Omega_H(\Gamma_{r+2+i+1})$ let bext(σ) be the non-empty set

$$\text{bext}(\sigma) = \{\sigma' \in \Omega_H(\Gamma_{r+2+i}) \mid \forall\, v \in V(\Gamma_{r+2+i+1}), \sigma(v) = \sigma'(v)\}.$$

For every $\sigma' \in$ bext(σ), let $\hat{w}(\sigma, \sigma') = w_{\sigma'}(\Gamma_{r+2+i})/w_\sigma(\Gamma_{r+2+i+1}) \ge 1$. The following experiment has mean $s_{r+2+i} = \varrho_{r+2+i}^{-1}$. Choose σ' from $\pi_{H,\Gamma_{r+2+i}}$. Let σ be the restriction of σ' to $V(\Gamma_{r+2+i+1})$. With probability $(\hat{w}(\sigma, \sigma') |\text{bext}(\sigma)|)^{-1}$, output 1. Otherwise, output 0. Since $Z_H(\Gamma_{r+2+i}) \le Z_H(P)Z_H(\Gamma_{r+2+i+1})$, we have $\varrho_i \le Z_H(P)$, so (vi) holds.

Note that the graph Γ_{2r+1} is the same as the graph Γ_0 except that the gadget $H(S)$ has been attached to $u(e_j)$ and the edge $(u(e_j), v(e_j))$ has been replaced with the path $u(e_j), f_{r,1}, \ldots, f_{r,2t-1}, w, v(e_j)$ and the gadget H' has been attached to w.

Finally, the graph Γ_{2r+2} is constructed from Γ_{2r+1} by deleting vertices $f_{r,1}, \ldots, f_{r,2t-1}$ (and the edges adjacent to theses vertices).

The proof that Property (v) and Property (vi) hold is similar to what we have just done. The new difficulty is showing that for every $\sigma \in \Omega_H(\Gamma_{2r+2})$, the set

$$\text{bext}(\sigma) = \{\sigma' \in \Omega_H(\Gamma_{2r+1}) \mid \forall\, v \in V(\Gamma_{2r+2}), \sigma(v) = \sigma'(v)\}$$

is non-empty.

Suppose that σ is a colouring $\in \Omega_H(\Gamma_{2r+2})$ in which $u(e_j)$ is coloured with colour a and w is coloured with colour b. We must show that there is a colouring of the path $u(e_j), f_{r,1}, \ldots, f_{r,2t-1}, w$ in which $u(e_j)$ is coloured a and w is coloured b. We will do this by looking at two cases.

Case 1: H is a loopless bipartite graph. Recall that (by construction) Γ_{2r+1} has at least one H-colouring. This means that Γ_{2r+1} is bipartite. Also, $u(e_j)$ and

w are in the same part of the vertex partition of Γ_{2r+1}. The graph Γ_{2r+2} is still connected (by construction) with $u(e_j)$ and w in the same part. This means that a and b are from the same side of H's vertex partition. Since H is connected, there is an even-length path from a to b of length at most $|V_H| - 1$. Thus, there is a walk of length $2t$ from a to b. (Take the path above and go back and forth on the last edge.)

Case 2: H is not a loopless bipartite graph, so it has an odd-length cycle of length at most $|V(H)|$. In this case, let c be some node on the cycle. We will construct an even-length path from a to b of length less than $2t$: First go from a to c using at most $|V(H)| - 1$ edges. Then go from c to b using at most $|V(H)| - 1$ edges. Finally, if the constructed path has odd length, then go around the odd-length cycle in the middle. The total number of edges is at most $3|V(H)| - 2 < 2t$. Once again, we can find a walk of length $2t$ from a to b by going back and forth on the last edge.

This completes the argument that Γ_{2r+2} is properly constructed and it completes the construction of $\Gamma_1, \ldots, \Gamma_{2r+2}$. Thus, we have constructed the sequence

$$\Gamma_0 = G_y, \Gamma_1 = G_{y+1}, \ldots, \Gamma_{2r+2} = G_{y+2r+2}$$

as required. Note that the graph Γ_{2r+2} is identical to $(V(G), \{e_1, \ldots, e_{j-1}\})$ except that every vertex $u \in V(G)$ may have some gadgets attached to u in Γ_{2r+2}. The gadgets that are present in Γ_{2r+2} which were not present in Γ_0 are the new gadget $H(S)$ (of size at most κ) which hangs off of $u(e_j)$ and the new gadget consisting of vertex w and the graph H' (of total size at most $\kappa + 1$) which hangs off of $v(e_j)$. If $j = n$ then we are finished and $y + 2r + 2 = p$. Otherwise, we start Part 1 again with $\Gamma_0 = G_{y'} = G_{y+2r+2}$.

Part 3: Computing $|\Omega_H(G_p)|$

We have now shown how to construct $G_0, \ldots, G_p$. We have shown that our construction satisfies (i), (iii), (iv), (v) and (vi). It remains to show that property (ii) is satisfied – namely, that we can compute $Z_H(G_p)$ in polynomial time (polynomial in n).

By construction, G_p is identical to the tree $T = (V(G), \{e_1, \ldots, e_{n-1}\})$ except that every vertex $u \in V(G)$ may have some gadgets attached to u in G_p. Each gadget is a graph H' of size at most $\kappa + 1$. One of the vertices of H' is identified with u. There are $2(m - n + 1)$ gadgets in all.

We can compute $Z_H(G_p)$ by dynamic programming. For each gadget (H', u') and each colour c, we first compute $Z_H(H')\{u' \to c\}$.

Now consider a rooted version of T. For each vertex $v \in V(G)$, let $G_p(v)$ denote the portion of G_p corresponding to the sub-tree rooted at v in T (including attached gadgets). We can calculate $Z_H(G_p(v))\{v \to c\}$ using the values of $Z_H(G_p(v'))\{v \to c'\}$ for all children v' of v in T and all colours $c' \in V(H)$ and all quantities $Z_H(H')\{u' \to c''\}$.

References

1. G.R. Brightwell and L.A. Goldberg, personal communication.

2. G.R. Brightwell and P. Winkler, Gibbs measures and dismantlable graphs, *J. Combin. Theory Ser. B* 78(1) 141–166 (2000)
3. C. Cooper, M. Dyer and A. Frieze, On Markov chains for randomly H-colouring a graph, *Journal of Algorithms,* **39(1)** (2001) 117–134.
4. M. Dyer and C. Greenhill, Random walks on combinatorial objects. In J.D. Lamb and D.A. Preece, editors, *Surveys in Combinatorics*, volume 267 of *London Mathematical Society Lecture Note Series*, pages 101–136. Cambridge University Press, 1999.
5. M.Dyer and C. Greenhill, The complexity of counting graph homomorphisms. *Random Structures and Algorithms,* **17** (2000) 260–289.
6. M. Dyer, M. Jerrum and E. Vigoda, Rapidly mixing Markov chains for dismantleable constraint graphs. In J. Nesetril and P. Winkler, editors, *Proceedings of a DIMACS/DIMATIA Workshop on Graphs, Morphisms and Statistical Physics*, March 2001, to appear.
7. L.A. Goldberg, Computation in permutation groups: counting and randomly sampling orbits. In J.W.P. Hirschfeld, editor, *Surveys in Combinatorics*, volume 288 of *London Mathematical Society Lecture Note Series*, pages 109–143. Cambridge University press, 2001.
8. L.A. Goldberg, M. Jerrum and M. Paterson, The computational complexity of two-state spin systems, Pre-print (2001).
9. L.A. Goldberg, S. Kelk and M. Paterson, The complexity of choosing an H-colouring (nearly) uniformly at random, To appear in STOC 2002.
10. O. Goldreich, *The Foundations of Cryptography - Volume 1,* (Cambridge University Press, 2001)
11. M. Jerrum, A very simple algorithm for estimating the number of k-colorings of a low-degree graph, *Random Structures and Algorithms,* **7** (1995) 157–165.
12. M. Jerrum, Sampling and Counting. Chapter 3 of *Counting, Sampling and Integrating: Algorithms and Complexity*, Birkhäuser, Basel. (In preparation.)
13. M.R. Jerrum, L.G. Valiant, and V.V. Vazirani, Random generation of combinatorial structures from a uniform distribution, *Theoretical Computer Science,* **43** (1986) 169–188.
14. C.H. Papadimitriou, *Computational Complexity,* (Addison-Wesley, 1994)

Rapidly Mixing Markov Chains for Dismantleable Constraint Graphs[*]

Martin Dyer[1], Mark Jerrum[2], and Eric Vigoda[3]

[1] School of Computing, University of Leeds, Leeds LS2 9JT, United Kingdom.
dyer@comp.leeds.ac.uk
[2] Laboratory for Foundations of Computer Science, University of Edinburgh, JCMB,
The King's Buildings, Edinburgh EH9 3JZ, United Kingdom.
mrj@dcs.ed.ac.uk
[3] Department of Computer Science, University of Chicago, 1100 E 58th Street,
Chicago, IL 60637, USA.
vigoda@cs.uchicago.edu

Abstract. If $G = (V_G, E_G)$ is an input graph, and $H = (V_H, E_H)$ a fixed constraint graph, we study the set Ω of homomorphisms (or colorings) from V_G to V_H, i.e., functions that preserve adjacency. Brightwell and Winkler introduced the notion of *dismantleable* constraint graph to characterize those H whose associated set Ω of homomorphisms is, for every G, connected under single vertex recolorings. Given fugacities $\lambda(c) > 0$ ($c \in V_H$) our focus is on sampling a coloring $\omega \in \Omega$ according to the *Gibbs distribution*, i.e., with probability proportional to $\prod_{v \in V_G} \lambda(\omega(v))$. The *Glauber dynamics* is a Markov chain on Ω which recolors a single vertex at each step, and leaves invariant the Gibbs distribution. We prove that, for each dismantleable H and degree bound Δ, there exist positive constant fugacities on V_H such that the Glauber dynamics has mixing time $O(n^2)$, for all graphs G whose vertex degrees are bounded by Δ.

1 Introduction

Graph homomorphisms provide a natural generalization of many well-studied combinatorial problems, including independent sets and colorings. Our focus is the computational complexity of randomly generating a homomorphism and computing the number of homomorphisms.

We consider an input graph $G = (V_G, E_G)$ with maximum degree Δ and a constraint graph $H = (V_H, E_H)$, where the latter may have loops. Let $n = |V_G|$, $h = |V_H|$, and let $x \sim y$ denote adjacency of a pair of vertices in G or H. Our interest is in the set Ω of H-*colorings* (or homomorphisms) $\sigma : V_G \to V_H$ where $\sigma(v) \sim \sigma(w)$ for all $v \sim w$. Just two of the numerous combinatorial structures that can be expressed as H-colorings are illustrated in Figure 1. It is easily

[*] This work was supported by EPSRC Research grant "Sharper Analysis of Randomised Algorithms: A Computational Approach" and in part by the ESPRIT Project RAND-APX. The last author was partially supported by a Koshland Scholar award from the Weizmann Institute of Science.

J.D.P. Rolim and S. Vadhan (Eds.): RANDOM 2002, LNCS 2483, pp. 68–77, 2002.

verified that when $H = H_{\mathrm{IS}}$ the set Ω of H-colorings of a graph G is in direct correspondence with the set of all independent sets in G; while when $H = H_{3\mathrm{Col}}$, the set Ω consists of all proper 3-colorings of G.

H_{IS} $H_{3\mathrm{Col}}$

Fig. 1. Independent sets (left) and proper 3-colorings (right).

Each color $c \in V_H$ is assigned a *fugacity* $\lambda(c)$. For convenience, we extend λ to colorings $\sigma \in \Omega$ by defining $\lambda(\sigma) = \prod_{v \in V_G} \lambda(\sigma(v))$. We are interested in the *Gibbs* (or "hard-core") probability distribution π over Ω given by $\pi(\sigma) = \lambda(\sigma)/Z$ for all $\sigma \in \Omega$, where the normalising factor $Z = \sum_{\sigma \in \Omega} \lambda(\sigma)$ is the *partition function* for H-colorings of G. The complexity of exactly computing Z has been investigated by Dyer and Greenhill [6] and is well-understood; in fact the problem is #P-complete for every non-trivial H even when we restrict attention to low-degree graphs G and uniform fugacity 1.

The picture appears (at the moment) more vague when we consider approximate computation of Z, though Dyer, Goldberg, Greenhill and Jerrum [7] discuss the case where all fugacities are equal, and uncover a part of the picture. In this paper, we examine the closely related problem of sampling colorings from the Gibbs distribution, or at least a close approximation to that distribution. We identify a class of constraint graphs H for which sampling can be done in polynomial time, for appropriately chosen non-zero fugacities. Since approximate counting is efficiently reducible to sampling for this class of constraint graphs, we obtain a polynomial-time approximation algorithm (technically an FPRAS) for this class of constraint graphs with their associated fugacities.

The typical approach to efficient sampling is to set up an ergodic Markov chain (X_t) on Ω whose stationary distribution is the desired distribution π. By simulating this Markov chain for sufficiently many steps, samples from a distribution arbitrarily close to π may be obtained. There are a number of ways of specifying appropriate transition probabilities for this Markov chain; a basic one is provided by the *Glauber dynamics*, which changes one vertex color at a time, according to the following experiment:

G1. Choose a vertex $v \in V_G$ uniformly at random.
G2. Let $X_{t+1}(w) := X_t(w)$ for all $w \neq v$.
G3. Let $S = \{c \in V_H : c \sim X_t(w) \text{ for all } w \sim v\}$ denote the set of valid colors for v.
G4. Choose the color $X_{t+1}(v)$ randomly from S with probability proportional to its fugacity.

In what circumstances might the Glauber dynamics provide an effective solution to the problem of sampling H-colorings? Certainly we require the state space Ω to be connected in some sense. A Markov chain with finite state space Ω and transition matrix $P : \Omega \times \Omega \rightarrow [0, 1]$ is called *ergodic* if the following conditions hold:

- *Irreducibility*: for all states (colorings) $\sigma, \tau \in \Omega$, there exists a time $t = t(\sigma, \tau)$ such that $P^t(\sigma, \tau) > 0$;
- *Aperiodicity*: for all states σ, $\gcd\{t : P^t(\sigma, \sigma) > 0\} = 1$.

It is a classical theorem from stochastic processes that a finite, ergodic Markov chain has a unique stationary distribution.

In many situations it is easy to verify that a candidate distribution is stationary. Specifically, a distribution π' on Ω that satisfies the so-called detailed balance conditions,

$$\pi'(\sigma)\, P(\sigma, \tau) = \pi'(\tau)\, P(\tau, \sigma) \quad \text{for all } \sigma, \tau \in \Omega, \tag{1}$$

is a stationary distribution. It is straightforward to verify that the Gibbs distribution π satisfies the detailed balance conditions for the Glauber dynamics on H-colorings. Nevertheless, we still need to check that π is the *unique* stationary distribution, i.e., that the Glauber dynamics meets the conditions of irreducibility and aperiodicity.

Since the Glauber dynamics satisfies $P(\sigma, \sigma) > 0$ for all colorings σ (it is always valid to recolor a vertex with the same color as before) it is immediate that the Markov chain specified by the Glauber dynamics is aperiodic. The question of irreducibility is more subtle. Brightwell and Winkler [1] characterized the constraint graphs H for which the Glauber dynamics on H-colorings is irreducible for every input graph G. They call such constraint graphs "dismantleable" and prove a host of equivalent conditions for dismantleability.

For efficient sampling, however, it is not sufficient that the Glauber dynamics converges eventually to the stationary (Gibbs) distribution; we need to know that the rate of convergence is rapid. Our main result is that for every dismantleable H and degree bound Δ there exists a set of non-zero fugacities such that the Glauber dynamics has polynomial mixing time[1] uniformly over graphs G of maximum degree at most Δ. The term *rapidly mixing* is often applied to a Markov chain, such as this one, whose mixing time is polynomial in some natural measure of input size, in this case the order of G.

If H is a complete graph with loops on all vertices then the Glauber dynamics is trivially rapidly mixing for any input graph G. Cooper, Dyer and Frieze [3] proved that, for any other H, there exists a set of fugacities (actually, one may set all fugacities equal to 1) and a sufficiently large degree Δ such that the Glauber dynamics on H-colorings has exponential mixing time, for some infinite

---

[1] A precise definition of mixing time is given in §2, but roughly it is the time t at which the t-step distribution of a Markov chain comes sufficiently close to the stationary distribution π in l_1 distance.

family of Δ-regular input graphs G. In contrast, we prove here that, provided H is dismantleable, there is another set of fugacities such that the Glauber dynamics on H-colorings has polynomial mixing time, on every bounded degree input graph G.

2   Definitions and Results

In graph-theoretic terms, a constraint graph H is said to be *dismantleable* if there exists an ordering $<$ of V_H for which the following holds:

> Let c^* be the smallest vertex (color) in the ordering $<$. Then, for all $c \in V_H \setminus \{c^*\}$, there exists $p(c) < c$ such that $c' \leq c$ and $c' \sim c$ entail $c' \sim p(c)$.

It is quite easy to show that the above condition is sufficient to ensure irreducibility. Fix an ordering on colors (i.e., vertices of H) for which the dismantleability condition holds, with c^* being the least color in the ordering. Starting from an arbitrary coloring $\sigma \in \Omega$, repeating the following procedure reaches the monochromatic coloring $(c^*)^{V_G}$: let c denote the greatest color appearing in σ; recolor all occurrences of c by $p(c)$. Note that the steps can be reversed to get from $(c^*)^{V_G}$ back to the original coloring.[2] That dismantleability is a necessary condition is a little trickier, and we refer the reader to [1].

We are interested in the asymptotic distance of the Glauber dynamics from stationarity. The traditional measure of distance in this context is (total) variation distance, defined as:

$$d_{\mathrm{TV}}(P^t(\sigma, \cdot), \pi) = \frac{1}{2} \sum_{\tau \in \Omega} |P^t(\sigma, \tau) - \pi(\tau)|,$$

where σ is the initial state (coloring). Our focus is the time to get close to stationarity, known as the mixing time. For an initial state $\sigma \in \Omega$ and $0 \leq \varepsilon < 1$, let

$$T_\sigma(\varepsilon) = \min\{t : d_{\mathrm{TV}}(P^t(\sigma, \cdot), \pi) \leq \varepsilon\}.$$

The *mixing time* is defined as

$$T(\varepsilon) = \max_{\sigma \in \Omega} T_\sigma(\varepsilon).$$

For the purposes of the analysis we do not consider the Glauber dynamics directly, but consider instead a variant dynamics with a more restricted set of possible transitions. (It is perhaps counterintuitive that the proof of rapid mixing might be simplified by limiting the available transitions!) In brief, we allow only

---

[2] Incidentally, the procedure just described provides a demonstration that $(c^*)^{V_G}$ is a valid H-coloring of G, provided *some* valid H-coloring exists. Thus, except in some trivial situations, the vertex c^* must be looped in H.

transitions in which a color c is replaced by its "parent" $p(c)$ or vice versa. If we were simply to restrict the available transitions in this way, while retaining the original fugacities $\lambda(c)$, then we would alter the stationary distribution. To correct for this, we assign to each color $c \in V_H$ a weight $\mu(c)$, which is in general different from $\lambda(c)$. The weight function μ influences transition probabilities just as λ does, but we avoid calling $\mu(c)$ a "fugacity", since the relation of the weights μ to the stationary distribution of the Markov chain is not quite the same as that of the fugacities λ to the Gibbs distribution.

Fix an ordering $<$ and parents $p(\cdot)$ consistent with the definition of dismantleability. Then our new Markov chain (X_t) has transitions $X_t \rightarrow X_{t+1}$ defined by the following experiment:

V1. Choose a vertex $v \in V_G$ uniformly at random. Let c denote the current color of v.
V2. Let $X_{t+1}(w) := X_t(w)$ for all $w \neq v$.
V3. Let $R = R(c) = \{c' \in V_H : c' = p(c), c' = c, \text{ or } c = p(c')\}$ denote the set of "relatives" of color c.
V4. Choose the color c' randomly from R with probability proportional to its weight $\mu(c')$. Provided it leads to a valid H-coloring, set $X_{t+1}(v) := c'$; otherwise set $X_{t+1}(v) := c$.

As before, the variant dynamics is an ergodic Markov chain provided H is dismantleable. The stationary distribution under the variant dynamics has probabilities $\pi(\sigma)$ proportional to $\prod_{v \in V_G} \mu(\sigma(v)) \, \mu(R(\sigma(v)))$, where $\mu(R(c)) = \sum_{c' \in R(c)} \mu(c')$. This may be verified from the detailed balance conditions (1). Suppose σ and τ agree at all vertices except v, at which $c = \sigma(v)$ and $c' = \tau(v)$. (Note that σ and τ must be of this form if $P(\sigma, \tau)$ is to be non-zero.) Then we have $P(\sigma, \tau) = \mu(c')/\mu(R(c))$ and $P(\tau, \sigma) = \mu(c)/\mu(R(c'))$. Hence

$$\frac{P(\sigma, \tau)}{P(\tau, \sigma)} = \frac{\mu(c') \, \mu(R(c'))}{\mu(c) \, \mu(R(c))} = \frac{\pi(\tau)}{\pi(\sigma)},$$

as required. Thus the stationary distribution is the Gibbs distribution with fugacities $\lambda(c) = \mu(c) \, \mu(R(c))$. We can now state our main theorem.

Theorem 1. *For every input graph $G = (V_G, E_G)$ with maximum degree Δ and dismantleable constraint graph $H = (V_H, E_H)$, there exists a set of weights (depending only on Δ and $h = |V_H|$) such that the variant dynamics (as defined in V1–V4 above) has mixing time $O(n \log n)$. Specifically, $T(\varepsilon) = O\big(n(\log n + \log \varepsilon^{-1})\big)$.*

Corollary 1. *Under the conditions of Theorem 1, there exists a set of fugacities (depending only on Δ and h) such that the Glauber dynamics (as defined in G1–G4 above) has mixing time $O(n^2)$. Specifically, $T(\varepsilon) = O\big(n(n + \log \varepsilon^{-1})\big)$.*

The corollary follows easily from the Diaconis and Saloff-Coste technique [4] for comparing the associated Dirichlet forms of the Markov chains. Indeed a slightly weaker bound on mixing time, with $n \log n + \log \varepsilon^{-1}$ replacing $n + \log \varepsilon^{-1}$, can be obtained simply by substituting the mixing time bound from Theorem 1 into a general comparison theorem of Randall and Tetali [9, Prop. 4]. However, by applying exactly the same method, but working from first principles, we can avoid the factor $\log n$. The argument is this. Applying Sinclair's [10, Prop. 1(ii)] to the mixing time bound of Theorem 1 with $\varepsilon = n^{-1}$, we discover that the spectral gap of the variant dynamics is $\Omega(n^{-1})$. Now the spectral gaps of the Glauber and variant dynamics are the same to within a constant factor: this can be seen by inspecting the variational characterisation of the second largest eigenvalue. Corollary 1 then follows by plugging the $\Omega(n^{-1})$ bound on spectral gap into [10, Prop. 1(i)].

Since the above comparison argument relies only on each possible transition in the variant dynamics being matched (in general with different probability) in the Markov chain under comparison, Corollary 1 holds for other single-site update rules, e.g., Metropolis. We conjecture that the true mixing time here is also $O\big(n(\log n + \log \varepsilon^{-1})\big)$ but the proof of this (if true) appears rather more complex than our proofs of Theorem 1 and Corollary 1.

As mentioned in the introduction, the existence of an efficient sampling procedure for certain structures usually entails the existence of an efficient approximation algorithm for the partition (or generating) function for those structures. The current situation is no exception.

Corollary 2. *Under the same conditions as Theorem 1, there is a fully polynomial randomized approximation scheme (FPRAS) for the partition function Z of H-colorings of G.*

For the definition of FPRAS, and also the reduction from approximate counting to sampling required to establish Corollary 2, see for example Jerrum's survey article [8], specifically §2. The reduction is given there in the context of usual (proper) colorings, but it works almost without change for H-colorings, when H is dismantleable.[3] To verify the reduction in the current context it is necessary to check that the ratios ρ_i appearing in that reduction are bounded away from 0. This can be done using a straightforward extension of the argument used to prove ergodicity of the Glauber dynamics.

Finally, note that many constraint graphs H are covered both by Theorem 1 and by the result of Cooper et al. [3]. In other words, there are graphs H — the simplest being K_2 with a single loop, which corresponds to independent sets in the input graph G — for which the mixing time of the Glauber dynamics is either polynomial or exponential, depending on the fugacities.

3 Coupling

We prove the main theorem via coupling. A coupling of a Markov chain is a joint evolution of two copies of the chain, designed to minimize the time till the

---

[3] It is in fact possible to reduce approximate counting to sampling without restriction on the graph H, but the reduction is then substantially more involved and, crucially in the context of Corollary 2, does not preserve the bound Δ on vertex degrees [5].

copies coalesce. In order to be a valid coupling we need that individually each copy behaves faithfully, and once the pair coalesce they evolve together. To be precise, a (Markovian) coupling is a Markov chain P' on $\Omega \times \Omega$ which satisfies the following conditions:

$$\sum_{\tau' \in \Omega} P'((\sigma, \tau), (\sigma', \tau')) = P(\sigma, \sigma') \quad \text{for all } \sigma, \sigma', \tau \in \Omega;$$

$$\sum_{\sigma' \in \Omega} P'((\sigma, \tau), (\sigma', \tau')) = P(\tau, \tau') \quad \text{for all } \sigma, \tau, \tau' \in \Omega;$$

and

$$P'((\sigma, \sigma), (\sigma', \sigma')) = P(\sigma, \sigma') \quad \text{for all } \sigma, \sigma' \in \Omega.$$

Our goal is to define a coupling which, in expectation, makes progress with respect to an appropriately defined distance metric after every coupled transition. Defining and analyzing a coupling for an arbitrary pair of states is typically a complicated task; however, the path coupling lemma of Bubley and Dyer simplifies matters. In particular, it suffices to focus on colorings which differ by a single transition of the dynamics, referred to as *adjacent* colorings. We write $\sigma \sim_v \tau$ to indicate that the pair of colorings σ, τ differ only at vertex v, and $\sigma \sim \tau$ if $\sigma \sim_v \tau$ for some vertex v.

We analyze this set of adjacent colorings with respect to the following distance metric. Each color $c \in V_H$ will be assigned a distance weight $d(c)$. For colorings $\sigma \sim_v \tau$, where without loss of generality $\tau(v) = p(\sigma(v))$, we will assign distance $d(\sigma, \tau) = d(\sigma(v))$. For any $\sigma, \tau \in \Omega$, let $\rho(\sigma, \tau)$ denote the collection of paths η such that $\sigma = \eta_0 \sim \eta_1 \sim \cdots \sim \eta_\ell = \tau$, where $\ell(\eta)$ is the length (in terms of number of transitions) of η. We define the distance between an arbitrary pair of states as the total distance along a shortest path:

$$d(\sigma, \tau) = \min_{\eta \in \rho(\sigma, \tau)} \sum_{0 \le i < \ell(\eta)} d(\eta_i, \eta_{i+1}).$$

For a pair of colorings σ, τ and a coupling (σ, τ), let σ', τ' denote the resulting pair of colorings after the coupled transition. The specialization of the path coupling lemma to our setting is as follows.

Lemma 1 (Bubley and Dyer [2]). *If there exists a $\beta < 1$ and a coupling for all $\sigma \sim \tau$ such that*

$$\mathbb{E}[d(\sigma', \tau')] < \beta\, d(\sigma, \tau),$$

then the mixing time is bounded by

$$T(\varepsilon) \le \frac{\log(D/\varepsilon)}{1 - \beta},$$

where $D = \max_{\sigma, \hat{\sigma} \in \Omega} d(\sigma, \hat{\sigma})$.

In our application, we shall discover that $\beta = 1 - \Theta(1/n)$ and $D = \Theta(n)$, leading easily to Theorem 1.

4   Proof of Theorem 1

Fix an ordering on the colors $c_1 < c_2 < \cdots < c_h$ for which the dismantleability condition holds. Set $d(c_i) = 2(\Delta + 2)^{i+1}$ and let $d_{\max} = d(c_h)$. Now set $\mu_i = \mu(c_i) = (d_{\max} + 1)^{-i}$, for $i = 1, \ldots, h$. Observe that the inequality

$$\frac{\sum_{k>i} \mu_k}{\mu_i} < \frac{1}{d_{\max}}$$

holds for all i.

Fix a pair of colorings $\sigma \sim_v \tau$ and once again let σ', τ' denote the resulting colorings after our coupled transition. To prove the theorem we need to demonstrate a coupling for which $\mathbb{E}[d(\sigma', \tau')] - d(\sigma, \tau)$ is negative and bounded away from zero.

We couple the two chains so that both chains attempt to modify the same vertex at every step. Therefore, for a vertex x, it is well defined to let

$$\mathbb{E}_x[d(\sigma', \tau')] = \mathbb{E}\left[d(\sigma', \tau') \mid \text{vertex } x \text{ is selected by the coupled process}\right].$$

Observe that the distance metric does not change when we recolor a vertex sufficiently far from v. Specifically, we only need to consider recolorings of v or a neighbor w of v. In summary, we have

$$n\left(\mathbb{E}[d(\sigma', \tau')] - d(\sigma, \tau)\right) = \left(\mathbb{E}_v[d(\sigma', \tau')] - d(\sigma, \tau)\right)$$
$$+ \sum_{w:w \sim v} \left(\mathbb{E}_w[d(\sigma', \tau')] - d(\sigma, \tau)\right). \tag{2}$$

Of all the colors that are valid for $\sigma'(x)$, let $c_\sigma^*(x)$ denote the one of greatest weight. Note that either $c_\sigma^*(x) = \sigma(x)$ or $c_\sigma^*(x) = p(\sigma(x))$, and that, in the latter case, $c_\sigma^*(x)$ has the greatest weight among the various colors that may be proposed. Either way, if vertex x is selected, then x will acquire color $c_\sigma^*(x)$ unless a color of lower weight than $c_\sigma^*(x)$ is proposed. Thus, conditioned on vertex x being selected, we have

$$\Pr[\sigma'(x) \neq c_\sigma^*(x)] \leq \frac{\sum_{c > c_\sigma^*(x)} \mu(c)}{\mu(c_\sigma^*(x))} < \frac{1}{d_{\max}}. \tag{3}$$

Defining $c_\tau^*(x)$ similarly, the analogous inequality holds with respect to τ as well.

In the light of inequality (3) and its analogue for τ we may complete the definition of the coupled process as follows. With probability $1 - 1/d_{\max}$ we couple $\sigma'(x) = c_\sigma^*(x)$ with $\tau'(x) = c_\tau^*(x)$. We call such a coupled move a *type A transition*. With the remaining probability of $1/d_{\max}$ the two chains independently recolor x (using the residual distributions), a *type B transition*. In the latter case, it is clear that $d(\sigma', \sigma) \leq d_{\max}$ and $d(\tau', \tau) \leq d_{\max}$ which implies

$$\mathbb{E}_x[d(\sigma', \tau')| \text{ type B transition}] \leq 2d_{\max} + d(\sigma, \tau). \tag{4}$$

We now proceed to bound the expected change in distance from a type A transition. It is clear that the distance after the type A transition on x is maximized when $c^*_\sigma(x) \neq c^*_\tau(x)$.

Recall that, by convention, $\tau(v) = p(\sigma(v))$, rather than vice versa. Consider the recoloring of a neighbor w of v. We begin by proving that $c^*_\sigma(w) \neq c^*_\tau(w)$ implies $\sigma(w) < \sigma(v)$. Suppose $\sigma(w) \geq \sigma(v)$. Vertex w has the same color in both chains, therefore colors $c^*_\sigma(w)$ and $c^*_\tau(w)$ are either $\sigma(w)$ or $p(\sigma(w))$. Since $\sigma(w) \sim \sigma(v)$ and $\sigma(w) \sim \tau(v)$, from the definition of dismantleability and our ordering on V_H we know $p(\sigma(w)) \sim \sigma(v)$ and $p(\sigma(w)) \sim \tau(v)$. Similarly if $u \neq v$ is any other neighbor of w, then dismantleability implies $p(\sigma(w)) \sim \sigma(u) = \tau(u)$. Therefore, recoloring vertex w to color $p(\sigma(w))$ is valid in both chains or neither and hence $c^*_\sigma(w) = c^*_\tau(w)$.

Now suppose $c^*_\sigma(w) \neq c^*_\tau(w)$. In addition to the fact $\sigma(w) < \sigma(v)$, it is clear that $c^*_\sigma(w) = \sigma(w)$ and $c^*_\tau(w) = p(\sigma(w))$. Following the type A transition for w we have $\sigma' = \sigma$ and $d(\tau', \tau) = d(\sigma(w)) \leq d(\sigma(v))/(\Delta + 2)$.

In summary, we have proven

$$\mathbb{E}_w[d(\sigma', \tau')| \text{ type A transition}] \leq d(\sigma, \tau) \left[1 + \frac{1}{\Delta + 2} \right]. \tag{5}$$

Combining (4) and (5) — recalling that a type B transition occurs with probability at most $1/d_{\max}$ — yields

$$\mathbb{E}_w[d(\sigma', \tau')] - d(\sigma, \tau) \leq 2 + \frac{d(\sigma, \tau)}{\Delta + 2}. \tag{6}$$

We complete the proof by considering the effect of recoloring v by a type A transition. Since $p(\sigma(v)) = \tau(v)$, it is clear that $p(\sigma(v))$ is a valid color for both $\sigma'(v)$ and $\tau'(v)$. Moreover, $c^*_\sigma(v) = p(\sigma(v))$ and $c^*_\tau(v)$ is either $p(\sigma(v))$ or $p(p(\sigma(v)))$. In other words, $d(\sigma', \tau') > 0$ implies $c^*_\tau(v) = p(p(\sigma(v)))$. In which case we have $d(\sigma', \tau') \leq d(p(\sigma(v))) \leq d(\sigma, \tau)/(\Delta + 2)$. Restating our bound:

$$\mathbb{E}_v[d(\sigma', \tau')| \text{ type A transition}] \leq \frac{d(\sigma, \tau)}{\Delta + 2}.$$

The above inequality together with (4) implies

$$\mathbb{E}_v[d(\sigma', \tau')] - d(\sigma, \tau) \leq 2 + d(\sigma, \tau) \left[\frac{1}{d_{\max}} + \frac{1}{\Delta + 2} - 1 \right]$$

$$\leq 3 - d(\sigma, \tau) \left[1 - \frac{1}{\Delta + 2} \right]. \tag{7}$$

Let $\Delta(v)$ denote the degree of vertex v. Putting inequalities (6) and (7) into (2), and recalling $d(\sigma, \tau) = d(\sigma(v))$, we obtain

$$n\big(\mathbb{E}[d(\sigma', \tau')] - d(\sigma, \tau) \big) \leq \Delta(v) \left[2 + \frac{d(\sigma, \tau)}{\Delta + 2} \right] + \left[3 - \frac{(\Delta + 1)\, d(\sigma, \tau)}{\Delta + 2} \right]$$

$$\leq (2\Delta + 3) - \frac{d(\sigma(v))}{\Delta + 2}$$

$$\leq (2\Delta + 3) - 2(\Delta + 2)$$

$$= -1.$$

Since $\sigma \sim \tau$ implies $d(\sigma, \tau) \leq d_{\max}$, we conclude

$$\mathbb{E}[d(\sigma', \tau')] \leq (1 - 1/nd_{\max})\, d(\sigma, \tau),$$

and hence we may take $\beta = 1 - 1/nd_{\max}$ in Lemma 1. If $\hat{\sigma} = (c_1)^{V_G}$, we know that $d(\sigma, \hat{\sigma}) \leq nhd_{\max}$ for all $\sigma \in \Omega$, by the path of transitions described in §2. Therefore, we may take $D = 2nhd_{\max}$ in Lemma 1. Theorem 1 now follows.

References

1. G. R. Brightwell and P. Winkler. Gibbs measures and dismantleable graphs. *J. Combin. Theory Ser. B*, 78(1):141–166, 2000.
2. R. Bubley and M. Dyer. Path coupling, Dobrushin uniqueness, and approximate counting. In *38th Annual Symposium on Foundations of Computer Science*, pages 223–231, Miami Beach, FL, October 1997. IEEE.
3. C. Cooper, M. Dyer, and A. Frieze. On Markov chains for randomly H-coloring a graph. *J. Algorithms*, 39(1):117–134, 2001.
4. P. Diaconis and L. Saloff-Coste. Comparison theorems for reversible Markov chains. *The Annals of Applied Probability*, 3(3):696–730, 1993.
5. M. Dyer, L.A. Goldberg and M. Jerrum. Counting and Sampling H-colourings. To appear in *Proceedings of the 6th International Workshop on Randomization and Approximation Techniques in Computer Science*, Springer-Verlag LNCS, September 2002.
6. M. Dyer and C. Greenhill. The complexity of counting graph homomorphisms. *Random Structures Algorithms*, 17(3-4):260–289, 2000.
7. M. E. Dyer, L. A. Goldberg, C. S. Greenhill, and M. R. Jerrum. On the relative complexity of approximate counting problems. In *Proceedings of APPROX 2000*, Lecture Notes in Computer Science vol. 1913, pages 108–119, Springer Verlag, 2000.
8. M. Jerrum. Mathematical foundations of the Markov chain Monte Carlo method. In *Probabilistic Methods for Algorithmic Discrete Mathematics* (M. Habib, C. Mc-Diarmid, J. Ramirez-Alfonsin & B. Reed, eds), Algorithms and Combinatorics vol. 16, Springer-Verlag, 1998, 116–165.
9. D. Randall and P. Tetali. Analyzing Glauber dynamics by comparison of Markov chains. *J. Math. Phys.*, 41(3):1598–1615, 2000.
10. A. Sinclair. Improved bounds for mixing rates of Markov chains and multicommodity flow. *Combinatorics, Probability and Computing*, 1(4):351–370, 1992.

On the 2-Colorability of Random Hypergraphs

Dimitris Achlioptas and Cristopher Moore*

[1] Microsoft Research, Redmond, Washington `optas@microsoft.com`
[2] Computer Science Department, University of New Mexico, Albuquerque and the
Santa Fe Institute, Santa Fe, New Mexico `moore@cs.unm.edu`

Abstract. A 2-coloring of a hypergraph is a mapping from its vertices to
a set of two colors such that no edge is monochromatic. Let $H_k(n, m)$ be
a random k-uniform hypergraph on n vertices formed by picking m edges
uniformly, independently and with replacement. It is easy to show that if
$r \geq r_c = 2^{k-1} \ln 2 - (\ln 2)/2$, then with high probability $H_k(n, m = rn)$
is not 2-colorable. We complement this observation by proving that if
$r \leq r_c - 1$ then with high probability $H_k(n, m = rn)$ is 2-colorable.

1 Introduction

For an integer $k \geq 2$, a k-*uniform hypergraph* H is an ordered pair $H = (V, E)$,
where V is a finite non-empty set, called the set of *vertices* of H, and E is a
family of distinct k-subsets of V, called the *edges* of H. For general hypergraph
terminology and background see [5]. A 2-*coloring* of a hypergraph $H = (V, E)$
is a partition of its vertex set V into two (color) classes so that no edge in E is
monochromatic. A hypergraph is *2-colorable* if it admits a 2-coloring.

The property of 2-colorability was introduced and studied by Bernstein [6] in
the early 1900s for infinite hypergraphs. The 2-colorability of finite hypergraphs,
also known as "Property B" (a term coined by Erdős in reference to Bernstein),
has been studied for about eighty years (e.g. [4,10,11,15,16,19,20]). For $k = 2$,
i.e. for graphs, the problem is well understood since a graph is 2-colorable if and
only if it has no odd cycle. For $k \geq 3$, though, much less is known and deciding
the 2-colorability of k-uniform hypergraphs is NP-complete [17].

In this paper we discuss the 2-colorability of random k-uniform hypergraphs
for $k \geq 3$. (For the evolution of odd cycles in random graphs see [12].) Let
$H_k(n, m)$ be a random k-uniform hypergraph on n vertices, where the edge
set is formed by selecting uniformly, independently and with replacement m
out of all possible $\binom{n}{k}$ edges. We will study asymptotic properties of $H_k(n, m)$
when $k \geq 3$ is arbitrary but fixed while n tends to infinity. We will say that
a hypergraph property A holds *with high probability* (w.h.p.) in $H_k(n, m)$ if
$\lim_{n \to \infty} \Pr[H_k(n, m) \text{ has } A] = 1$. The main question in this setting is:

As m is increased, when does $H_k(n, m)$ stop being 2-colorable?

---

* Supported by NSF grant PHY-0071139, the Sandia University Research Program,
and Los Alamos National Laboratory.

J.D.P. Rolim and S. Vadhan (Eds.): RANDOM 2002, LNCS 2483, pp. 78–90, 2002.
© Springer-Verlag Berlin Heidelberg 2002

It is popular to conjecture that the transition from 2-colorability to non-2-colorability is *sharp*. That is, it is believed that for each $k \geq 3$, there exists a constant r_k such that if $r < r_k$ then $H_k(n, m = rn)$ is w.h.p. 2-colorable, but if $r > r_k$ then w.h.p. $H_k(n, m = rn)$ is not 2-colorable. Determining r_k is a challenging open problem, closely related to the satisfiability threshold conjecture for random k-SAT. Although r_k has not been proven to exist, we will take the liberty of writing $r_k \geq r^*$ to denote that for $r < r^*$, $H_k(n, rn)$ is 2-colorable w.h.p. (and analogously for $r_k \leq r^*$).

A relatively recent result of Friedgut [13] supports this conjecture as it gives a *non-uniform* sharp threshold for hypergraph 2-colorability. Namely, for each $k \geq 3$ there exists a *sequence* $r_k(n)$ such that if $r < r_k(n) - \epsilon$ then w.h.p. $H_k(n, rn)$ is 2-colorable, but if $r > r_k(n) + \epsilon$ then w.h.p. $H_k(n, rn)$ is not 2-colorable. We will find useful the following immediate corollary of this sharp threshold.

Corollary 1. *If*

$$\liminf_{n \to \infty} \Pr[H_k(n, r^*n) \text{ is 2-colorable}] > 0 \ ,$$

then for $r < r^*$, $H_k(n, rn)$ is 2-colorable w.h.p.

Alon and Spencer [3] were the first to give bounds on the potential value of r_k. In particular, they observed that the expected number of 2-colorings of $H_k(n, m = rn)$ is $o(1)$ if $2(1 - 2^{1-k})^r < 1$, implying

$$r_k < 2^{k-1} \ln 2 - \frac{\ln 2}{2} \ . \tag{1}$$

Their main contribution, though, was providing a lower bound on r_k. Specifically, by applying the Lovász Local Lemma, they were able to show that if $r = c\, 2^k/k^2$ then w.h.p. $H_k(n, rn)$ is 2-colorable, for some small constant $c > 0$.

In [1], Achlioptas, Kim, Krivelevich and Tetali reduced the asymptotic gap between the upper and lower bounds of [3] from order k^2 to order k. In particular, they proved that there exists a constant $c > 0$ such that if $r \leq c\, 2^k/k$ then a simple, linear-time algorithm w.h.p. finds a 2-coloring of $H_k(n, rn)$. Their algorithm was motivated by algorithms for random k-SAT due to Chao and Franco [7] and Chvátal and Reed [8]. In fact, those algorithms give a similar $\Omega(2^k/k)$ lower bound on the random k-SAT threshold which, like r_k, can also be easily bounded as $O(2^k)$.

Very recently, the authors eliminated the gap for the random k-SAT threshold, determining its value within a factor of two [2]. The proof amounts to applying the "second moment" method to the set of satisfying truth assignments whose complement is also satisfying. Alternatively, one can think of this as applying the second moment method to the number of truth assignments under which every k-clause contains at least one satisfied literal *and* at least one unsatisfied literal, i.e. which satisfy the formula when interpreted as a random instance of Not-All-Equal k-SAT (NAE k-SAT).

Here we extend the techniques of [2] and apply them to hypergraph 2-colorability. This allows us to determine r_k within a small additive constant.

Theorem 1. *For every $\epsilon > 0$ and all $k \geq k_0(\epsilon)$,*

$$r_k \geq 2^{k-1} \ln 2 - \frac{\ln 2}{2} - \frac{1 + \epsilon}{2} .$$

Our method actually yields an explicit lower bound for r_k for each value of k as the solution to a simple equation (yet one without a pretty closed form, hence Theorem 1). Below we compare this lower bound to the upper bound of (1) for small values of k. The gap converges to $1/2$ rather rapidly.

Table 1. Upper and lower bounds for r_k

k	3	4	5	7	9	11	12
Lower bound	3/2	49/12	9.973	43.432	176.570	708.925	1418.712
Upper bound	2.409	5.191	10.740	44.014	177.099	709.436	1419.219

2   Second Moment and NAE k-SAT

We prove Theorem 1 by applying the following version of the second moment method (see Exercise 3.6 in [18]) to the number of 2-colorings of $H_k(n, m = rn)$.

Lemma 1. *For any non-negative integer-valued random variable X,*

$$\Pr[X > 0] \geq \frac{\mathbf{E}[X]^2}{\mathbf{E}[X^2]} . \tag{2}$$

In particular, if X is the number of 2-colorings of $H_k(n, m = rn)$, we will prove that for all $\epsilon > 0$ and all $k \geq k_0(\epsilon)$, if $r = 2^{k-1} \ln 2 - \ln 2/2 - (1 + \epsilon)/2$ then there exists some constant $C = C(k)$ such that

$$\mathbf{E}[X^2] < C \times \mathbf{E}[X]^2 .$$

By Lemma 1, this implies $\Pr[X > 0] = \Pr[H_k(n, rn)$ is 2-colorable$] > 1/C$. Theorem 1 follows by invoking Corollary 1.

This approach parallels the one taken recently by the authors for random NAE k-SAT [2]. Naturally, what differs is the second-moment calculation which here is prima facie significantly more involved.

We start our exposition by outlining the NAE k-SAT calculation of [2]. This serves as a warm up for our calculations and allows us to state a couple of useful lemmata from [2]. We then proceed to outline the proof of our main result, showing the parallels with NAE k-SAT and reducing the proof of Theorem 1 to the proof of three independent lemmata.

The first such lemma is specific to hypergraph 2-colorability and expresses $\mathbf{E}[X^2]$ as a multinomial sum. The second one is a general lemma about bounding multinomial sums by a function of their largest term and is perhaps of independent interest. It generalizes Lemma 2 of [2], which we state below. After applying these two lemmata, we are left to maximize a three-variable function parameterized by k and r. This is analogous to NAE k-SAT, except that there we only have to deal with a one-variable function, similarly parameterized. That simpler maximization, in fact, amounted to the bulk of the technical work in [2]. Luckily, here we will be able to get away with much less work: a convexity argument will allow us to reduce our three-dimensional optimization precisely to the optimization in [2].

2.1 Proof Outline for NAE k-SAT

Let Y be the number of satisfying assignments of a random NAE k-SAT formula with n variables and $m = rn$ clauses. It is easy to see that $\mathbf{E}[Y] = 2^n (1 - 2^{1-k})^{rn}$. Then $\mathbf{E}[Y^2]$ is the sum, over all ordered pairs of truth assignments, of the probability that both assignments in the pair are satisfying. It is not hard to show that if two assignments assign the same value to $z = \alpha n$ variables, then the probability that both are satisfying is

$$p(\alpha) = 1 - 2^{1-k} \left(2 - \alpha^k - (1-\alpha)^k \right) .$$

Since there are $2^n \binom{n}{z}$ pairs of assignments sharing z variables, we have

$$\frac{\mathbf{E}[Y^2]}{\mathbf{E}[Y]^2} = \sum_{z=0}^{n} \binom{n}{z} \left[\frac{1}{2} \left(\frac{p(z/n)}{(1 - 2^{1-k})^2} \right)^r \right]^n .$$

To bound such sums within a constant factor, we proved the following in [2].

Lemma 2. *Let f be any real positive analytic function and let*

$$S = \sum_{z=0}^{n} \binom{n}{z} f(z/n)^n .$$

Define $0^0 \equiv 1$ and let g on $[0,1]$ be

$$g(\alpha) = \frac{f(\alpha)}{\alpha^\alpha (1-\alpha)^{1-\alpha}} .$$

If there exists $\alpha_{\max} \in (0,1)$ such that $g(\alpha_{\max}) \equiv g_{\max} > g(\alpha)$ for all $\alpha \neq \alpha_{\max}$, and $g''(\alpha_{\max}) < 0$, then there exist constants B and C such that for all sufficiently large n

$$B \times g_{\max}^n \leq S \leq C \times g_{\max}^n .$$

Thus, using Lemma 2, bounding $\mathbf{E}[X^2]/\mathbf{E}[X]^2$ reduces to maximizing

$$\ell_r(\alpha) = \frac{1}{2\,\alpha^\alpha\,(1-\alpha)^{1-\alpha}} \left(\frac{p(\alpha)}{(1-2^{1-k})^2}\right)^r . \tag{3}$$

Note now that $\ell_r(1/2) = 1$ for all r and that our goal is to find r such that $\ell_r(\alpha) \le 1$ for all $\alpha \in [0,1]$. Indeed, in [2] we showed that

Lemma 3. [2] *For every $\epsilon > 0$, and all $k \ge k_0(\epsilon)$, if*

$$r \le 2^{k-1}\ln 2 - \frac{\ln 2}{2} - \frac{1+\epsilon}{2}$$

then $\ell_r(1/2) = 1 > \ell_r(\alpha)$ for all $\alpha \ne 1/2$ and $\ell_r''(1/2) < 0$.

Thus, for all r, k, ϵ as in Lemma 3, we see that Lemma 2 implies $\mathbf{E}[Y^2]/\mathbf{E}[Y]^2 < C \times \ell_r(1/2)^n = C$, concluding the proof.

3 Proof Outline for Hypergraph 2-Colorability

Let X be the number of 2-colorings of $H_k(n, rn)$. Let $q = 1 - 2^{1-k}$ and

$$p(\alpha, \beta, \gamma) = 1 - \alpha^k - (1-\alpha)^k - \beta^k - (1-\beta)^k + \gamma^k + (\alpha-\gamma)^k + (\beta-\gamma)^k + (1-\alpha-\beta+\gamma)^k.$$

We will prove that

Lemma 4. *There exists a constant A such that*

$$\frac{\mathbf{E}[X^2]}{\mathbf{E}[X]^2} \le \frac{1}{A^2} \sum_{z_1+\cdots+z_4=n} \binom{n}{z_1, z_2, z_3, z_4} \left(\frac{1}{4}\left(\frac{p\left(\frac{z_1+z_2}{n}, \frac{z_1+z_3}{n}, \frac{z_1}{n}\right)}{q^2}\right)^r\right)^n .$$

Similarly to NAE k-SAT we would like to bound this sum by a function of its maximum term. To do this we will establish a multidimensional generalization of the upper bound of Lemma 2.

Lemma 5. *Let f be any real positive analytic function and let*

$$S = \sum_{z_1+\cdots+z_d=n} \binom{n}{z_1, \cdots, z_d} f(z_1/n, \cdots, z_{d-1}/n)^n .$$

Let $Z = \left\{(\zeta_1, \ldots, \zeta_{d-1}) : \zeta_i \ge 0 \text{ for all } i, \text{ and } \sum \zeta_i \le 1\right\}$. Define g on Z as

$$g(\zeta_1, \ldots, \zeta_{d-1}) = \frac{f(\zeta_1, \ldots, \zeta_{d-1})}{\zeta_1^{\zeta_1} \cdots \zeta_{d-1}^{\zeta_{d-1}}(1 - \zeta_1 - \cdots - \zeta_{d-1})^{1-\zeta_1-\cdots-\zeta_{d-1}}} .$$

If i) there exists $\zeta_{\max}$ in the interior of Z such that for all $\zeta \in Z$ with $\zeta \ne \zeta_{\max}$, we have $g(\zeta_{\max}) \equiv g_{\max} > g(\zeta)$, and ii) the determinant of the $(d-1) \times (d-1)$ matrix of second derivatives of g is nonzero at $\zeta_{\max}$, then there exists a constant D such that for all sufficiently large n

$$S < D \times g_{\max}^n .$$

Applying Lemma 5 to the sum in Lemma 4 we see that we need to maximize

$$g_r(\alpha, \beta, \gamma) = \frac{\left(\dfrac{p(\alpha, \beta, \gamma)}{q^2}\right)^r}{4\,\gamma^\gamma\,(\alpha - \gamma)^{\alpha - \gamma}\,(\beta - \gamma)^{\beta - \gamma}\,(1 - \alpha - \beta + \gamma)^{1 - \alpha - \beta + \gamma}}\ , \qquad (4)$$

where for convenience we defined g_r in terms of α, β, γ instead of $\zeta_1, \zeta_2, \zeta_3$. We will show that g_r has a unique maximum at

$$\zeta^* = (1/2, 1/2, 1/4)\ .$$

Lemma 6. *For every $\epsilon > 0$, and all $k \geq k_0(\epsilon)$ if*

$$r \leq 2^{k-1}\ln 2 - \frac{\ln 2}{2} - \frac{1 + \epsilon}{2}$$

then $g_r(\zeta^*) = 1 > g_r(\zeta)$ for all $\zeta \in Z$ with $\zeta \neq \zeta^*$. Moreover, the determinant of the matrix of second derivatives of g_r at ζ^* is nonzero.

Therefore, for all r, k, ϵ as in Lemma 6

$$\frac{\mathbf{E}[X^2]}{\mathbf{E}[X]^2} < \frac{D}{A^2} \times g_r(\zeta^*)^n = D/A^2\ ,$$

completing the proof of Theorem 1 modulo Lemmata 4, 5 and 6.

The proof of Lemma 4 is a straightforward probabilistic calculation. The proof of Lemma 5 is somewhat technical but follows standard asymptotic methods. To prove Lemma 6 we will rely very heavily on Lemma 3. In particular, we will show that all local maxima of g_r occur within a one-dimensional subspace, in which g_r coincides with the function ℓ_r of (3). Specifically, we prove

Lemma 7. *If (α, β, γ) is a local extremum of g_r, then $\alpha = \beta = 1/2$.*

This reduces our problem to the one-dimensional maximization for NAE k-SAT, allowing us to easily prove Lemma 6.

Proof of Lemma 6. Observe that

$$g_r(1/2, 1/2, \gamma) = \ell_r(2\gamma)\ ,$$

where ℓ_r is the function defined in (3) for NAE k-SAT. Thus, the inequality $g_r(\zeta^*) > g_r(\zeta)$ for $\zeta \neq \zeta^*$ follows readily from Lemma 3, giving the first part of the lemma.

To prove the condition on the determinant of the 3×3 matrix of second derivatives, a little arithmetic shows that at ζ^* it is equal to

$$\frac{256\,(2^k - 2 - 2kr + 2k^2 r)^2}{2^{4k} q^4}\,(4k(k-1)\,r - 2^{2k}\,q^2)\ .$$

Thus, the determinant is negative whenever

$$4k(k-1)\,r < 2^{2k}\,q^2\ .$$

For $k = 3, 4$ this is true for $r < 3/2$ and $r < 49/12$ respectively, while for $k \geq 5$ it is true for all $r < \ln 2 \times 2^{k-1}$. $\qquad\qquad\Box$

4 Proof of Lemma 4

Recall that X denotes the number of 2-colorings of $H_k(n, m = rn)$.

4.1   First Moment

Recall that

$$q = 1 - 2^{1-k} \ .$$

The probability that a 2-coloring with $z = \alpha n$ black vertices and $n - z = (1-\alpha)n$ white vertices makes a random hyperedge of size k bichromatic is

$$s(\alpha) = 1 - \alpha^k - (1 - \alpha)^k \le q \ .$$

Summing over the 2^n colorings gives

$$\mathbf{E}[X] = \sum_{z=0}^{n} \binom{n}{z} s(z/n)^{rn} \ .$$

To bound this sum from below we apply the lower bound of Lemma 2 with $f(\alpha) = s(\alpha)^r$. In particular, it is easy to see that for all $r > 0$

$$g(\alpha) = \frac{s(\alpha)^r}{\alpha^\alpha (1 - \alpha)^{1-\alpha}} = \frac{(1 - \alpha^k - (1 - \alpha)^k)^r}{\alpha^\alpha (1 - \alpha)^{1-\alpha}}$$

is maximized at $\alpha = 1/2$ and that $g(1/2) = 2q^r$. Moreover, for any $k > 1$

$$g''(1/2) = -8 \left(1 - 2^{1-k}\right)^{r-1} \left(1 + 2^{1-k}(k(k-1)r - 1)\right) < 0 \ .$$

Therefore, we see that there exists a constant A such that

$$\mathbf{E}[X] \ge A \times (2q^r)^n \ . \tag{5}$$

4.2   Second Moment

We first observe that $\mathbf{E}[X^2]$ equals the expected number of ordered pairs S, T of 2-partitions of the vertices such that both S and T are 2-colorings. Suppose that S and T have αn and βn black vertices respectively, while γn vertices are black in both. By inclusion-exclusion a random hyperedge of size k is bichromatic under both S and T with probability $p(\alpha, \beta, \gamma)$, i.e.

$$1 - \alpha^k - (1 - \alpha)^k - \beta^k - (1 - \beta)^k + \gamma^k + (\alpha - \gamma)^k + (\beta - \gamma)^k + (1 - \alpha - \beta + \gamma)^k \ .$$

The negative terms above represent the probability that the hyperedge is monochromatic under either S or T, while the positive terms represent the probability that it is monochromatic under both (potentially with different colors). Since the $m = rn$ hyperedges are chosen independently and with replacement, the probability that all $m = rn$ hyperedges are bichromatic is $p(\alpha, \beta, \gamma)^{rn}$.

If z_1, z_2, z_3 and z_4 vertices are respectively black in both assignments, black in S and white in T, white in S and black in T, and white in both, then $\alpha = (z_1 + z_2)/n$, $\beta = (z_1 + z_3)/n$ and $\gamma = z_1/n$. Thus,

$$\mathbf{E}[X^2] = \sum \binom{n}{z_1, z_2, z_3, z_4} p\left(\frac{z_1 + z_2}{n}, \frac{z_1 + z_3}{n}, \frac{z_3}{n}\right)^{rn} . \qquad (6)$$

$\square$

5   Proof of Lemma 7

We wish to show that at any extremum of g_r we have $\alpha = \beta = 1/2$. We start by proving that at any such extremum $\alpha = \beta$. Note that since, by symmetry, we are free to flip either or both colorings, we can restrict ourselves to the case where $\alpha \leq 1/2$ and $\gamma \leq \alpha/2$.

Let $h(x_1, x_2, x_3, x_4) = \sum_i x_i \ln x_i$ denote the entropy function, and let us define the shorthand $(\partial/\partial x - \partial/\partial y)f$ for $\partial f/\partial x - \partial f/\partial y$. Also, recall that $q = 1 - 2^{1-k}$ and that $p(\alpha, \beta, \gamma) \equiv p$ is

$$1 - \alpha^k - (1-\alpha)^k - \beta^k - (1-\beta)^k + \gamma^k + (\alpha - \gamma)^k + (\beta - \gamma)^k + (1 - \alpha - \beta + \gamma)^k.$$

We will consider the gradient of $\ln g_r$ along a vector that increases α while decreasing β. We see

$$\left(\frac{\partial}{\partial \alpha} - \frac{\partial}{\partial \beta}\right) \ln g_r(\alpha, \beta, \gamma)$$

$$= \left(\frac{\partial}{\partial \alpha} - \frac{\partial}{\partial \beta}\right) \left(h(\gamma, \alpha - \gamma, \beta - \gamma, 1 - \alpha - \beta + \gamma) - \ln 4 + r(\ln p - 2\ln q)\right)$$

$$= -\ln(\alpha - \gamma) + \ln(\beta - \gamma)$$

$$+ \frac{kr}{p}\left(-\alpha^{k-1} + (1-\alpha)^{k-1} + (\alpha - \gamma)^{k-1} + \beta^{k-1} - (1-\beta)^{k-1} - (\beta - \gamma)^{k-1}\right)$$

$$\equiv \phi(\alpha) - \phi(\beta) , \qquad (7)$$

where

$$\phi(x) = -\ln(x - \gamma) + \frac{kr}{p}\left(-x^{k-1} + (1-x)^{k-1} + (x - \gamma)^{k-1}\right) .$$

Here we regard p as a constant in the definition of $\phi(x)$.

Observe now that if (α, β, γ) is an extremum of g_r then it is also an extremum of $\ln g_r$. Therefore, it must be that $(\partial/\partial \alpha) \ln g_r = (\partial/\partial \beta) \ln g_r = 0$ at (α, β, γ) which, by (7), implies $\phi(\alpha) = \phi(\beta)$. This, in turn, implies $\alpha = \beta$ since $\phi(x)$ is monotonically decreasing in the interval $\gamma < x < 1$:

$$\frac{d\phi}{dx} = -\frac{1}{x - \gamma} - \frac{k(k-1)r}{p}\left(x^{k-2} + (1-x)^{k-2} - (x - \gamma)^{k-2}\right) < 0 .$$

Next we wish to show that in fact $\alpha = \beta = 1/2$. Setting $\alpha = \beta$, we consider the gradient of $\ln g_r$ along a vector that increases α and γ simultaneously (using a similar shorthand for $\partial g/\partial\alpha + \partial g/\partial\gamma$):

$$\left(\frac{\partial}{\partial\alpha} + \frac{\partial}{\partial\gamma}\right)\ln g_r(\alpha, \alpha, \gamma)$$

$$= \left(\frac{\partial}{\partial\alpha} + \frac{\partial}{\partial\gamma}\right)\left(h(\gamma, \alpha - \gamma, \alpha - \gamma, 1 - 2\alpha + \gamma) - \ln 4 + r\left(\ln p - 2\ln q\right)\right)$$

$$= -\ln\gamma + \ln(1 - 2\alpha + \gamma)$$

$$+ \frac{kr}{p}\left(-2\alpha^{k-1} + 2(1 - \alpha)^{k-1} - (1 - 2\alpha + \gamma)^{k-1} + \gamma^{k-1}\right)$$

$$\equiv \psi(\alpha) \ .$$

Clearly, $\psi(\alpha) = 0$ when $\alpha = 1/2$. To show that $1/2$ is the only such α, we show that ψ decreases monotonically with α by showing that if $0 < \alpha < 1/2$ and $\gamma \leq \alpha/2$, all three terms below are negative for $k \geq 3$.

$$\frac{\partial\psi}{\partial\alpha} = -\frac{2}{1 - 2\alpha + \gamma}$$

$$+ \frac{2k(k-1)r}{p}\left(-\alpha^{k-2} - (1 - \alpha)^{k-2} + (1 - 2\alpha + \gamma)^{k-2}\right)$$

$$+ \frac{2k^2 r}{p^2} \times \left(\gamma^{k-1} - (1 - 2\alpha + \gamma)^{k-1} - 2\alpha^{k-1} + 2(1 - \alpha)^{k-1}\right)$$

$$\times \left(-(\alpha - \gamma)^{k-1} + \alpha^{k-1} + (1 - 2\alpha + \gamma)^{k-1} - (1 - \alpha)^{k-1}\right) \ .$$

The first and second terms are negative since $1 - \alpha > 1 - 2\alpha + \gamma > 0$, implying $(1 - \alpha)^{k-2} > (1 - 2\alpha + \gamma)^{k-2}$. The second factor of the third term is positive since $f(z) = z^{k-1}$ is convex and $(1 - \alpha) - \alpha = (1 - 2\alpha + \gamma) - \gamma$ (the factor of 2 on the last two terms only helps us since $1 - \alpha \geq \alpha$). Similarly, the third factor is negative since $(1 - \alpha) - (1 - 2\alpha + \gamma) = \alpha - \gamma \geq \alpha - (\alpha - \gamma) = \gamma$.

Thus, $\partial\psi/\partial\alpha < 0$ and $\alpha = 1/2$ is the unique solution to $\psi(\alpha) = 0$. Therefore, if (α, α, γ) is an extremum of g_r we must have $\alpha = 1/2$.   □

6   Proof of Lemma 5

6.1   Preliminaries

We will use the following form of Stirling's approximation for $n!$, valid for $n > 0$

$$\sqrt{2\pi n}\, n^n\, e^{-n}\left(1 + \frac{1}{12n}\right) < n! < \sqrt{2\pi n}\, n^n\, e^{-n}\left(1 + \frac{1}{6n}\right) \ . \tag{8}$$

We will also use the following crude lower bound for $n!$, valid for $n \geq 0$

$$n! \geq (n/e)^n \ , \tag{9}$$

using the convention $0^0 \equiv 1$.

Let $z_1, \ldots, z_d$ be such that $\sum_{i=1}^d z_i = n$. Let $\zeta_i = z_i/n$. Let $\boldsymbol{\zeta} = (\zeta_1, \ldots, \zeta_{d-1})$.

- If $z_i > 0$ for all i, then using the upper and lower bounds of (8) for $n!$ and $z_i!$ respectively, and reducing the denominator further by changing the factor $1 + 1/(12z_i)$ to 1 for $i \neq 1$, we get

$$\binom{n}{z_1, \cdots, z_d} < (2\pi n)^{-(d-1)/2} \left(\prod_{i=1}^{d} \zeta_i^{-1/2}\right) \left(\prod_{i=1}^{d} \zeta_i^{-\zeta_i}\right)^n \times \frac{1 + 1/(6n)}{1 + 1/(12z_1)}$$

$$\leq (2\pi n)^{-(d-1)/2} \left(\prod_{i=1}^{d} \zeta_i^{-1/2}\right) \left(\prod_{i=1}^{d} \zeta_i^{-\zeta_i}\right)^n , \qquad (10)$$

where for (10) we assumed w.l.o.g. that $z_1 \leq n/2$. Thus,

$$\binom{n}{z_1, \cdots, z_d} f(z_1/n, \ldots, z_{d-1}/n)^n \leq (2\pi n)^{-(d-1)/2} \left(\prod_{i=1}^{d} \zeta_i^{-1/2}\right) g(\zeta)^n . \qquad (11)$$

- For any $z_i \geq 0$, the upper bound of (8) and (9) give

$$\binom{n}{z_1, \cdots, z_d} < \frac{7}{6} \sqrt{2\pi n} \left(\prod_{i=1}^{d} \zeta_i^{-\zeta_i}\right)^n ,$$

implying a cruder bound

$$\binom{n}{z_1, \cdots, z_d} f(z_1/n, \ldots, z_{d-1}/n)^n \leq \frac{7}{6} \sqrt{2\pi n}\, g(\zeta)^n . \qquad (12)$$

6.2 The Main Proof

Our approach is a crude form of the Laplace method for asymptotic integrals [9] which amounts to approximating functions near their peak as Gaussians.

We wish to approximate $g(\zeta)$ in the vicinity of $\zeta_{\max}$. We will do this by Taylor expanding $\ln g$, which is analytic since g is analytic and positive. Since $\ln g$ increases monotonically with g, both g and $\ln g$ are maximized at $\zeta_{\max}$. Furthermore, at $\zeta_{\max}$ the matrix of second derivatives of $\ln g$ is that of g divided by a constant, since

$$\left.\frac{\partial^2 \ln g}{\partial \zeta_i \partial \zeta_j}\right|_{\zeta=\zeta_{\max}} = \frac{1}{g_{\max}} \frac{\partial^2 g}{\partial \zeta_i \partial \zeta_j} - \frac{1}{g_{\max}^2} \frac{\partial g}{\partial \zeta_i} \frac{\partial g}{\partial \zeta_j}$$

and at $\zeta_{\max}$ the first derivatives of g are all zero. Therefore, if the matrix of second derivatives of g at $\zeta_{\max}$ has nonzero determinant, so does the matrix of the second derivatives of $\ln g$.

Note now that since the matrix of second derivatives is by definition symmetric, it can be diagonalized, and its determinant is the product of its eigenvalues. Therefore, if its determinant is nonzero, all its eigenvalues are smaller than some $\lambda_{\max} < 0$. Thus, Taylor expansion around $\zeta_{\max}$ gives

$$\ln g(\zeta) \leq \ln g_{\max} + \frac{1}{2}\lambda_{\max} |\zeta - \zeta_{\max}|^2 + O(|\zeta - \zeta_{\max}|^3)$$

or, exponentiating to obtain g,

$$g(\zeta) \leq g_{\max} \exp\left(\frac{1}{2}\lambda_{\max}|\zeta - \zeta_{\max}|^2\right) \times \left(1 + O(|\zeta - \zeta_{\max}|^3)\right) .$$

Therefore, there is a ball of radius $\rho > 0$ around $\zeta_{\max}$ and constants $Y > 0$ and $g_* < g_{\max}$ such that

$$\text{If } |\zeta - \zeta_{\max}| \leq \rho, \ g(\zeta) \leq g_{\max} \exp\left(-Y|\zeta - \zeta_{\max}|^2\right) , \tag{13}$$

$$\text{If } |\zeta - \zeta_{\max}| > \rho, \ g(\zeta) \leq g_* . \tag{14}$$

We will separate S into two sums, one inside the ball and one outside:

$$\sum_{\zeta \in Z : |\zeta - \zeta_{\max}| \leq \rho} \binom{n}{\zeta_1 n, \cdots, \zeta_d n} f(\zeta)^n \ + \ \sum_{\zeta \in Z : |\zeta - \zeta_{\max}| > \rho} \binom{n}{\zeta_1 n, \cdots, \zeta_d n} f(\zeta)^n .$$

For the terms inside the ball, first note that if $|\zeta - \zeta_{\max}| \leq \rho$ then

$$\prod_{i=1}^{d} \zeta_i^{-1/2} \leq W \text{ where } W = \left(\min_i \zeta_{\max,i} - \rho\right)^{-d/2} .$$

Then, since $|\zeta - \zeta_{\max}|^2 = \sum_{i=1}^{d-1}(\zeta_i - \zeta_{\max,i})^2$, using (11) and (13) we have

$$\sum_{\zeta \in Z : |\zeta - \zeta_{\max}| \leq \rho} \binom{n}{\zeta_1 n, \cdots, \zeta_d n} f(\zeta)^n$$

$$\leq (2\pi n)^{-(d-1)/2} W g_{\max}^n \ \times \ \sum_{z_1, \cdots, z_{d-1} = -\infty}^{\infty} \exp\left(-nY \sum_{i=1}^{d-1}(\zeta_i - \zeta_{\max,i})^2\right)$$

$$= (2\pi n)^{-(d-1)/2} W g_{\max}^n \times \prod_{i=1}^{d-1}\left(\sum_{z_i=-\infty}^{\infty} \exp\left(-nY\,(z_i/n - \zeta_{\max,i})^2\right)\right) .$$

Now if a function $\phi(z)$ has a single peak, on either side of which it is monotonic, we can replace its sum with its integral with an additive error at most twice its largest term:

$$\left|\sum_{z=-\infty}^{\infty} \phi(z) - \int_{-\infty}^{\infty} \phi(z)\, dz\right| \leq 2\max_z \phi(z)$$

and so

$$\sum_{z_i=-\infty}^{\infty} \exp\left(-nY(z_i/n - \zeta_{\max,i})^2\right) \leq 2 + \int_{-\infty}^{\infty} \exp\left(-nY(z_i/n - \zeta_{\max,i})^2\right) dz$$

$$= \sqrt{\pi n/Y} + 2 < \sqrt{2\pi n/Y}$$

where the last inequality holds for sufficiently large n. Multiplying these $d-1$ sums together gives

$$\sum_{\zeta \in Z : |\zeta - \zeta_{\max}| \leq \rho} \binom{n}{\zeta_1 n, \cdots, \zeta_d n} f(\zeta)^n \leq W Y^{-(d-1)/2} g_{\max}^n . \tag{15}$$

Outside the ball, we use (12), (14) and the fact that the entire sum has at most n^{d-1} terms to write

$$\sum_{\zeta \in Z: |\zeta - \zeta_{\max}| > \rho} \binom{n}{\zeta_1 n, \cdots, \zeta_d n} f(\zeta)^n \leq n^{d-1} \times \frac{7}{6} \sqrt{2\pi n}\, g_*^n < g_{\max}^n \qquad (16)$$

where the last inequality holds for sufficiently large n. Combining (16) and (15) gives

$$S < (WY^{-(d-1)/2} + 1)\, g_{\max}^n \equiv D \times g_{\max}^n$$

which completes the proof. (We note that the constant D can be optimized by replacing our sums by integrals and using Laplace's method [2,9].)

7   Conclusions

We have shown that the second moment method yields a very sharp estimate of the threshold for hypergraph 2-colorability. It allows us not only to close the asymptotic gap between the previously known bounds but, in fact, to get the threshold within a small additive constant. Yet:

• While the second moment method tells us that w.h.p. an exponential number of 2-colorings exist for $r = \Theta(2^k)$, it tells us nothing about how to find a single one of them efficiently. The possibility that such colorings actually cannot be found efficiently is extremely intriguing.

• While we have shown that the second moment method works really well, we'd be hard pressed to say why. In particular, we do not have a criterion for determining a constraint satisfaction problem's amenability to the method. The fact that the method fails spectacularly for random k-SAT suggests that, perhaps, rather subtle forces are at play.

Naturally, one can always view the success of the second moment method in a particular problem as an aposteriori indication that the satisfying solutions of the problem are "largely uncorrelated". This viewpoint, though, is hardly predictive. (Yet, it might prove useful to the algorithmic question above).

The solution-symmetry shared by NAE k-SAT and hypergraph 2-colorability but not by k-SAT, i.e. the property that the complement of a solution is also a solution, explains why the method gives a nonzero lower bound for these two problems (and why it fails for k-SAT). Yet symmetry alone does not explain why the bound becomes essentially tight as k grows. In any case, we hope (and, worse, consider it natural) that an appropriate notion of symmetry is present in many more problems.

References

1. ACHLIOPTAS, D., KIM, J.H., KRIVELEVICH, M. AND TETALI, P. Two-coloring random hypergraphs. *Random Structures Algorithms*, 20(2):249–259, 2002.

2. ACHLIOPTAS, D., AND MOORE, C. The asymptotic order of the random k-SAT threshold. To appear in *43rd Annual Symposium on Foundations of Computer Science (Vancouver, BC, 2002)*.

3. ALON, N., AND SPENCER, J. A note on coloring random k-sets. Unpublished manuscript.

4. BECK, J. On 3-chromatic hypergraphs. *Discrete Math. 24*, 2 (1978), 127–137.

5. BERGE, C. *Hypergraphs.* North-Holland Publishing Co., Amsterdam, 1989. Combinatorics of finite sets, Translated from the French.

6. BERNSTEIN, F. Zur theorie der trigonometrische reihen. *Leipz. Ber. 60*, (1908), 325–328.

7. CHAO, M.-T., AND FRANCO, J. Probabilistic analysis of a generalization of the unit-clause literal selection heuristics for the k-satisfiability problem. *Inform. Sci. 51*, 3 (1990), 289–314.

8. CHVÁTAL, V., AND REED, B. Mick gets some (the odds are on his side). In *33rd Annual Symposium on Foundations of Computer Science (Pittsburgh, PA, 1992)*. IEEE Comput. Soc. Press, Los Alamitos, CA, 1992, pp. 620–627.

9. DE BRUIJN, N.G. *Asymptotic Methods in Analysis.* North-Holland, 1958.

10. ERDŐS, P. On a combinatorial problem. *Nordisk Mat. Tidskr. 11* (1963), 5–10, 40.

11. ERDŐS, P., AND LOVÁSZ, L. Problems and results on 3-chromatic hypergraphs and some related questions. 609–627. Colloq. Math. Soc. János Bolyai, Vol. 10.

12. FLAJOLET, P., KNUTH, D. E., AND PITTEL, B. The first cycles in an evolving graph. *Discrete Math. 75*, 1-3 (1989), 167–215. Graph theory and combinatorics (Cambridge, 1988).

13. FRIEDGUT, E. Necessary and sufficient conditions for sharp thresholds of graph properties, and the k-SAT problem. *J. Amer. Math. Soc. 12* (1999), 1017–1054.

14. FRIEZE, A. AND MCDIARMID, C. Algorithmic theory of random graphs. *Random Struct. Alg. 10*, (1997), 5–42.

15. KAROŃSKI, M. AND ŁUCZAK, T. Random hypergraphs. Combinatorics, Paul Erdős is eighty, Vol. 2 (Keszthely, 1993), 283–293, Bolyai Soc. Math. Stud., 2, Janos Bolyai Math. Soc., Budapest, 1996.

16. KRIVELEVICH, M. AND SUDAKOV, B. The chromatic numbers of random hypergraphs. *Random Struct. Alg. 12*, (1998), 381–403.

17. LOVÁSZ, L. Coverings and coloring of hypergraphs. *Proceedings of the Fourth Southeastern Conference on Combinatorics, Graph Theory, and Computing* (Boca Raton, Florida, 1973), 3–12.

18. MOTWANI, R. AND RAGHAVAN, P. *Randomized Algorithms.* Cambridge University Press, Cambridge, 1995.

19. RADHAKRISHNAN, J., AND SRINIVASAN, A. Improved bounds and algorithms for hypergraph 2-coloring. *Random Structures Algorithms 16*, 1 (2000), 4–32.

20. SCHMIDT-PRUZAN, J., AND SHAMIR, E., AND UPFAL, E. Random hypergraph coloring algorithms and the weak chromatic number. *Journal of Graph Theory 8*, (1985), 347–362.

Percolation on Finite Cayley Graphs

Christopher Malon[1]* and Igor Pak[1]**

Massachusetts Institute of Technology, Cambridge MA 02139, USA,
{malon,pak}@math.mit.edu

Abstract. In this paper, we study percolation on finite Cayley graphs. A conjecture of Benjamini says that the critical percolation p_c of such a graph can be bounded away from one, for any Cayley graph satisfying a certain diameter condition. We prove Benjamini's conjecture for some special classes of groups. We also establish a reduction theorem, which allows us to build Cayley graphs for large groups without increasing p_c.

Introduction

Percolation on finite graphs is a new subject with a classical flavor. It arose from two important and, until recently, largely independent areas of research: Percolation Theory and Random Graph Theory. The first is a classical Bernoulli percolation on a lattice, initiated as a mathematical subject by Hammersley and Morton in 1950s, and which now became a major area of research. A fundamental albeit elementary observation that the critical percolation p_c is bounded away from 1 on $\mathbb{Z}^2$ has led to a number of advanced results and quests for generalizations. Among those most relevant to this work, let us mention the Grimmett Theorem regarding the 'smallest' possible region under a graph in $\mathbb{Z}^2$, for which one still has $p_c < 1$. Similarly, a percolation in finite boxes has become crucially important as a source of new questions, as well as a tool (see [12] for references and major results in the area.)

In the past decade, within the subject of percolation, there has been much attention devoted to study of percolation on Cayley graphs, and, more generally, vertex-transitive graphs. As envisioned by Benjamini and Schramm [6] in a series of conjectures, it concentrated on the interplay between Probability Theory and Group Theory, when the probabilistic properties of the (bond or site) percolation depend heavily on the algebraic properties of the underlying (infinite) group, and not on a particular generating set. We refer to [7] for a recent progress report on the subject.

Motivated by the study of percolation on infinite Cayley graphs, Benjamini in [5] (see also [2]) extends the notion of critical probability to finite graphs by asking at which point the resulting graph has a large (constant proportion size) connected component. He conjectured that one can prove a new version of $p_c < 1 - \varepsilon$, under a weak diameter condition. (Here and everywhere in the

---

* NSF Graduate Research Fellow
** Partially supported by an NSF Grant

J.D.P. Rolim and S. Vadhan (Eds.): RANDOM 2002, LNCS 2483, pp. 91–104, 2002.

introduction, $\varepsilon > 0$ is a universal constant independent of the size of the graph.) In this paper we present a number of positive results toward this unexpected, and, perhaps, overly optimistic conjecture.

Our main results are of two types. First, we concentrate on special classes of groups and establish $p_c < 1 - \varepsilon$ for these. We prove Benjamini's conjecture for all abelian groups, and then for nilpotent groups, with Hall bases as generating sets. We also prove that $p_c < 1 - \varepsilon$ for groups with small disjoint sets of relations, a notion somewhat similar to that in [4]. Our most important, and technically most difficult result is the Reduction Theorem, which enables us in certain cases to obtain sharp bounds for p_c of a Cayley graph of a group G depending on those of a normal subgroup $H \lhd G$ and a quotient group G/H. While the full version of Benjamini's conjecture remains wide open, the Reduction Theorem allows us in certain cases to concentrate on finite simple groups (a sentiment expressed in [5]). By means of classification of finite simple groups [11], and recent series of probabilistic results relying on classification (see e.g. [17]) one can hope that our results will lead to further progress towards understanding percolation on finite Cayley graphs.

Let us also describe a connection to Random Graph Theory, already mentioned above. First introduced by Erdős and Rényi in a pioneer paper [10], the authors considered random graphs either as random subgraphs of a complete graph K_n, or as a result of a random graph process, when edges are added one at a time. Although one needs probability p of an edge to be roughly $\log n/n$ for the graph to become connected, already $p = (1 + \varepsilon)/n$, suffices for creation of a 'giant' ($c(\varepsilon)n$ size) connected component.

The work of Erdős and Rényi generated the whole study of properties of random graphs, and more recently random subgraphs of finite graphs (see e.g. [8, 14,1]) In the past years, connectivity and Hamiltonicity have remained the most studied properties, ever since the celebrated Margulis' Lemma, rediscovered later by Russo (see e.g. [15,16].) One can view our work as a new treatment of the existence of a giant component in a large class of vertex-transitive graphs.

1   Definitions and Main Results

In a *p–percolation* process on a finite graph Γ, every edge $e \in \Gamma$ is deleted with probability $1-p$, independently. Such a process defines a probability distribution on subgraphs of Γ, in which each subgraph $H \subset \Gamma$ is assigned the probability $p^{|H|}(1 - p)^{|\Gamma|-|H|}$. Later we informally refer to edges of H as 'p–percolated'.

For constants ρ, α, and p between zero and one, we let $\mathcal{L}(\rho, \alpha, p)$ denote the collection of finite graphs Γ, such that a random subgraph $H \subset \Gamma$ as above will have a connected component joining $\rho|\Gamma|$ of their vertices, with probability at least α.

Let ρ and α be fixed, and let Γ be a finite graph. Define the *critical probability* $p_c(\Gamma)$ as follows:

$$p_c(\Gamma) = p_c(\Gamma; \rho, \alpha) := \inf\{p : \Gamma \in \mathcal{L}(\rho, \alpha, p)\}.$$

From monotonicity of the percolation, this implies that $\Gamma(G, S) \in \mathcal{L}(\rho, \alpha, p)$ for all $1 \geq p > p_c$.

We are interested in conditions which bound the critical probability away from 1, as the size of graph Γ grows. Benjamini conjectured in [5] that $p_c(\Gamma; \rho, \alpha) < 1 - \varepsilon(\rho, \alpha)$ for *every* vertex-transitive graph Γ with n vertices, provided $\operatorname{diam}(\Gamma) < n / \log n$.

It is well known and easy to see that all finite vertex-transitive graphs are Schreier graphs of finite groups, and thus can be obtained as projections of the Cayley graphs. Even in the domain of infinite vertex-transitive graphs, no known example is shown to be *not* quasi-isometric to a Cayley graph [9]. From this point on, we consider only finite Cayley graphs.

Let G be a finite group and let $S = S^{-1}$ be a symmetric set of generators. A graph with vertices $g \in G$ and edges $(g, g \cdot s)$, $s \in S$ is called the *Cayley graph* of the group G.

Definition 1 *Suppose that generators* $s_1, \ldots, s_n$ *for a finite abelian group* G *are given so that the products* $s_1^{i_1} \cdots s_n^{i_n}$ *are distinct for* $0 \leq i_k < a_k$, *where* a_k *is the order of* s_k. *Then we say that* $\{s_1, \ldots, s_n\}$ *is a Hall basis for* G.

The following result establishes Benjamini's conjecture for all abelian groups whose generating sets are Hall bases:

Theorem 1 *For any constants* ρ *and* α, *there is a constant* $\varepsilon = \varepsilon(\rho, \alpha) > 0$, *such that for every Cayley graph* $\Gamma = \Gamma(G, S)$ *of any finite abelian group* G *and Hall basis* S *satisfying* $\operatorname{diam}(\Gamma) < \frac{|G|}{\log|G|}$, *we have* $p_c(G; \rho, \alpha) < 1 - \varepsilon$.

Let G be a finite nilpotent group. Consider a lower central series

$$G = G_0 \triangleright G_1 \triangleright \ldots \triangleright G_\ell = \{1\},$$

where each $G_i = [G, G_{i-1}]$. A *Hall basis* is a generating set $S = S_1 \sqcup S_2 \sqcup \cdots \sqcup S_\ell$, such that $S_i \subset G_{i-1}$ and the image $\gamma_i(S_i)$ of S_i onto the quotient $H_i = G_i/G_{i-1}$ is a Hall basis for the abelian group H_i. These bases were introduced by Philip Hall [13], and recently appeared in a probabilistic context of random walks on nilpotent groups [3].

We say that Hall basis satisfies *diameter condition* $(*)$, if

$$(*) \quad \operatorname{diam}(\Gamma(H_i, \gamma_i(S_i))) < \frac{|H_i|}{\log|H_i|}, \quad \text{for all } 1 \leq i \leq \ell.$$

Theorem 2 *For any constants* ρ *and* α, *there is a constant* $\varepsilon = \varepsilon(\rho, \alpha) > 0$, *such that for every Cayley graph* $\Gamma = \Gamma(G, S)$ *of finite nilpotent group* G *with a composition series* $\{G_i\}$ *and a Hall basis* $S = \sqcup_i S_i$ *that satisfies the diameter condition* $(*)$, *we have* $p_c(G; \rho, \alpha) < 1 - \varepsilon$.

One can view Theorem 2 as a generalization of Theorem 1, since the former becomes the latter when G is abelian, i.e. when $\ell = 1$.

Let $\Gamma_n = \Gamma(G_n, R_n)$ be a sequence of Cayley graphs with diameters $d_n = \operatorname{diam}(\Gamma_n)$. For each $s \in R_n$, let $T_n(s) = \{r \in R_n : [r, s] = 1\}$.

Theorem 3 *If $d_n \to \infty$ as $n \to \infty$, and each $|T_{n,s}| \geq 4 \log d_n$, then there exists $\varepsilon > 0$ such that $p_c(\Gamma(G_n, R_n); \frac{2}{3}, \frac{1}{2}) \leq 1 - \varepsilon$ for all n.*

Without information about the structure or the critical probability of G/H, it still may be possible to bound the critical probability of G if the index of H in G is not too large.

Theorem 4 (Reduction Theorem) *Let $\Gamma = \Gamma(G, S)$ be a Cayley graph of a finite group, let $H \triangleleft G$ be a normal subgroup. and let $0 < \rho, \alpha < 1$ be any constants. Suppose that $R = H \cap S$ generates H, and $p > \max\{\frac{1}{\sqrt{2}}, p_c(\Gamma(H, R); \rho, \alpha)\}$. There exist constants $\beta = \beta(\rho) < 1$ and $N = N(\rho, \alpha)$, so that if $\alpha > \beta$ and $[G : H] > N$, and*

$$\left(\ln \frac{3}{1 - \alpha} + \ln [G : H] \right) [G : H] \leq \left(\rho - \frac{1}{2} \right) |H| \tag{1}$$

we have $p_c(\Gamma(G, S); \rho, \alpha) \leq p$.

We prove these theorems in the sections that follow, and conclude with a few examples and open problems.

2 Basic Results

Large components in finite graphs are the analogues of infinite clusters in infinite graphs. The Benjamini conjecture appears to be inspired by Grimmett's Theorem (see, *e.g.*, [12], pages 304–309), which guarantees the existence of infinite clusters in certain subsets of the square lattice.

Theorem 5 (Grimmett) *Let f be a function so that $\frac{f(x)}{\log x} \to a$ as $x \to \infty$, for some positive constant a. Let $G(f)$ denote the region in the positive quadrant of the square lattice under the function $f(x)$. There exists $p < 1$ so that this region has an infinite component after p–percolation almost surely.*

The following lemma follows easily from this theorem. We will use it in our proof of Theorem 1.

Lemma 6 *Let Γ be an $m \times n$ box within the square grid, and let $\rho, \alpha < 1$ be constants. Then there exists $\varepsilon = \varepsilon(\rho, \alpha) > 0$ such that if $n \geq m > \log n$, we have $p_c(\Gamma; \rho, \alpha) < 1 - \varepsilon$.*

We conclude this section with a counting lemma that provides a way to bound the critical probability of any vertex transitive graph.

Proposition 7 *Let Γ be a vertex transitive graph undergoing p–percolation. Distinguish a vertex z. Suppose that there are constants $0 < \tau, \rho < 1$ such that for every vertex $v \in \Gamma$, the probability that z lies in the same connected component as v after percolation is at least $\tau + \rho - \tau\rho$. Then the probability that z belongs to a configuration of size at least ρ is at least τ.*

Proof: We prove the contrapositive: If the probability that z is in a component of size smaller than ρ is at least $1 - \tau$, then there exists a vertex x whose probability of being in a different component than z is at least $1 - \tau - \rho + \tau\rho$.

For each vertex $v \in \Gamma$, let $m(v)$ denote the probability that v is connected to z after percolation. Then

$$\sum_{v \in \Gamma} m(v) \leq \tau|\Gamma| + (1 - \tau)\rho|\Gamma| \qquad (2)$$

Indeed, even if all the graphs with ρ–size connected component were entirely connected, they would not contribute more than $\tau|\Gamma|$ to the sum, because such graphs occur with probability no more than τ. This gives the first term. The remaining graphs contribute to $m(v)$ for no more than ρ fraction of the vertices. This gives the second term.

Therefore, some vertex v must have $m(v) \leq \tau + \rho - \tau\rho$. $\square$

3   Commuting Generators

In this section, we prove Theorem 9, which generalizes Theorem 3 from the introduction. For clarity, we start with the following motivating example.

Let $\Gamma = \Gamma(S_n, R_n)$ be the Cayley graph for the symmetric group, with $R_n = \{(1\ 2), (2\ 3), \dots, (n - 1\ n)\}$ the Coxeter transpositions. We may bound the critical probability of this Cayley graph using an idea that applies to any family of groups with lots of generators and lots of short disjoint relations.

Proposition 8 *There exists $\varepsilon > 0$ such that for all n, $p_c(\Gamma(S_n, R_n); \frac{2}{3}, \frac{1}{2}) \leq 1 - \varepsilon$.*

Proof: By Proposition 7, it suffices to show that every element $g \in S_n$ remains connected to the identity 1 with probability at least $5/6$.

Let d be the diameter of $\Gamma(S_n, R_n)$; we have $d = O(n^2)$. Fix a path from 1 to g of length no more than d. By the Chernoff bound, with probability $1 - e^{-\frac{p^2 d}{2}}$, no more than δd edges of this path are deleted under p–percolation, where $\delta = 1 - p + p^2$.

Let's consider how to get around a deleted edge $(i\ i + 1)$. Any of the $n - 4$ generators $(j\ j+1)$ with $\{j, j+1\}$ disjoint from $\{i, i+1\}$ commutes with $(i\ i+1)$, in which case we can replace the edge $(i\ i + 1)$ by the three–edge sequence $(j\ j+1)(i\ i+1)(j\ j+1)$. Each three–edge sequence is unbroken with probability p^3, and since they are disjoint from each other, the probability that all $n - 4$ detours break is $(1 - p^3)^{n-4}$.

Assuming that no more more than δd edges of the path are deleted, we can find unbroken detours around all the deleted edges with probability at least $1 - \delta d(1 - p^3)^{n-4}$. Therefore, if $e^{-\frac{p^2 d}{2}} + \delta d(1 - p^3)^{n-4} < \frac{1}{6}$, the proposition is proven. Since $d = O(n^2)$, we can find a $p = 1 - \varepsilon$ to satisfy this inequality for large n. $\square$

We can generalize this result for other sequences of Cayley graphs as follows.

If $\Gamma(G, R)$ is a Cayley graph and $r \in R$, a *relation for* r is an identity of the form $r = s_1 \cdots s_m$, where $s_i \in R$. Its length is $m + 1$. We say that two relations $r = s_1 \cdots s_m$ and $r = t_1 \cdots t_n$ for r are *disjoint* if $s_1 \cdots s_j \neq s_1 \cdots s_k$ for any $j, k > 0$, except when $j = m$ and $k = n$.

Theorem 9 *Let* $\Gamma_n = \Gamma(G_n, R_n)$ *be a sequence of Cayley graphs with diameters* $d_n = \text{diam}(\Gamma_n)$. *Suppose that* $d_n \to \infty$, *and that there is a constant* C *such that for all* n *and all* $s \in R_n$, *there are at least* $2 \log d_n$ *disjoint relations for* s, *each having length no more than* C. *There exists* $\varepsilon > 0$ *such that for all* n, $p_c(\Gamma(G_n, R_n); \frac{2}{3}, \frac{1}{2}) \leq 1 - \varepsilon$.

Proof: As above, we count disjoint detours around an edge $\{a, b\} \in \Gamma$. For simplicity, we may assume $a = 1$ so that $b \in S$.

For each relation $b = s_1 \cdots s_n$, we consider the detour that replaces the edge $\{1, b\}$ with the edges $\{1, s_1\}$, $\{s_1, s_1 s_2\}$, $\ldots$, $\{s_1 s_2 \cdots s_{n-1}, b\}$, and apply Proposition 7 to obtain our result. Consider a path of length at most d_n from 1 to x. With probability $1 - e^{-\frac{p^2 d_n}{2}}$, at most $\delta = 1 - p + p^2$ fraction of its edges are broken. For each of these deleted edges $\{a, ar\}$, we have constructed at least $2 \log d_n$ disjoint detours of C edges. The probability that all of these are broken is no more than $(1 - p^C)^{2 \log d_n}$. Thus, the total probability we cannot patch the path from 1 to x with our detours is no more than

$$e^{-\frac{p^2 d_n}{2}} + \delta d_n (1 - p^C)^{2 \log d_n} \tag{3}$$

If pick p so that $p^C > \frac{1}{2}$, then $\delta d_n (1 - p^C)^{2 \log d_n} < \frac{\delta d_n}{d_n^2}$. Since $d_n \to \infty$ as $n \to \infty$, $e^{-\frac{p^2 d_n}{2}} \to 0$. Increase p so that $\Gamma(G_n, R_n)$ has a large component in each of the finitely many graphs where the expression (3) is greater than $\frac{1}{6}$. $\square$

Proof: Pruning $T_n(s)$ so that $r \in T_n(s) \Rightarrow rs \notin T_n(s)$ cuts the size of $T_n(s)$ by at most half, and ensures that the commutation relations between s and the elements of $T_n(s)$ are disjoint. Each commutation relation has length $C = 3$. $\square$

4   Semidirect Products

Recall the construction of a semidirect product. Let H and K be groups, with an action of K on H whereby $k \in K$ sends $h \in H$ to an element denoted $h^k \in H$. The *semidirect product* of H and K, denoted $H \rtimes K$, is the set of ordered pairs $(h, k) \in H \times K$ equipped with the operation

$$(h_1, k_1) \cdot (h_2, k_2) = (h_1 \cdot h_2^{k_1}, k_1 \cdot k_2) .$$

We can regard H and K as subgroups of G via the inclusions $h \to (h, 1)$ and $k \to (1, k)$.

Theorem 10 *Let constants ρ and α be given, with $\rho^2(2 - \rho) > \alpha$ and α sufficiently close to one. There exists a constant C so that if $|G| > C^2$, $G = H \rtimes K$, $p > 0$, and H and K have generating sets R and S for which $\Gamma(H, R)$ and $\Gamma(K, S)$ belong to $\mathcal{L}(\rho, \alpha, p)$, then $\Gamma(G, R \cup S) \in \mathcal{L}(\rho, \alpha, p)$.*

Proof: We may write the elements of G uniquely as $g = hk$ where $h \in H$ and $k \in K$. For $h \in H$, let K_h be the subgraph $\{hk : k \in K\}$, with edges joining hk to hks for $s \in K$. For $k \in K$, let H_k be the subgraph $\{hk : h \in H\}$, with edges joining hk to hkr for $r \in H$. The product structure in $H \rtimes K$ is given by

$$h_1 k_1 \cdot h_2 k_2 = (h_1(k_1 h_2 k_1^{-1}))(k_1 k_2)$$

Examining this product when $k_2 = e$ or $h_2 = e$, we see that the sets H_k and K_h are closed under right multiplication by elements of H or K, respectively.

The graph of K_h clearly is isomorphic to the Cayley graph of K. The graph of H_k is isomorphic to the Cayley graph of H under the isomorphism $(khk^{-1})k = kh \to h$, for if $h_1 r = h_2$, then $(kh_1)r = k(h_1 r) = kh_2$. Thus, each K_h and each H_k has a component of size at least $\rho|K|$ with probability at least α independently.

We regard the sets K_h as the "rows" and the sets hK as the "columns" of the Cayley graph G. If some column H_k (or row K_h) has a connected component of size ρ considering only the generators in R (or in S), we call the column (or row) "good," and its "good part" comprises the vertices of this component.

Either H or K has order at least C; we assume that K does (in the argument that follows, K and H can be reversed if necessary). If $\delta > 0$, the probability q that the number of good columns H_k is less than $\alpha\delta|K|$ satisfies

$$q < e^{-\frac{\alpha|K|(1-\delta)^2}{2}} \tag{4}$$

by the Chernoff bound.

Denote by a and b the fractions of the columns and rows that are good. Assuming $a, b > \frac{1}{2}$, all good parts of good rows lie in the same component of G, because they intersect $a|K|$ good columns, and therefore any pair of good rows must hit $(2a - 1)|K|$ of the same good columns. As $|K|$ goes to infinity, this means they almost surely both hit the good part of some good column.

Some of the good columns may not be joined to the large component of G. In the worst case, $|H| = 2$. Since $b > \frac{1}{2}$, at least two rows are good. We expect $1 - (1 - \rho)^2 = \rho(2 - \rho)$ fraction of the good columns to hit the good parts of one of these two good rows; the good parts of these columns belong to the large component. Let M be the expected elements of G that belong either to the good part of a good row or to the good part of a good column. Since membership in a good row is independent from membership in a good column, we have

$$M = 1 - (1 - a\rho^2(1 - \rho))(1 - b\rho) . \tag{5}$$

All such vertices belong to the large component of G. We wish to choose a and b to make expression (5) bigger than ρ. This occurs when

$$b > \frac{1 - a\rho(2 - \rho)}{1 - a\rho^2(2 - \rho)} .$$

The mean number of good columns is $\alpha |K|$. The Chernoff bound shows that the probability that there are fewer than $a = \alpha \delta |K|$ good columns or $b = \alpha \epsilon |H|$ good rows is no more than

$$\xi = e^{-\alpha |K|(1-\delta)^2} + e^{-\alpha |H|(1-\epsilon)^2} .$$

To satisfy the conclusion of the theorem, we need this failure probability to be $\xi < 1 - \alpha$. To make expression (5) greater than ρ, set

$$\epsilon = \frac{1 - \alpha \delta \rho(2 - \rho)}{2\alpha(1 - \alpha \delta \rho^2(2 - \rho))} .$$

For any $\delta < 1$, we can require K (the larger of the subgroups) to be big enough so that $e^{-\alpha |K|(1-\delta)^2} < \frac{1-\alpha}{2}$. Choose δ so that $\delta \rho^2(2 - \rho) > \alpha$ (possible by the hypothesis on ρ and α) and $\frac{1}{\alpha} + \delta \rho(2 - \rho) > 2$ (possible since $\rho(2 - \rho) > \rho^2(2 - \rho) > \alpha > 2 - \frac{1}{\alpha}$). We have

$$
\begin{aligned}
e^{-\alpha |H|(1-\epsilon)^2} &\le e^{-2\alpha(1-\epsilon)^2} \\
&= e^{-2\alpha\left(1 - \frac{1-\alpha\delta\rho(2-\rho)}{2\alpha(1-\alpha\delta\rho^2(2-\rho))}\right)^2} \\
&= e^{-2\alpha\left(1 - \frac{\frac{1}{2}\left(\frac{1}{\alpha}+\delta\rho(2-\rho)\right)}{1-\alpha\delta\rho^2(2-\rho)}\right)^2} \\
&< e^{-2\alpha\left(1 - \frac{1}{1-\alpha^2}\right)^2} \\
&< \frac{1-\alpha}{2}
\end{aligned}
$$

for α sufficiently close to 1.

Thus, $\Gamma(G, R \cup S)$ has a connected component of size $\rho |G|$ after p–percolation with probability at least α. $\square$

Proof of Theorem 2:   Let G be a nilpotent basis with Hall basis given as in Section 1. By Theorem 1, there exists $\varepsilon = \varepsilon(\rho, \alpha)$ independent of i, such that $p_c(\Gamma(H_i, \gamma_i(S_i)); \rho, \alpha) < 1 - \varepsilon$ for every i. As each $G_i = G_i \rtimes H_{i+1}$, Theorem 10 shows $p_c(\Gamma(G, S); \rho, \alpha) < 1 - \varepsilon$. $\square$

5   Proof of Reduction Theorem

Our proof has three steps. Roughly, we show:

1. Most of the cosets in G/H, considered as subgraphs, have big (size $\rho |H|$) connected components
2. Most of these components are connected to each other, forming a pretty big (but not yet size $\rho |G|$) component of G
3. Enough additional vertices are attached to the big component so that it has size $\rho |G|$.

We formalize these ideas below.

Let $n = [G : H]$, and write $G/H = \{Hg_i\}_{i=1}^n$. Consider each coset as a subgraph of G. Prior to percolation, each is isomorphic as a graph to H. When we p–percolate G, our assumption implies that each has a connected component of size at least $\rho|H|$ with probability at least α independently, just by considering the generators $R \subset S$. Call a coset with such a large component a "good coset," and call its large component the "good part" of the good coset. Applying the Chernoff bound, with probability $1 - e^{-\frac{(\alpha-\alpha^2)n}{2}}$, there are at least $\alpha^2 n$ good cosets.

Fix a spanning tree on the Cayley graph $\Gamma(G/H, \pi(S))$, and choose the root to be a good coset. Although the parent of a good coset need not be a good coset, we can take parents recursively until we find reach one that is good. All the good cosets will be connected if each good coset is connected to its good parent. With high probability, say $1 - \epsilon$, every good parent is no more than $2 \log n$ levels higher in the tree.

Suppose Hg_j is the good parent of Hg_i. Then $g_j = g_i s_{i_1} \cdots s_{i_r}$ for some string of generators $s_{i_1}, \dots, s_{i_r} \in S$ where $Hg_i s_{i_1}$ is the parent of Hg_i, etc. We have $r \leq m$. Right multiplication by $s_{i_1} \cdots s_{i_r}$ gives a bijection from Hg_i to Hg_j. By the inclusion–exclusion principle, the image of the good part of Hg_i must hit at least $\rho - \frac{1}{2}$ fraction of the elements of Hg_j. Therefore, in order for the good part of Hg_i to fail to be connected to the good part of Hg_j, we would need each of the $(\rho - \frac{1}{2})|H|$ paths of the form $x, xs_{i_1}, \dots, xs_{i_1} \cdots s_{i_r}$ to break. Since these paths are all disjoint, the probability that they all break is no more than $(1 - p^r)^{(\rho-\frac{1}{2})|H|}$. Thus, the probability that some good coset fails to have its good part connected to that of its good parent is no more than

$$n(1 - p^m)^{(\rho-\frac{1}{2})|H|} \leq n(e^{-p^m})^{(\rho-\frac{1}{2})|H|}$$
$$\leq n(e^{-p^{2\log n}})^{(\rho-\frac{1}{2})|H|}$$
$$= n(e^{-(\rho-\frac{1}{2})|H|p^{2\log n}})$$
$$\leq ne^{-\frac{(\rho-\frac{1}{2})|H|}{n}}$$

applying the hypothesis $p > \frac{\sqrt{2}}{2}$. This failure probability is less than $\frac{1-\alpha}{3}$ if

$$\left(\ln \frac{3}{1-\alpha} + \ln n\right) n \leq \left(\rho - \frac{1}{2}\right)|H|$$

which is implied by our hypothesis relating $[G : H]$ and $|H|$.

For every good coset Hg_i, choose a $\rho|H|$–size connected subset of Hg_i uniformly at random, and call it A_i. Let B_i be the complement of A_i in Hg_i.

Call a coset Hg_i "nearly good" if there exists $s \in S$ so that $Hg_i s$ is a good coset. Put $g_j = g_i s$.

Lemma 11 *We expect*

$$|B_j \cap A_i s| \geq \left(\frac{1}{2}\rho - \frac{1}{4}\right)(1 - \rho)|H| . \tag{6}$$

Furthermore, $|B_j \cap A_i s| > \frac{1}{4}(1-\rho)|H|$ with probability at least $2\rho - 1$.

Proof: Before percolation, Hg_i and $Hg_j = Hg_i s$ are each isomorphic to the Cayley graph $\Gamma(H, R)$, and the percolation on G affects these two subgraphs independently.

Let $\mathcal{A}$ be the set of connected subsets of H having cardinality exactly $\rho|H|$. Observe that if $X \in \mathcal{A}$ and $g \in H$, then $Xg \in \mathcal{A}$. Thus, we can partition $\mathcal{A}$ into equivalence classes of $|H|$ elements each, by identifying $X \in \mathcal{A}$ with all its right translates $\{Xg\}_{g \in H}$. Percolation on $\Gamma(H, R)$ defines a probability distribution on $\mathcal{A}$ that assigns to $X \in \mathcal{A}$ the probability that $A_i = Xg_i$, and this distribution is independent of i. The expected intersection $|A_j \cap A_i s|$ is the same as the expected intersection $|X \cap Y|$ where X and Y are chosen from $\mathcal{A}$ according to this distribution.

Fix any set $X \in \mathcal{A}$. We want to show that the intersection $|X \cap Y|$ where Y is chosen from $\mathcal{A}$ is expected to be small. Because Cayley graphs are vertex transitive, we have $\text{Pr}_Y(A_i = Yg_i) = \text{Pr}_Y(A_i = (Yg)g_i)$ for any $g \in H$. Therefore, if we find a constant M such that the expectation value $\text{E}_{g \in H}|X \cap Yg| \le M$ for all fixed $X, Y \in \mathcal{A}$, choosing g uniformly at random from H, we can conclude that $\text{E}_{Y \in \mathcal{A}}|X \cap Y| \le M$ when Y is chosen randomly according to our distribution on $\mathcal{A}$.

We proceed to show that $M = \left(\frac{1}{4} + \frac{1}{4}\rho + \frac{1}{2}\rho^2\right)|H|$ has the desired property. Replace Y by its right translate that maximizes $|X \cap Y|$. Let $K = H - Y$ and $K' = H - X$. Evidently, $|K| = |K'| = (1-\rho)|H|$. As

$$|X \cap Y| = |X| - (|K'| - |K \cap K'|) \tag{7}$$

maximizing $|X \cap Y|$ is the same as maximizing $|K \cap K'|$. If $|K \cap K'| \le \frac{3}{4}(1-\rho)|H|$, we are done because equation (7) implies $|X \cap Y| \le \frac{5}{4}\rho - \frac{1}{4}$, and $\frac{5}{4}\rho - \frac{1}{4} \le \frac{1}{4} + \frac{1}{4}\rho + \frac{1}{2}\rho^2 = M$ always.

Suppose, on the other hand, that $|K \cap K'| > \frac{3}{4}(1-\rho)|H|$. If $g \in H$ has the property that $|Kg \cap K'| > \frac{3}{4}(1-\rho)|H|$ as well, then $|K \cap Kg| > \frac{1}{2}(1-\rho)|H|$. But at most $2|K| = 2(1-\rho)|H|$ values of g can have this property. Indeed, for $g \in G$, let $K_g = \{x \in K : xg \in K\}$. Then

$$\sum_{g \in H} |K_g| = |K|^2$$

(each element of K has $|K|$ translates that lie inside K), which implies that at most $2|K|$ values of g have $|K_g| > \frac{1}{2}|K|$. This is what we desire, as $|K_g| = |K \cap Kg|$.

For the values of g with $|K \cap Kg| \le \frac{1}{2}|K|$ and hence $|Kg \cap K'| \le \frac{3}{4}|K|$ we have by equation (7)

$$|X \cap Yg| \le |H| - \left(2 - \frac{3}{4}\right)|K| \tag{8}$$

$$= |H|\left(1 - \frac{5}{4}(1-\rho)\right) \tag{9}$$

For the remaining values of g (of which there are no more than $2|K|$), we bound $|X \cap Yg| \leq |X|$. This yields

$$E_{g \in H}(|X \cap Yg|)$$
$$\leq \frac{2|K||X| + (|H| - 2|K|)|H|(1 - \frac{5}{4}(1 - \rho))}{|H|}$$
$$= \frac{2(1 - \rho)\rho|H|^2 + (2\rho - 1)(\frac{5}{4}\rho - \frac{1}{4})|H|^2}{|H|}$$
$$= \left(\frac{1}{4} + \frac{1}{4}\rho + \frac{1}{2}\rho^2\right)|H|$$

This shows that $M = (\frac{1}{4} + \frac{1}{4}\rho + \frac{1}{2}\rho^2)|H|$ has the desired property.

Thus we expect

$$|B_j \cap A_i s| = |A_i s| - |A_j \cap A_i s|$$
$$\geq \rho|H| - (\frac{1}{4} + \frac{1}{4}\rho + \frac{1}{2}\rho^2)|H|$$
$$= \left(\frac{1}{2}\rho - \frac{1}{4}\right)(1 - \rho)|H|$$

which yields the first statement of the lemma. In the same way, equation (9) yields the second statement. $\square$

Fix a good neighbor $Hg_{j_i} = Hg_i s_i$ where $s_i \in S$ for as many nearly good cosets Hg_i as possible (which is at least $\alpha^2 n$). Let $r = j_i$. The sizes of the intersections $B_r \cap A_i s_i$ and $B_{j_r} \cap A_r s_{j_r}$ are independent unless $j_r = i$. We may arrange for this not to be the case. Applying the Chernoff bound to the second statement of the lemma, there remain at least $(\rho - \frac{1}{2})\alpha^2 n$ great cosets Hg_i with $|B_{j_i} \cap A_i s_i| > \frac{1}{4}(1 - \rho)|H|$, with probability at least $1 - e^{-\frac{(\rho - \frac{1}{2})\alpha^2 n}{2}}$.

The $\frac{1}{4}(1 - \rho)$ vertices of these nearly good cosets attach to the large connected component of size $\rho\alpha^2$ we found in step 2. Altogether, we have a connected component of size

$$\rho\alpha^2 + (\rho - \frac{1}{2})\alpha^2 \frac{1}{4}(1 - \rho) \tag{10}$$

with probability at least

$$1 - \epsilon - e^{-\frac{\alpha^2 n}{2}} - ne^{-\frac{(\rho - \frac{1}{2})|H|}{n}} - e^{-\frac{(\rho - \frac{1}{2})\alpha^2 n}{2}} \tag{11}$$

Write $\omega = \frac{1}{4}(\rho - \frac{1}{2})(1 - \rho)$. If $\alpha > \sqrt{\frac{\rho + \frac{\omega}{2}}{\rho + \omega}}$, expression 10 describes a component of size at least ρ. Requiring $n > \frac{8 \ln \frac{3}{1 - \alpha}}{(2\rho - 1)(2\alpha^2 - 1)}$ and applying the hypothesis relating n and $|H|$, we find that expression 11 is at least α. This proves Theorem 4. with $\beta = \sqrt{\frac{\rho + \frac{\omega}{2}}{\rho + \omega}}$ and $N = \frac{8 \ln \frac{3}{1 - \alpha}}{(2\rho - 1)(2\alpha^2 - 1)}$. $\square$

6   Proof of Theorem 1

We wish to embed the Cayley graph $\Gamma = \Gamma(G, S)$ of an abelian group G into a two–dimensional box, so that we can apply Grimmett's Theorem (Lemma 6).

The generators $s_1, \ldots, s_n$ in the Hall basis define a homomorphism $\varphi : \mathbb{Z}^n \to G$, given by

$$\varphi(a_1, \ldots, a_n) = s_1^{a_1} \cdots s_n^{a_n} .$$

The homomorphism φ maps the box $B = [0, a_1 - 1] \times \cdots \times [0, a_n - 1]$ bijectively onto G, To flatten B into a two–dimensional box, we will select k dimensions and choose Hamiltonian paths in the Cayley graphs of a section and a cross section. Unwrapping these Hamiltonian paths each into one dimension will produce the desired two–dimensional box.

We claim that there exists $I \subset \{1, \ldots, n\}$ such that $\prod_{i \in I} a_i > \frac{\log |G|}{2}$ and $\prod_{i \notin I} a_i > \frac{\log |G|}{2}$. Indeed, choose the smallest k such that $a_1 \cdots a_k > \frac{\log |G|}{2}$. If $a_1 \cdots a_k < \frac{2|G|}{\log |G|}$, then we may take $I = \{1, \ldots, k\}$, and we are done. If $k = 1$, this inequality is assured by the diameter assumption, since $\mathrm{diam}\Gamma = (a_1 + \cdots + a_n)/2$. If $k > 1$ and yet $a_1 \cdots a_k > \frac{2|G|}{\log |G|}$, then $a_k > \frac{2|G|}{\log^2 |G|} > 2 \log |G|$, assuming $|G|$ is large enough. The diameter condition implies that $a_1 \cdots a_{k-1} a_{k+1} \cdots a_n > \frac{\log |G|}{2}$, so $I = \{k\}$ has the desired property.

Now choose Hamiltonian paths $g : [0, x - 1] \to \prod_{i \in I}[0, a_i]$ and $h : [0, y - 1] \to \prod_{i \notin I}[0, a_i]$. Let A be the box $[0, x - 1] \times [0, y - 1]$. Because $\varphi \circ (g, h)$ is a graph homomorphism mapping A bijectively onto G, it suffices to show that $A \in \mathcal{L}(\rho, \alpha p)$. This follows immediately from Lemma 6, since x and y are each at least $\frac{\log |G|}{2}$.

7   Examples

1. Our first example is a hypercube C_n, which is a Cayley graph of the group $\mathbb{Z}_2^n$ with the usual set of generators $R = \{r_1, \ldots, r_n\}$. In this case, $\mathrm{diam}(C_n) = n = o(\frac{2^n}{n})$. Therefore, $p_c(C_n) < 1 - \varepsilon$ for some $\varepsilon > 0$, by Theorem 1. Of course, this bound is much weaker than $p_c = (1 + o(1))/n$ established in [1].

2. Consider $G_n = S_n \ltimes \mathbb{Z}_2^n$, with the generating set $R_n = \{((i \ \ i + 1), \overline{0}), (\mathrm{id}, r_j); i = 1, \ldots, n - 1; j = 1, \ldots, n\}$. From the previous example, Proposition 8, and Theorem 10, $p_c(\Gamma(G_n, R_n)) < \max\{p_c(C_n), p_c(\Gamma(S_n, R_n))\} < 1 - \varepsilon$ for some $\varepsilon > 0$.

3. Fix a prime power q. Let $G_n = U(n, \mathbb{F}_q)$ be the group of $n \times n$ upper triangular matrices over the finite field with q elements, with ones along the diagonal. Consider the set $L_n = \{E_{i,j}^{\pm} : 1 \leq i < j \leq n\}$ of all elementary transvections $E_{i,j}^{\pm}$ with ± 1 in position (i, j), ones along the diagonal, and zeros elsewhere. Clearly, L_n generates G_n. The subgroup H generated by the $E_{i,j}^{\pm}$ with $j > i+1$ (consisting of matrices with zero on the first superdiagonal) is isomorphic to G_{n-1}, and the quotient $K = G_n/H$ is isomorphic to $\mathbb{F}_q^{n-1}$. By Theorems 1 and 10 and induction on n, there exists a constant ε so that $p_c(\Gamma(G_n, L_n)) < 1 - \varepsilon$.

4. Let $G_n = B(n, \mathbb{F}_q)$, and $H_n = U(n, \mathbb{F}_q)$. Let R_n be any generating set for the diagonal subgroup. Then $R_n \cup L_n$ generates G_n, and equation (1) is satisfied for large n. The Reduction Theorem 4 gives $p_c(\Gamma(G_n, R_n \cup L_n)) < 1 - \varepsilon$.

5. Let $G_n = U(n, \mathbb{F}_q)$ and $R_n = \{E_{i,i+1}^{\pm} : i = 1, \ldots, n-1\}$. Because R_n is no longer a Hall basis in G_n, we cannot apply Theorem 2. However, Theorem 3 applies in the same manner as in Proposition 8.

6. Consider S_n with the star transpositions $R_n = \{r_i = (1\ i) : i = 2, \ldots, n\}$. None of these generators commute, so we cannot apply Theorem 3. However, the short relations $(r_i\ r_j)^3 = 1$ can be used in Theorem 9 to obtain $p_c(\Gamma(S_n, R_n); \frac{2}{3}, \frac{1}{2}) < 1 - \varepsilon$.

8   Concluding Remarks

We are unable to prove the Benjamini conjecture in its full generality, even for abelian groups. It would be nice to prove the Benjamini conjecture for all generating sets of finite abelian groups.

In the notation of Theorem 2, we have

$$\mathrm{diam}(\Gamma(H_1, \gamma_1(S_1))) \leq \mathrm{diam}(G, S)$$

$$\leq \sum_{i=1}^{\ell} \mathrm{diam}(\Gamma(H_i, \gamma_i(S_i))) .$$

Therefore, the condition $(*)$ is formally neither weaker nor stronger than Benjamini's conjectured condition.

In view of the Reduction Theorem, it is important to study simple groups with small generating sets. For example, any simple group can be generated by two elements, one of which is an involution (see [11]). The corresponding Cayley graph may provide interesting test cases for Benjamini's conjecture.

It is well known that all Cayley graphs Γ_n of the symmetric group S_n have a diameter $e^{o(\sqrt{n \log n})} = o\left(\frac{n!}{n \log n}\right)$. Proving Benjamini's conjecture in these cases is the ultimate challenge for the reader. Even for the generating set $\{(1\ 2), (1\ 2\ \cdots n)^{\pm 1}\}$, we are unable to bound p_c away from 1.

References

[1] AJTAI, M., KOMLÓS, J., AND SZEMERÉDI, E. Largest random component of a k-cube. *Combinatorica 2*, 1 (1982), 1–7.

[2] ALON, N., BENJAMINI, I., AND STACEY, A. Percolation on finite graphs and isoperimetric inequalities. In preparation.

[3] ASTASHKEVICH, A., AND PAK, I. Random walks on nilpotent groups. *http://www-math.mit.edu/~pak/research.html* (2001).

[4] BABSON, E., AND BENJAMINI, I. Cut sets and normed cohomology with applications to percolation. *Proc. Amer. Math. Soc. 127*, 2 (1999), 589–597.

[5] BENJAMINI, I. Percolation on finite graphs. *math.PR/0106022* (2001).

[6] BENJAMINI, I., AND SCHRAMM, O. Percolation beyond $\mathbf{Z}^d$, many questions and a few answers. *Electron. Comm. Probab. 1* (1996), no. 8, 71–82 (electronic).

[7] BENJAMINI, I., AND SCHRAMM, O. Recent progress on percolation beyond $\mathbf{Z}^d$. *http://www.wisdom.weizmann.ac.il/~schramm/ papers/pyond-rep* (2000).

[8] BOLLOBÁS, B. *Random graphs*, second ed. Cambridge University Press, Cambridge, 2001.

[9] DIESTEL, R., AND LEADER, I. A conjecture concerning a limit of non-Cayley graphs. *J. Algebraic Combin. 14*, 1 (2001), 17–25.

[10] ERDŐS, P., AND RÉNYI, A. On the evolution of random graphs. *Magyar Tud. Akad. Mat. Kutató Int. Közl. 5* (1960), 17–61.

[11] GORENSTEIN, D. *Finite simple groups*. Plenum Publishing Corp., New York, 1982.

[12] GRIMMETT, G. *Percolation*, second ed. Springer-Verlag, Berlin, 1999.

[13] HALL, P. *The Edmonton notes on nilpotent groups*. Mathematics Department, Queen Mary College, London, 1969.

[14] JANSON, S., ŁUCZAK, T., AND RUCINSKI, A. *Random graphs*. Wiley-Interscience, New York, 2000.

[15] KRIVELEVICH, M., SUDAKOV, B., AND VU, V. H. Sharp threshold for network reliability. *Combinatorics, Probability and Computing*, To appear.

[16] PAK, I., AND VU, V. H. On mixing of certain random walks, cutoff phenomenon and sharp threshold of random matroid processes. *Discrete Appl. Math. 110*, 2-3 (2001), 251–272.

[17] SHALEV, A. Simple groups, permutation groups, and probability. In *Proceedings of the International Congress of Mathematicians, Vol. II (Berlin, 1998)* (1998), no. Extra Vol. II, pp. 129–137 (electronic).

Computing Graph Properties by Randomized Subcube Partitions

Ehud Friedgut[1], Jeff Kahn[2], and Avi Wigderson[3]

[1] The Hebrew University
ehudf@math.huji.ac.il
[2] Rutgers University
jkahn@math.rutgers.edu
[3] Institute for Advanced Study
and The Hebrew University
avi@ias.edu

Abstract. We prove a new lower bound on the randomized decision tree complexity of monotone graph properties. For a monotone property $\mathcal{A}$ of graphs on n vertices, let $p = p(\mathcal{A})$ denote the threshold probability of $\mathcal{A}$, namely the value of p for which a random graph from $G(n, p)$ has property $\mathcal{A}$ with probability $1/2$. Then the expected number of queries made by any decision tree for $\mathcal{A}$ on such a random graph is at least $\Omega(n^2 / \max\{pn, \log n\})$.

Our lower bound holds in the subcube partition model, which generalizes the decision tree model. The proof combines a simple combinatorial lemma on subcube partitions (which may be of independent interest) with simple graph packing arguments. Our approach motivates the study of packing of "typical" graphs, which may yield better lower bounds.

1 Introduction

A decision tree is the most basic computational device. It has unlimited power, and attempts to compute a function of an input string by successively (and adaptively) querying its values in different coordinates. The complexity of such an algorithm is the maximum number of queries used for a worst case input.

A basic problem studied for this model is the complexity of monotone graph properties. Here the input is a graph (on n vertices), and the function is the indicator of a monotone (i.e. upward closed) family of such graphs $\mathcal{A}$. A fundamental result of Rivest and Vuillemin [RV76] is that *every* such function requires $\Omega(n^2)$ queries. In other words, the trivial algorithm of querying the whole input is not worse than the best algorithm by more than a constant factor. (We assume throughout that properties $\mathcal{A}$ are nontrivial, meaning $\mathcal{A}$ contains some, but not all graphs.) Moreover, further results [KSS84,CKS02] show that for some values of n every such function is *evasive*, namely the trivial algorithm is best possible—no algorithm can save even one query.

When randomization enters, the story becomes more interesting. A randomized decision tree for a given function is simply a probability distribution over

J.D.P. Rolim and S. Vadhan (Eds.): RANDOM 2002, LNCS 2483, pp. 105–113, 2002.
© Springer-Verlag Berlin Heidelberg 2002

(deterministic) decision trees for that function. Thus, we deal here only with errorless algorithms. The complexity of such an algorithm is the maximum over inputs of the expectation (with respect to the given distribution) of the number of queries. Randomized decision trees can be much more efficient than their deterministic counterparts. The largest gap known is for a function on N bits, which is evasive (hence its deterministic complexity is exactly N), while its randomized complexity is $\Theta(N^{.753..})$. Moreover, this function is transitive (that is, invariant under some group acting transitively on the coordinates).

The question of the randomized complexity of monotone graph properties was raised by several people, and a conjecture attributed to Karp is that for these functions randomization *does not* help—namely that every such property of graphs on n vertices has randomized complexity $\Omega(n^2)$. For some properties, e.g. that of having an isolated vertex, it is easy to see that the randomized complexity is about $n^2/4$, roughly half the trivial bound, but no better upper bound (even for algorithms that are allowed to make a small error) is known for any property. For some specific properties a quadratic lower bound is known.

The first nontrivial lower bounds on the randomized complexity of general graph properties—$\Omega(n(\log n)^{1/12})$—were given by Yao [Y87]. The basic methods introduced by Yao were significantly improved by [K88,H91,CK01] to yield the current best lower bound $\Omega(n^{4/3}(\log n)^{1/3})$, still a far cry from the (conjectured tight) upper bound. A slightly better bound of $\Omega(n^{3/2})$ has been proved for the class of all properties defined by a single minterm (that is, the property holds iff the input graph contains a (copy of a) specified graph as a subgraph) [G92].

In this paper we give a lower bound that depends on the threshold probability of the property. Let $\mathcal{A}$ be a graph property, and let p be such that

$$\Pr[G(n,p) \in \mathcal{A}] = 1/2,$$

where $G(n,p)$ is the usual random graph on n vertices with edge probability p. Then our lower bound is $\Omega(\min\{n^2/\log n, n/p\})$. Note that this is nearly quadratic for many known properties (e.g. connectivity, Hamiltonicity, absence of isolated vertices, containing a triangle), and improves the best bounds above as long as $p << n^{-1/3}$. On the other hand, we get no improvement for general properties.

The above bound is actually stronger than stated in two senses. One is that it holds not just in the worst case, but on average for $G(n,p)$ (this is typical of almost all lower bounds). The other, more interesting aspect, is that our lower bound holds in a seemingly stronger model. Note that a decision tree for any function f partitions the inputs into subcubes (determined by its paths), each of which contain inputs with a constant f value. Our model allows *any* such partition, not just one derived from a decision tree. The complexity on a given input is simply the co-dimension of the subcube containing that input (corresponding to the number of positions "read" by the algorithm, or equivalently, to the path length in a decision tree). Then our lower bound on the expected complexity (of an input drawn from $G(n,p)$) holds for any partition as above.

The result is obtained using a simple lemma about such partitions. Given any partition of the Boolean cube $\{0,1\}^N$ to subcubes, pick an input at random using the product distribution assigning a 1 to each coordinate independently with probability p. Then the ratio of the expected number of positions "read" containing 1, and the expected number of those containing 0, is exactly $p/(1-p)$. Like the packing lemmas typically used in such lower bounds, this lemma is used to show (roughly) that if the minterms of the function are small, then the maxterms are large. However, here this trade-off improves as p shrinks. Combining this lemma with standard packing arguments gives our lower bound.

Before going into the proof we wish to raise a natural question, which is probably not relevant to improving the present bounds, but is nonetheless of some theoretical interest. We would like to know whether the random subcube partition model we present is essentially different than the random decision tree model.

Question 1. Are there graph properties, or properties of other families of sets, for which the random decision tree complexity has a different asymptotic behavior then the random subcube partition complexity (called RAND(f) below). Are there any "natural" examples of this phenomenon?

The paper is organized as follows. In section 2 we present the subcube partition model, define basic notions and prove the basic combinatorial theorem for it mentioned above. In section 3 we derive the lower bound on computing graph properties in this model. In section 4 we propose some questions and conjectures on packing "typical" graphs, which seem potentially relevant to improving the current bounds, and also of independent interest.

2   Preliminaries

Consider the Boolean cube $C = \{0,1\}^N$. Let $C = \bigcup C_i$ be a partition of C into subcubes. Every such subcube C_i can be associated with a characteristic vector in $\{0,1,*\}^N$. The 0's and 1's are the coordinates that are fixed in C_i and the *'s are the free coordinates. For every such C_i let $X_1(C_i)$ be the number of 1's in the corresponding vector, and $X_0(C_i)$ the number of 0's. We also set $X(C_i) = X_0(C_i) + X_1(C_i)$, the co-dimension of the subcube C_i.

The deterministic complexity of a partition is simply the maximum value of $X(C_i)$. We will say that a partition of C computes a function f if f is constant on each C_i. Let $DET(f)$ denote the minimum deterministic complexity over all partitions that compute f. Note that this is a lower bound on the deterministic decision tree complexity of f.

The distributional complexity over some probability distribution D on C is defined as follows. Let $x \in C$ be a point chosen at random according to D. Let $i(x)$ be the index of the subcube containing x, i.e. $x \in C_{i(x)}$. Define the random variables

$$X_1 = X_1(C_{i(x)}), \ X_0 = X_0(C_{i(x)}), \ X = X_0 + X_1.$$

Now the distributional complexity for this partition is the expectation of X. The distributional complexity of a function f, denoted $DIST(f, D)$, is the minimum of this value over all partitions computing f. Again, this is a lower bound on the distributional complexity of a decision tree for f on D.

Finally, a randomized partition is a distribution on partitions, and we may consider the expectation of X over such a distribution, for a worst case input x. We let $RAND(f)$ be the minimum of this value over distributions supported on partitions which compute f. Again, this is a lower bound on the randomized decision tree complexity of f.

As observed by Yao for decision trees, the von-Neumann MiniMax theorem [vN28] applies in this context, giving, for every f,

$$RAND(f) = \max_D DIST(f, D).$$

As is common, we will choose a particular D to prove a lower bound on $RAND(f)$. We consider only product distributions $D = \mu_p = \mu$ given by by $\mu(x) = p^{\sum x_i} q^{N - \sum x_i}$, where (here and throughout the paper) we set $q = 1 - p$.

Our approach is based on the following simple observation.

Lemma 1. *For any partition $\{C_i\}$ of C and x drawn from μ_p, $\frac{E(X_1)}{E(X_0)} = \frac{p}{q}$.*

We give two proofs of this simple fact.

First proof. It is tempting to try to prove this easy lemma by induction on the number of subcubes. However this does not work, since not every partition into subcubes can be achieved from a coarser one by splitting one of the cubes involved into two; indeed this is what makes the partition model stronger than the decision tree model, in which such a coarser partition always exists—simply merge two maximal paths which diverge only at the last query).

Instead, we will work our way up by reverse induction, starting with the partition into 2^n subcubes and showing that merging two neighboring subcubes does not change $\frac{E(X_1)}{E(X_0)}$.

To begin notice that the lemma is trivial in the case when there are 2^n subcubes (=points). In this case X_1 is the number of 1's in a point chosen at random, $E(X_1) = np$ and $E(X_0) = nq$ so the lemma holds.

Now assume the lemma holds for a given partition: Let $E(X_1)/E(X_0) = p/q$. Let C_1 and C_0 be two neighboring subcubes, i.e. their characteristic vectors are identical except for one coordinate where the C_1-vector has a 1 and the C_0-vector has a 0. Merging the two subcubes into one results in a subcube whose characteristic vector is the same as the two aforementioned vectors except for a * replacing the 1 and the 0 at the coordinate where they disagreed. Note that $\mu(C_1)/\mu(C_0) = p/q$. Let X_1' and X_0' be the new random variables corresponding to the new partition. It is easy to see that $E(X_i') = E(X_i) - \mu(C_i)$ for $1 = 1, 2$ so that

$$\frac{E(X_1')}{E(X_0')} = \frac{E(X_1) - \mu(C_1)}{E(X_0) - \mu(c_0)} = \frac{p}{q}.$$

$\square$

Second proof. For $b \in \{0, 1\}$ and $j \in [N]$ let X_b^j be 1 if the jth coordinate of the characteristic vector of $C_{i(x)}$ is b and 0 otherwise (it is $1 - b$ or *). Note that

$X_b = \sum_{j\in[N]} X_b^j$, so by additivity of expectation, all we have to prove is that for every j, $E(X_1^j)/E(X_0^j) = p/q$. But in fact this holds even if we condition on an arbitrary setting of all coordinates other than x_j: either this setting determines the subcube containing our input, in which case there is no contribution to either numerator or denominator; or it does not, in which case the subcube containing the input is determined by the jth bit, and there is a contribution of p to the numerator and q to the denominator. □

3   Graph Properties

A graph is associated with its characteristic vector of 1's (edges) and 0's (non-edges). Graphs on n vertices are thus represented by binary vectors of length $N = \binom{n}{2}$.

Let $\mathcal{A}$ be a (nontrivial) monotone graph property of graphs on n vertices. "Graph property" means that as a Boolean function on $C = \{0,1\}^N$, $\mathcal{A}$ is invariant under the action induced on edges of the symmetric group S_n (acting on the vertices). Slightly abusing notation, we use $\mathcal{A}$ to denote the set of graphs that have the property, and also the corresponding characteristic function: $\mathcal{A}(G)$ is 1 or 0 depending on whether or not $G \in \mathcal{A}$.

The product measure μ_p on C now becomes the standard measure defining $G(n,p)$, the random graph on n vertices with edge probability p.

Let $p = p(\mathcal{A})$ be the threshold probability for $\mathcal{A}$; that is,

$$\Pr[G(n,p) \in \mathcal{A}] = 1/2.$$

Note that there exists a (unique) such p because $\mu_{\mathcal{A}}(p) = \Pr[G(n,p) \in \mathcal{A}]$ is a continuous, strictly increasing function of $p \in [0,1]$ with $\mu_{\mathcal{A}}(0) = 0, \mu_{\mathcal{A}}(1) = 1$. By replacing $\mathcal{A}$ by its dual (the family of graphs whose complements are not in $\mathcal{A}$) if necessary, we may assume that $p(\mathcal{A}) \leq 1/2$. We will in fact assume in what follows that $p(\mathcal{A}) = o(1)$, since for constant p our result is much weaker than known bounds.

Let C be partitioned into subcubes that compute $\mathcal{A}$ (recall this means that the graphs represented by the vectors in each subcube either all belong to $\mathcal{A}$ or all do not). Let X_1, X_0 and X be defined as before with respect to the given partition and the product measure μ_p with $p = p(\mathcal{A})$.

Theorem 1. $\max\{E(X_0), E(X_1)\} \geq (1 - o(1)) \min\{\frac{n}{64p}, \frac{n^2}{256 \log n}\}.$

Let MIN_1 denote the number of edges in a smallest graph with property $\mathcal{A}$ (i.e. the size of a smallest minterm of the function $\mathcal{A}$). Let MIN_0 denote the number of edges in the complement of the largest graph not in $\mathcal{A}$ (size of a smallest maxterm of $\mathcal{A}$).

Observation 1. $E(X_0) \geq MIN_0/2$

Proof: Note first that if $\mathcal{A}(G) = 0$ for (all) $G \in C_i$ then $X_0(C_i) \geq MIN_0$. Consequently, $E(X_0|x \notin \mathcal{A}) \geq MIN_0$. The result now follows since $\Pr(x \notin \mathcal{A}) = 1/2$) □

Proof of Theorem 1: If $E(X_1) \geq \frac{n}{64}$ then by Lemma 1 $E(X_0) \geq \frac{qn}{64p}$ and we are done. So let us assume $E(X_1) \leq \frac{n}{64}$. Writing Δ for maximum degree, let

$$\mathcal{B} = \{G : \Delta(G) \leq np + \sqrt{4np\log n}\ \}.$$

According to the Chernoff bound, the probability of a given vertex of $G(n, p)$ having degree more than $np + \sqrt{4np\log n}$ is $O(1/n^2)$; hence $\mu_p(\mathcal{B}) = 1 - O(1/n)$. Let $\mathcal{C} = \mathcal{B} \cap \mathcal{A}$. Then $\mu_p(\mathcal{C}) = 1/2 - O(1/n)$, so that

$$E(X_1|x \in \mathcal{C}) \leq (1/2 - O(1/n))^{-1}E(X_1) \leq (1 + o(1))\frac{n}{32}.$$

In particular, this implies the existence of a graph $G^* \in \mathcal{A}$ on n vertices with at most $(1 + o(1))\frac{n}{32}$ edges and $\Delta(G^*) \leq np + \sqrt{4np\log n}$.

Graphs G and H on vertex set $[n]$ are said to *pack* if there is some permutation $\sigma \in S_n$ for which $\sigma(H) \subseteq \overline{G}$.

Lemma 2. *Let G and H be n-vertex graphs with $|G| < (1+o(1))n/32$, $\Delta(G) \leq np + \sqrt{4np\log n}$, and $|H| < \frac{n^2}{16(np+\sqrt{4np\log n})}$. Then G and H pack.*

Corollary 1. *If $E(X_1) \leq \frac{n}{64}$ then*

(a) $MIN_0 \geq \frac{n^2}{16(np+\sqrt{4np\log n})}$,

(b) $MIN_0 \geq \left(\min\{\frac{n}{32p}, \frac{n^2}{128\log n}\}\right)$, *and*

(c) $E(X_0) \geq \left(\min\{\frac{n}{64p}, \frac{n^2}{256\log n}\}\right)$.

Of course this gives Theorem 1.

Proof of corollary. Since $G^* \in \mathcal{A}$, no maxterm H of $\mathcal{A}$ can pack with G; so Lemma 2 gives (a), which immediately gives (b); and (c) then follows from Observation 1. □

Proof of Lemma 2. We will use the following packing lemma due to Catlin [C74] and Sauer and Spencer [SS78].

Lemma 3. *For n-vertex graphs G and H, if*

$$\Delta(H)\Delta(G) \leq n/2,$$

then H and G pack.

Now let G, H be as in Lemma 2. Notice that since G has at most $(1 + o(1))\frac{n}{32}$ edges it has at most $(1 + o(1))\frac{n}{16}$ nonisolated vertices. It thus suffices to pack (say) some $(n/2)$-vertex subgraph G' of G containing all nonisolated vertices of G, with any spanned subgraph of H on $n/2$ vertices. Let H' be the subgraph of

H spanned by the (some) $n/2$ vertices of lowest degree. Since the average degree in H is at most $\frac{n}{8(np+\sqrt{4np\log n})}$, we have $\Delta(H') \leq \frac{n}{4(np+\sqrt{4np\log n})}$. Thus

$$\Delta(H')\Delta(G') < n/4 = \frac{n/2}{2},$$

and by Lemma 3 G' and H' do indeed pack. $\square$

4   Packing Families of Random Graphs

The approach we follow in this paper gives rise to a line of questions that seems to be of great theoretical interest, both since it is essential for exploiting this technique further, and because it is a natural extension of the much-studied topic of graph packing.

As the reader may have noticed, at the end of the previous section we guaranteed the packing of two graphs G^* and H, using the fact that G^* was something like a subgraph of a random graph. The canonical examples showing that the classical graph packing theorems are tight use examples such as the impossibility of packing a complete matching with a star of degree $n - 1$. But in our setting—involving graphs that have a reasonable chance of showing up in a random graph—one does not find such high degree vertices, and average degree becomes a more relevant parameter.

Let us begin with a rather bold conjecture that, using the technique in this paper, would immediately imply, up to log factors, the $\Omega(n^2)$ lower bound for randomized decision tree complexity.

Let $\mathcal{A}$ be a monotone graph property, and G a graph. We wish to find a small witness to the fact that $G \in \mathcal{A}$ or $G \notin \mathcal{A}$. Let

$$Witness(G, \mathcal{A}) = \begin{cases} \min\{|H| : H \subseteq G, H \in \mathcal{A}\} & \text{if } G \in \mathcal{A} \\ \min\{|H| : H \subseteq \overline{G}, H \notin \mathcal{A}\} & \text{if } G \notin \mathcal{A} \end{cases}$$

Recall that $p(\mathcal{A})$ is that (threshold) probability p for which

$$\Pr(G(n,p) \in \mathcal{A}) = 1/2.$$

Our conjecture is that at the threshold for $\mathcal{A}$, in the random graph $G = G(n,p)$ typically either a smallest witness for membership in $\mathcal{A}$ uses $\Omega(1/\log(n))$ of the edges of G or a smallest witness for non-membership uses $\Omega(1/\log(n))$ of the edges of $\overline{G}$ (or both):

Conjecture 1. *Let $\mathcal{A}$ be a monotone graph property of graphs on n vertices and $p = p(\mathcal{A})$. Then for $G = G(n,p)$ either*

$$E[Witness(G, \mathcal{A})|G \in \mathcal{A}] = \Omega(n^2 p/\log(n))$$

or

$$E[Witness(G, \mathcal{A})|G \notin \mathcal{A}] = \Omega(n^2(1 - p)/\log(n)).$$

As a start, this conjecture seems quite fascinating even for balanced graph properties, those for which $p(\mathcal{A}) = 1/2$. In fact in this case we believe the $\log(n)$ factor may be unnecessary, as, for example, we can show to be true when $\mathcal{A}$ is the property of containing a clique of size k ($\approx 2\log_2(n)$). In this case each minimal witness for membership in $\mathcal{A}$ has size of order $\log^2(n)$, but it turns out that a typical witness for non-membership is of size $\Omega(n^2)$; namely we can prove that with high probability *every* subgraph H of G of size less than $n^2/100$ edges packs with a clique of size k. This illustrates the difference between packing arbitrary graphs and packing subgraphs of random graphs. We know there is a Turán-type graph of size $O(n^2/\log n)$ that does not pack with a clique of size k; but such a graph almost never occurs as a subgraph of G.

Though packing remains in some sense the heart of the matter, real progress (on randomized complexity) probably cannot be based on new packing theorems for pairs of random-like graphs; rather, we need to understand something about packing two *families* of graphs.

Definition 1. *Let $\mathcal{A}$, $\mathcal{B}$ be families of graphs. We say $\mathcal{A}$ and $\mathcal{B}$ pack if there exist $A \in \mathcal{A}$ and $B \in \mathcal{B}$ such that A and B pack.*

The relevance of this notion derives from the fact that a family $\mathcal{A}$ and its dual cannot pack. For example Conjecture 1 implies the following statement, which seems to be of independent interest. (Apply the conjecture to the property obtained from $\mathcal{A}$ by adding all graphs with at least $n^2/4$ edges whose complements are not in $\mathcal{B}$.)

Conjecture 2. *Let $\mathcal{A}$ and $\mathcal{B}$ be monotone graph properties on n vertices generated by minterms of size $o(n^2/\log n)$, and for which $p(\mathcal{A}), p(\mathcal{B}) > \Omega(1)$. Then $\mathcal{A}$ and $\mathcal{B}$ pack.*

We close with one more question in a similar vein, though now just for pairs of graphs. For a graph G on n vertices, set $\mu(G) = \Pr(G \subset G(n, 1/2))$ (where the containment is up to isomorphism). It is easy to see that if each of $\mu(G), \mu(H)$ is more than $1/2$, then G and H pack; but we expect more is true:

Conjecture 3. *For each $\alpha > 0$, if n is sufficiently large and G, H are n-vertex graphs with $\mu(G), \mu(H) \geq \alpha$, then G and H pack.*

(And of course one may ask how quickly $\alpha = \alpha(n)$ can go to zero without falsifying the conjecture.)

Though we have given only a small sample, we hope it is enough to indicate the wealth of possibilities suggested by the present point of view.

References

[B78]   B. Bollobás, Extremal Graph Theory, Academic Press, Chapter 8, 1978.

[C74]   P. A. Catlin, *Subgraphs of graphs I*, Discrete Math, 10, pp. 225–233, 1974.

[CKS02]   A. Chakrabarti, S. Khot and Y. Shi, *Evasiveness of subgraph containment and related properties*, SIAM J. on Computing, 31, pp. 866–875, 2002.

[CK01]   A. Chakrabarti and S. Khot, *Improved lower bounds on the randomized complexity of graph properties*, Proc. of the 28th ICALP conference, Lecture Notes in Computer Science 2076, pp. 285–296, 2001.

[G92]   D. H. Groġer, *On the randomized complexity of monotone graph properties*, Acta Cybernetica, 10, pp. 119–127, 1992.

[H91]   P. Hajnal, *An $\Omega(n^{4/3})$ lower bound on the randomized complexity of graph properties*, Combinatorica, 11, pp. 131–143, 1991.

[K88]   V. King, *Lower bounds on the complexity of graph properties*, Proc. of the 20th STOC conference, pp. 468–476, 1988.

[KSS84] J. Kahn, M. Saks and D. Sturtevant, *A topological approach to evasiveness*, Combinatorica 4, pp. 297–306, 1984.

[RV76]   R. Rivest and J. Vuillemin, *On recognizing graph properties from adjacency matrices*, TCS 3, pp. 371–384, 1976.

[SS78]   N. Sauer and J. Spencer, *Edge-disjoint placement of graphs*, J. of Combinatorial Theory Ser. B, 25, pp. 295–302, 1978.

[SW86]   M. Saks and A. Wigderson, *Probabilistic Boolean Decision Trees and the Complexity of Evaluating Game Trees*, Proc. of the 27th FOCS conference, pp. 29–38, 1986.

[vN28]   J. von Neumann, *Zur Theorie der Gesellschaftspiele*, Matematische Annalen, 100, pp. 295–320, 1928. English translation *On the theory of games and strategy*, in "Contributions to the theory of Games", IV, pp.13–42, 1959.

[Y87]   A.C. Yao, *Lower bounds to randomized algorithms for graph properties*, Proc. of the 28th FOCS conference, pp. 393–400, 1987.

Bisection of Random Cubic Graphs[*]

J. Díaz[1], N. Do[2], M.J. Serna[1], and N.C. Wormald[2]

[1] Dept. Llenguatges i Sistemes, Universitat Politecnica de Catalunya,
Jordi Girona Salgado 1–3, 08034 Barcelona, Spain
{diaz,mjserna}@lsi.upc.es,
[2] Department of Mathematics and Statistics, University of Melbourne,
VIC 3010, Australia
nick@ms.unimelb.edu.au, ducdo@smart.net.au.

Abstract. We present two randomized algorithms to bound the bisection width of random n-vertex cubic graphs. We obtain an asymptotic upper bound for the bisection width of $0.174039n$ and a corresponding lower bound of $1.325961n$. The analysis is based on the differential equation method.

1 Introduction

Given a graph $G = (V, E)$ with $|V| = n$ and n even, a *bisection* of V is a partition of V into two parts each of cardinality $n/2$, and its *size* is the number of edges crossing between the parts. A *minimum bisection* is a bisection of V with minimal size. The size of a minimum bisection is called the *bisection width* and the *min bisection problem* consists of finding a minimum bisection in a given G. In the same manner, we can also consider a *maximum bisection*, i.e. a bisection that maximizes the number of crossing edges. A related problem is that of finding the largest bipartite subgraph of a graph, i.e. a bipartite subgraph with as many edges as possible. This problem is known as the *Max Cut Problem* (see for example [6]). Given a graph, the size of a maximum bisection is clearly a lower bound on the size of a Max Cut in the graph.

The min bisection problem has received a lot of attention, as the bisection width plays an important role in finding lower bounds to the routing performance of a network. The decisional version of the problem is known to be NP-complete [6], even for cubic graphs [3]. On the other hand, several exact and heuristic positive results are known (see for example [4]). In this paper, we deal with the problem of estimating the typical size of minimum and maximum bisections of random cubic graphs.

It is shown in [10] that all cubic graphs have bisection width at most $\frac{n}{4} + O(\sqrt{n}\log n)$, and there are cubic graphs with bisection width of at least $\frac{n}{9.9}$.

[*] The first and third author are partially supported by the FET programme of the EU under contract IST-1999-14186 (ALCOM-FT). The last author is supported by the Australian Research Council and by the Centre de Recerca Matematica, Bellaterra, Spain.

J.D.P. Rolim and S. Vadhan (Eds.): RANDOM 2002, LNCS 2483, pp. 114–125, 2002.
© Springer-Verlag Berlin Heidelberg 2002

Our first result is an asymptotic bound on the bisection width of random cubic graphs. We refer the reader to [8], for the definitions of u.a.r. (uniformly at random) and a.a.s. (asymptotically almost surely). For such statements, $n \to \infty$ and we restrict n to even integers.

Theorem 1 *The bisection width of a random cubic graph on n vertices is a.a.s. smaller than $0.174039n$.*

We actually give two quite different proofs of approximately equal upper bounds; the other is $0.17451n$ (see Theorem 3).

Regarding the size of the maximum bisection, we are not aware of any non-trivial lower bounds. Our second result provides an asymptotic bound on the maximum bisection of random cubic graphs.

Theorem 2 *The maximum bisection of a random cubic graph with n vertices is a.a.s. greater than $1.325961n$.*

Notice that, as the number of edges in a cubic graph is 1.5n, then we have a 1.131255 randomized approximation to the Max Cut and Max Bisection problems on cubic graphs. For Max Bisection the best known approximation ratio is 1.4313 [16] and for Max Cut the best known approximation ratio is 1.1383 [7].

We conjecture that the largest balanced bipartite subgraph of a random cubic graph is a.a.s. almost the same size as the largest bipartite subgraph. We can state this even more strongly, as follows.

Conjecture 1 *For every $\epsilon > 0$, a.a.s. the largest bipartite subgraph of a random cubic graph has a 2-coloring with the difference in the numbers of vertices of the two colors less than ϵn.*

2   Greedy Algorithms for Minimum Bisection

In this section we prove Theorems 1 and 2. Given a random cubic graph, and given a partial assignment of colors red (R) and blue (B) to its vertices, we classify the non-colored vertices according with the number of their colored neighbors:

A vertex is of **Type** (r, b) if it has r neighbors colored R and b neighbors colored B.

We will consider the greedy procedure Simple greedy given in Figure 1 to find a.a.s. a balanced partition (R, B) with small bisection. The algorithm colors vertices in pairs to maintain balancedness. It repeatedly uses three operations. Op1 consists of choosing one vertex of type $(1, 0)$ and one of type $(0, 1)$ u.a.r., and coloring each the same as its colored neighbor. Op2 consists of choosing one vertex of type $(2, 0)$ and one of type $(0, 2)$ u.a.r., and coloring each the same as its colored neighbors. Op3 consists of choosing two non-adjacent vertices of type $(1, 1)$ u.a.r., and coloring one with R and one with B.

Initial step: select two non-adjacent vertices u.a.r., color one with R
and the other with B

Phase 1: **repeat**
    if there are vertices of both types (2,0) and (0,2) **then** perform Op2;
    else if there are vertices of both types (1,0) and (0,1) **then** perform Op1;
    until no new vertex is colored

Phase 2: **repeat**
    if there are vertices of both types (1,0) and (0,1)**then** perform Op1;
    else if there are at least two vertices of type (1,1) **then** perform Op3;
    until no new vertex is colored

Phase 3: **repeat**
    if there are vertices of both types (3,0) and (0,3)
        then choose one of each type at random, and color each
        the same as its colored neighbor;
    if there are vertices of both types (2,1) and (1,2)
        then choose one of each type at random, and color each
        with the majority of its colored neighbors;
    until no new vertex is colored
color any remaining uncolored vertices, half of them R and half B,
in any manner, and output the bisection R, B.

Fig. 1. Algorithm simple min greedy for Min Bisection

Note that the size of the bisection is the number of *bicolored* edges, with one vertex of each color.

One method of analyzing the performance of a randomized algorithm is to use a system of differential equations to express the expected changes in variables describing the state of the algorithm during its execution. An exposition of this method is in [13], which includes various examples of graph-theoretic optimization problems. For purposes of exposition, we continue for the present to discuss Algorithm simple min greedy, without giving full justification. After this, in order to reduce the complexity of the justification, it is in fact a different but related algorithm which we will analyze to yield our claimed bounds. We call this variation of algorithm a *deprioritized* algorithm as in [15], where this technique was first used, though Achiloptas [1] used a related idea to different effect.

We use the pairing model to analyze n-vertex cubic graphs, generated u.a.r. Briefly, to generate such a random graph, it is enough to begin with $3n$ *points* in n *cells*, and choose a random perfect matching of the points, which we call a *pairing*. The corresponding pseudograph (possibly with loops or multiple edges) has the cells as vertices and the pairs as edges. Any property a.a.s. true of the random pseudograph is also a.a.s. true of the restriction to random graphs, with no loops or multiple edges, and this restricted probability space is uniform (see for example [2,14] for a full description).

Without loss of generality, when stating such asymptotic results, we restrict n to being even to avoid parity problems. We consider Algorithm simple min greedy applied directly to the random pairing. As discussed in [13], the random pairing can be generated pair by pair, and at each step a point p can be chosen by any rule whatsoever, as long as the other point in the pair is chosen u.a.r. from the remaining unused points. We call this step *exposing* the pair containing p.

At each point in the algorithm, let Z_{rb} represent the number of uncolored vertices of type (r, b). To analyze the algorithm, when a vertex is colored we immediately expose all pairs involved in that vertex. In this way, the numbers Z_{rb} are always determined.

At any time, let W denote the number of points not yet involved in exposed pairs. These are the points available for the pairs that will be exposed during the next step. Then $W = 3Z_{00} + 2Z_{01} + 2Z_{10} + Z_{02} + Z_{20} + Z_{11}$.

Consider what happens when a vertex u is newly colored R and one of the pairs containing a point p in that cell is exposed. The other point will lie in some vertex v. Let d_{rb} denote the expected contribution to the increment $\Delta(Z_{rb})$ in Z_{rb} due to the change in the status of v. Then, up to terms $O(1/W)$,

$$d_{00} = -\frac{3Z_{00}}{W}, \; d_{01} = -\frac{2Z_{01}}{W}, \; d_{02} = -\frac{Z_{02}}{W}, \; d_{03} = 0, \; d_{11} = \frac{2Z_{01} - Z_{11}}{W},$$

$$d_{12} = \frac{Z_{02}}{W}, \; d_{10} = \frac{3Z_{00} - 2Z_{10}}{W}, \; d_{20} = \frac{2Z_{10} - Z_{20}}{W}, \; d_{30} = \frac{Z_{02}}{W}, \; d_{21} = \frac{Z_{11}}{W}.$$

The error term $O(1/W)$ is due to adjustments occurring when v happens to be the same as u (and also saves us from specifying whether the variables refer to the graph before or after coloring u).

The corresponding equations when a vertex is colored B form a symmetric set with these: they are the same but with the index pair on all variables swapped. Therefore, the expected increments due to a dual step, consisting of one new pair from a vertex of each color, is $\bar{d}_{r,b}$:

$$\bar{d}_{00} = -\frac{6Z_{00}}{W}, \; \bar{d}_{01} = \frac{3Z_{00} - 4Z_{01}}{W}, \; \bar{d}_{02} = \frac{2Z_{01} - 2Z_{02}}{W}, \; \bar{d}_{11} = \frac{2Z_{01} + 2Z_{10} - 2Z_{11}}{W},$$

$$\bar{d}_{12} = \frac{Z_{02} + Z_{11}}{W}, \text{ where } W = 3Z_{00} + 2Z_{01} + 2Z_{10} + Z_{02} + Z_{20} + Z_{11}. \tag{1}$$

Symmetrically corresponding variables have symmetrically corresponding equations. Note that $\bar{d}_{03}$ and its symmetric mate are not required, since vertices of type $(0, 3)$ are just colored in phase 3 with the color of all their neighbors and therefore are not incident with any bicolored edges.

The rest of our discussion, until considering the deprioritized algorithm, is mainly motivation but also includes some derivations of formulae used later. The difficulty of analysis is caused by the prioritization in phases 1 and 2. In phase 1 the algorithm performs one of two types of operations on a pair of vertices, in each case coloring them the same as their neighbors. If vertices of types $(0, 2)$ and $(2, 0)$ exist, these have priority (Op2), while if they don't, but at least one of each of types $(0, 1)$ and $(1, 0)$ exist, then these are treated (Op1). In practice, for a random graph there are never very many vertices of types $(0, 2)$ and $(2, 0)$; as soon as there are some of each, they are processed, and the number of new ones arising tends to be less than the number used up. Which leaves a few of one type or the other, waiting for the matching ones to be created. To proceed with the discussion, we assume that a given iteration in phase 1 performs Op1 with probability ϕ, and therefore Op2 with probability $1 - \phi$.

Define

$$\theta = \frac{2Z_{01} - 2Z_{02}}{W}.$$

Due to Op1, there are *two* pairs exposed from each vertex. Thus the expected number of new vertices of type $(0,2)$ arising from this is $2\bar{d}_{02} = \frac{4Z_{01} - 4Z_{02}}{W} = 2\theta$. At the moment, we make the assumption that we have the *rb-symmetry*: for all i and j, $Z_{ij} = Z_{ji}$ (later we will see how to remove it). With this last assumption, there also are 2θ new vertices of type $(2,0)$ it is also 2θ. So each iteration of phase 1 performing Op1 gives rise on average to 2θ steps performing Op2 in order to keep Z_{02} and Z_{20} constant.

However, these iterations with Op2 cause further vertices of types $(0,2)$ and $(2,0)$. In Op2, one pair is exposed from each vertex, which gives expected increase of θ to Z_{02} and to Z_{20}. Following from this, we expect θ extra steps of Op2. For $\theta < 1$, the expected number of iterations of Op2 executed by the algorithm for each Op1, is $\sum_{i=0}^{\infty} 2\theta(\theta)^i = \frac{2\theta}{1-\theta}$.

As the probability of Op2 will be $\frac{2\theta}{1-\theta}\phi = 1 - \phi$, and we can conclude that

$$\phi = \frac{1 - \theta}{1 + \theta}. \tag{2}$$

Using this, we can compute the expected increments of the random variables Z_{ij} in each iteration in phase 1.

$$\mathbf{E}\left[\Delta(Z_{ij})\right] = \phi(2\bar{d}_{ij} - \delta_{01}) + (1 - \phi)(\bar{d}_{ij} - \delta_{02}) = (1 + \phi)\bar{d}_{ij} - \phi\delta_{01} - (1 - \phi)\delta_{02} \tag{3}$$

for any i,j with $i \leq j$ and $i + j = 3$, where $-\delta_{pq} = 1$ if $(p,q) = (i,j)$, and 0 otherwise. The equations for $i > j$ are the symmetric ones.

In (3) we may use $\bar{d}_{ij}$ as given in (1) but with $\bar{d}_{11} = (4Z_{01} - 2Z_{11})/W$ by rb-symmetry. Without justification at this point, we may express the above expected increments as a set of differential equations, where each $\mathbf{E}\left[\Delta(Z_{ij})\right]$ is expressed as the differential Z'_{ij} (all as functions of the number t of iterations). If we scale both time and the variables by dividing by n, and denote Z_{ij}/n by z_{ij} and W/n by w, then the equations are

$$z'_{00} = -(1 + \phi)\frac{6z_{00}}{w}, \quad z'_{01} = (1 + \phi)\frac{3z_{00} - 4z_{01}}{w} - \phi, \quad z'_{02} = (1 + \phi)\frac{2z_{01} - 2z_{02}}{w} + (\phi - 1),$$

$$z'_{11} = (1 + \phi)\frac{4z_{01} - 2z_{11}}{w}, \quad z'_{12} = (1 + \phi)\frac{z_{02} + z_{11}}{w}, \quad \text{where } w = 3z_{00} + 4z_{01} + 2z_{02} + z_{11},$$

by rb-symmetry, $\phi = \frac{1-\theta}{1+\theta}$ and $\theta = \frac{2z_{01} - 2z_{02}}{w}$. The initial conditions at $x = t/n = 0$ are

$$z_{00}(0) = 1, \quad z_{01}(0) = 0, \quad z_{02}(0) = 0, \quad z_{03}(0) = 0, \quad z_{11}(0) = 0, \quad z_{12}(0) = 0. \tag{4}$$

Note that by the way we defined ϕ in (2), $z'_{02} \equiv 0$ and hence $z_{02} \equiv 0$. We are interested in the point that z_{01} first goes negative, which by numerical solution occurs when

$$x = x_1 \approx 0.41178, \ z_{00} \approx 0.002405, \ z_{11} \approx 0.046633, \ z_{12} \approx 0.063700. \quad (5)$$

The whole algorithm "takes off" at the start because the derivative of z_{01} is strictly positive, so a.a.s. phase 1 does not quickly use up all vertices to be processed.

At this point, since z_{01} and z_{02} are both 0, phase 2 is entered. The situation is similar to phase 1, but now the operation with highest priority is Op1. The other operation, Op3, is such that two vertices of type $(1,1)$ are randomly chosen and colored, one B and one R. Following the discussion above, let $\theta_2 = \frac{3Z_{00}-4Z_{01}}{W}$. Then, due to Op3, there is one pair exposed from each of two vertices of type $(1,1)$, and the expected number of new vertices of type $(0,1)$ arising from this is $2\bar{d}_{01} = \frac{4Z_{01}-4Z_{02}}{W} = 2\theta_2$, where *two* vertices of type $(1,1)$ are used in this operation. In Op1, the expected number is $2\theta_2$. Letting ϕ_2 denote the probability that at a given time Op3 is performed in phase 2, the probability of Op1 will be $\frac{\theta_2}{1-2\theta_2}\phi_2 = 1 - \phi_2$, giving

$$\phi_2 = \frac{1-2\theta_2}{1-\theta_2}.$$

In place of (3) we now have

$$\mathbf{E}\left[\Delta(Z_{ij})\right] = \phi_2(\bar{d}_{ij} - \delta_{11}) + (1 - \phi_2)(2\bar{d}_{ij} - \delta_{01}) = (2 - \phi_2)\bar{d}_{ij} - \phi_2\delta_{11} - (1 - \phi_2)\delta_{01}. \quad (6)$$

To set the differential equations, let Y be a random variable that keeps track of the number of times Op3 is performed, and let $y = Y/n$. This is needed for record-keeping because each such operation causes two bicolored edges. Then

$$z'_{00} = -(2-\phi_2)\frac{6z_{00}}{w}, \ z'_{01} = (2-\phi_2)\frac{3z_{00}-4z_{01}}{w} + (\phi_2 - 1), \ z'_{02} = (2-\phi_2)\frac{2z_{01}-2z_{02}}{w},$$

$$z'_{11} = (2-\phi_2)\frac{4z_{01}-2z_{11}}{w} - 2\phi_2, \ z'_{12} = (2-\phi_2)\frac{z_{02}+z_{11}}{w}, \ y' = \phi_2,$$

with w as before and with initial conditions given by (5) and $z_{01} = z_{02} = 0$.

By the choice of ϕ_2, $z_{01} \equiv 0$, so $z_{02} \equiv 0$. The point of interest is

$$x_2 = \sup\{x : z_{11} > 0, \ w > 0, \ \theta_2 < 1\}. \quad (7)$$

Numerically, we find that θ_2 does not reach 1 before z_{11} reaches 0, which clearly must therefore hold at x_2. This corresponds to the beginning of phase 3. During phase 3 the number of bicolored edges created is 2 for every pair of vertices of types $(1,2)$ and $(2,1)$ (using rb-symmetry) and at most 6 for every other pair colored except types $(0,3)$ and $(3,0)$, which give none. Since $z_{01} = z_{02} = z_{11} = 0$ at x_2, our upper bound for the size of the bisection is thus $(6z_{00} + 2z_{12} + 2y)n$ where the variables are evaluated at x_2. Solving numerically, we find

$$z_{00}(x_2) + 2z_{12}(x_2) + 2y(x_2) < 0.1740381057, \quad (8)$$

where the constant is correct to ten decimal places.

Now we are in position to carry out the formal analysis via a deprioritized algorithm. For a given sufficiently small $\epsilon > 0$, consider the deprioritized algorithm in Figure 2. Notice that pre-phase 1 ensures a good supply of vertices of types $(0,1)$, $(1,0)$, $(0,2)$ and $(2,0)$.

> *Pre-phase 1:* **do** the following $\lfloor \epsilon n \rfloor$ times:
>
> select two non-adjacent type $(0,0)$ vertices u.a.r.,
> color one with R and the other with B;
>
> *Phase 1:* **while** all of Z_{01}, Z_{10}, Z_{02} and Z_{20} are non-zero
>
> let $\theta = \frac{2Z_{01} - 2Z_{02}}{W}$ and $\phi = \frac{1-\theta}{1+\theta}$;
> **with probability** ϕ perform Op1;
> **otherwise** perform Op2;
>
> *Pre-phase 2:* **do** $\lfloor \epsilon n \rfloor$ steps as in Pre-phase 1;
>
> *Phase 2:* **while** $Z_{01} > 0$, $Z_{10} > 0$ and $Z_{11} > 1$
>
> let $\theta_2 = \frac{3Z_{00} - 4Z_{01}}{W}$ and $\phi_2 = \frac{1-2\theta_2}{1-\theta_2}$;
> **with probability** ϕ_2 perform Op3;
> **otherwise** perform Op1;
>
> *Phase 3:* as for Algorithm simple min greedy.

Fig. 2. Algorithm deprioritized min greedy for Min Bisection

The expected changes in the variables Z_{ij} for each edge exposed are given in (1). In pre-phase 1, the derivation of (2) applies, but with ϕ redefined 1 at all times, and with different terms in the equations for z'_{00} and z'_{01} due to the fact that the vertex being processed is type $(0,0)$ rather than $(0,1)$ or $(0,2)$. At this stage, we entirely avoid using the rb-symmetry assumption. Referring back to (1), this requires only the adjustment of the formulae for z'_{11} and w. The result is

$$z'_{00} = 1 - \frac{12z_{00}}{w}, \quad z'_{01} = 2\frac{3z_{00} - 4z_{01}}{w}, \quad z'_{02} = 2\frac{2z_{01} - 2z_{02}}{w},$$

$$z'_{11} = 2\frac{2z_{01} + 2z_{10} - 2z_{11}}{w}, \quad z'_{12} = 2\frac{z_{02} + z_{11}}{w}, \quad w = 3z_{00} + 2z_{01} + 2z_{10} + z_{02} + z_{20} + z_{11}.$$

Other derivatives z'_{ji} are the symmetric versions of z'_{ij} (with indices swapped); z'_{03} and z'_{30} are not needed. It follows that the unique solution must be the symmetric one, which satisfies $z_{ij}(t) = z_{ji}(t)$ for all i, j and t, as well as the stated equations.

Let $z_{ij}(t/n)$ denote Z_{ij}/n after t steps. Thus the previous equations give the expected one-step change in the variables Z_{ij} with error $O(1/n)$. This error is due to the changing value of the variables between when one vertex of type $(0,0)$ is chosen and the next. This applies with initial conditions given in (4). We write $\tilde{z}_{ij}(x)$ for the (unique) solutions of this initial value problem, $0 \leq x \leq \epsilon$. We may now apply the differential equation method (using, for example, [12, Theorem 1] or [13, Theorem 5.1]) to deduce that during pre-phase 1, we have a.a.s.

$$Z_{ij}(t) = n\tilde{z}_{ij}(t/n) + o(n) \tag{9}$$

for each i and j, where $Z_{ij}(t)$ is the value of Z_{ij} after t steps. This applies until either $t = \lfloor \epsilon n \rfloor$ or one of the derivatives approaches a singularity, which we can prevent by restricting to a domain in which $\theta > -1 + \epsilon$ and $w > \epsilon$, or the differential equations no longer apply for some other reason, which in this case only occurs if Z_{00} reaches 0. Note that the derivatives are all $O(1)$, so $\tilde{z}_{00}(x)$ stays close to 1 for $x < \epsilon$ assuming that $\epsilon > 0$ is sufficiently small. We conclude that a.a.s.

$$Z_{ij}(t_0) = n\tilde{z}_{ij}(t_0/n) + o(n), \qquad t_0 := \lfloor \epsilon n \rfloor. \tag{10}$$

We also note that z'_{01} must be strictly positive, and so $\tilde{z}_{01}$ and hence $\tilde{z}_{02}$ are strictly positive on $(0, \epsilon)$. Thus, in particular, for sufficiently small $\epsilon_1 = \epsilon_1(\epsilon) > 0$,

$$\tilde{z}_{01}(\epsilon) \geq \epsilon_1, \quad \tilde{z}_{02}(\epsilon) \geq \epsilon_1. \tag{11}$$

Now consider phase 1. Arguing as above, the expected changes in the Z_{ij} are given, with error $O(1/W)$, by the right hand sides of the equations in (2), with

$$w = 3z_{00} + 2z_{01} + 2z_{10} + z_{02} + z_{20} + z_{11}, \quad \phi = \frac{1-\theta}{1+\theta}, \quad \theta = \frac{2z_{01} - 2z_{02}}{w},$$

and the replacement equation

$$z'_{11} = (1 + \phi)\frac{2z_{01} + 2z_{10} - 2z_{11}}{w}$$

to avoid the rb-symmetry assumption. Again the symmetrically reversed functions have symmetrically reversed equations (except that θ stays the same). Continue the definition of the functions $\tilde{z}_{ij}(x)$ for $x > \epsilon$ by the solution of these equations with initial conditions given by the values of these functions at $x = \epsilon$ as determined above.

Note that setting $z_{ij} = z_{ji}$ for all i and j in the equations, except in the definition of θ, again makes the formulae for z'_{ij} and z'_{ji} identical, despite the asymmetrical definition of θ. It follows that again the unique solution is symmetric, with $\tilde{z}_{ij} = \tilde{z}_{ji}$. We deduce that the equations (2) are satisfied, with the symmetric definitions of w and z_{11}, and we may restrict attention to the variables appearing there.

Again applying the differential equation method, we deduce that (9) holds a.a.s. as long as the solution set $\tilde{z}_{ij}$ stays within a predefined closed domain which does not contain singularities of the derivatives, and also the variables Z_{01} and Z_{02} stay positive. We may select the domain D satisfying $\tilde{z}_{01} \geq \epsilon_1$, $\tilde{z}_{02} \geq \epsilon_1$, $w > \epsilon$ and $\theta > 1 - \epsilon$. By (11), the first two of these inequalities hold at $x = \epsilon$, and the other two also hold for ϵ sufficiently small by boundedness of derivatives. Arguing by continuity, z'_{01} as given in (2) is strictly positive for $x < \delta$, where $\delta > 0$ is an absolute constant independent of ϵ. By definition of

θ, z'_{02} is identical to 0, and so $\tilde{z}_{02} \equiv \tilde{z}_{02}(\epsilon)$. Thus, for ϵ sufficiently small, the solution set $\tilde{z}_{ij}$ stays within D for $x < \delta$, and can only leave D when, for $x > \delta$,

$$\tilde{z}_{01} = \epsilon_1, \ w = \epsilon \text{ or } \theta = 1 - \epsilon. \tag{12}$$

Note that for ϵ and ϵ' sufficiently small, the initial conditions for $\tilde{z}_{ij}$ are arbitrarily close to (4). Let us denote the solutions with initial conditions (4) by $\bar{z}_{ij}$. By standard theory of first order systems of differential equations, it follows that the functions $\tilde{z}_{ij}$ can be made arbitrarily close to $\bar{z}_{ij}$ in the domain D, by taking ϵ and ϵ_1 sufficiently small. By numerical computation, the conditions corresponding to (12) are not reached by the solution $\bar{z}_{ij}$, until x approaches x_1 given in (5), at which point $\bar{z}_{01}$ reaches 0. It follows that, as ϵ and $\epsilon_1 \to 0$, the exit point of the $\tilde{z}_{ij}$ from D also tends towards x_1, and the values are given in the limit by the values of $\bar{z}_{ij}$ in (5).

A similar argument applies for phase 2, and at this point we also introduce the variable y, to keep track of the number Y of times Op3 is performed. The conclusion is that there is a deprioritized algorithm in which the values of the variables Z_{ij} are a.a.s. $\tilde{z}_{ij}n + o(n)$, where the functions $\tilde{z}_{ij}$ and $\tilde{y}$ solve (2). They can be made arbitrarily close to $\bar{z}_{ij}$ for all $x < x_2 - \delta'$, where x_2 is given in (7) and δ' is an arbitrary positive quantity. Additionally, $Y = yn + o(n)$ a.a.s. Note that $|\bar{z}_{ij}(x_2 - \delta') - \bar{z}_{ij}(x_2)| = O(\delta')$ since the derivatives are all bounded. Examining phase 3 as for Algorithm simple min greedy, it follows that the size of the bisection produced is a.a.s.

$$(z_{00}(x_2) + 2z_{12}(x_2) + 2y(x_2))n + O(\delta'n),$$

where δ' can be chosen arbitrarily small. By (8), this completes the proof of Theorem 1.

3   Maximum Bisection

Let us consider the maximization version of the bisection problem, Max Bisection, i.e. given a connected cubic graph with n vertices, for n even, find a bisection which maximizes the number of crossing edges. In general, the problem is also known to be NP-complete, even for cubic graphs, moreover for the particular case of cubic graphs, the problem can be approximated with an approximation ratio of 0.847 [9]. For motivation on the problem, see [5].

Let us consider the variation Simple max greedy of the Algorithm 1 Simple min greedy, obtained by changing the meaning of Op1 and Op2. In the Simple max greedy algorithm, Op1 consists of choosing one vertex of type $(1, 0)$ and one of type $(0, 1)$ u.a.r., and coloring each opposite to its colored neighbor. Op2 consists of choosing one vertex of type $(2, 0)$ and one of type $(0, 2)$ u.a.r., and coloring each opposite to its colored neighbors. Op3, as before, consists of choosing two non-adjacent vertices of type $(1, 1)$ u.a.r., and coloring one with R and one with B. Also, in phase 3, change equal and majority, by different and minority.

Let us say that an edge is *fully colored* when both its ends are finally colored. A fully colored edge is *monocolored* if both ends have the same color and *bicolored* if both ends have different color. So the monocolored edges by Min greedy get bicolored by Max greedy and vice versa, whenever the vertices of the graph are treated in the same order (which happens with the same probability, in both cases). That is, every edge that counts in the bisection for one algorithm does not count in the other and vice versa. Therefore, taking into account that the total number of edges in a cubic graph is $1.5n$, we have proved Therorem 2.

4 The Comb Swapping Algorithm

In this section we include a different approach that provides an independent verification of approximately the same bound as in the previous section. Its proof is just sketched, since it gives a slightly weaker result.

Let us recall that Robinson and Wormald proved that asymptotically almost all random cubic graphs are Hamiltonian [11]. The proof gave a contiguity result which permits us, for the purpose of proving any statement to be true a.a.s., to assume that a random cubic graph with n vertices (n even) is given as the union of a Hamiltonian cycle and perfect matching chosen u.a.r. (See [14].) Given such a graph G we can construct an initial bisection, by placing $n/2$ consecutive vertices in the cycle on one side, say the R side, and the remaining $n/2$ on the B side.

Given an arbitrary bisection (R, B) of G, define an R-comb to be a maximal path of vertices in R such that each is adjacent to at least one vertex in B, and such that this property also holds in the initial bisection. In a symmetric way we can define a B-comb. The *length* of a comb is the length of the corresponding path. An R-comb and a B-comb of length k are *compatible* if there is no edge joining a vertex in one to a vertex in the other. To *swap* a compatible pair of combs, swap each vertex in the B-comb into R, and each vertex in the R-comb into B. Notice that by swapping two compatible combs of length k, we decrease the bisection size by $k - 2$. We will consider the algorithm in Fig. 3, where $K > 2$ is a constant to be fixed later. At any step of the algorithm, let Y_i be a random

Given a random cubic graph as a hamiltonian cycle + a perfect matching. Construct an initial bisection (R, B), by breaking the Hamiltonian cycle in two disjoint paths of length $n - 1$. Set $k = K$.
while $k \geq 2$
    while there is a compatible R-comb and B-comb of length k
        swap the two combs.
    endwhile
    $k = k - 1$
endwhile

Fig. 3. The comb swapping algorithm for Min Bisection

variable counting the number of B-combs of length i ($i \leq k$). We will consider only the B point of view, as R has the same equations.

Initially, the expected bisection size is $n/4$, and the expected number of B-combs of length i is asymptotically $\mathbf{E}[Y_i] \sim 2^{-i-3}n$. So, at this point, the expected contribution to the bisection of combs with length bigger than d is $(1 + o(1)) \sum_{i=d+1}^{n} i 2^{-i-3} n$. This tends quickly to 0 as $d \to \infty$, showing that our bounds will not improve much by considering very long combs.

Let k be fixed. During phase k, the maximum comb length is k. Let $s = \sum_{i=1}^{k} i Y_i$. This quantity corresponds (roughly) to the number of vertices in combs of one color. We compute the expected increment in Y_i by first considering the contribution due to one edge from a vertex in the red comb. This edge, e say, may subdivide, or split, a B-comb. Then the probability that e hits a B-comb of length i is $\frac{iY_i}{s}$ (arguing about randomness without justification in this sketch). For each B-comb of length $j > i$ there are two positions giving a splitting with length $i < j$, except when $j = 2i + 1$, but in such a case we have two equally sized split portions. So, due to e, the expected increase in the number of combs with length i is

$$d_i = -\frac{iY_i}{s} + \sum_{j=i+1}^{k} \frac{2Y_j}{s}.$$

The total increase due to the $2k$ edges involved in one swap is therefore

$$\mathbf{E}[\Delta(Y_i)] = -\delta_{ik} + kd_i,$$

where δ denotes the Kronecker delta function, and the expected bisection decrease is $4 - 2k$. Phase k finishes when all pairs of combs of different colors are incompatible. It turns out that at this point, we may assume that the number of combs of length k is negligible.

We may write down an associated differential equation (as with the other method) and solve it numerically (we use a Runge-Kutta method), with initial conditions $y_i = 2^{-i-3}$, $1 \leq i \leq K$, $z = 1/4$ for the bisection width, and $k = K$ initially. When y_k hits 0, we can move to the next phase and make the initial values of phase $k - 1$ equal to the final values of phase k. We select the value K such that the possible gain by treating the longer combs (as calculated above) is not significant. Taking $K = 24$ the solution of this system of equations, provides a bisection with size less than $0.17451n$ (and treating longer combs cannot reduce to $0.174507n$). So we have

Theorem 3 *The bisection width of a random cubic graph on n vertices is a.a.s. less than $0.17451n$.*

The same argument can be used to derive a lower bound for the maximum bisection. In this case we start with a partition obtained from the Hamiltonian cycle, but this time putting consecutive vertices in different parts of the bipartition. This provides an initial bisection of expected size $1.25n$. Define an equivalent notion of *anti-comb*, a maximal alternating path so that for every vertex v its third neighbor is in the same side as v. Then by swapping anti-combs with the same length we get the same gain (now positive) as when swapping combs. Furthermore the equations for both systems are the same, so we improve the initial bisection by the same amount, which gives the following.

Theorem 4 *The maximum bisection of a random cubic graph on n vertices is a.a.s. greater than $1.32549n$.*

References

1. D. Achlioptas. Setting 2 variables at a time yields a new lower bound for random 3-sat. In *32nd Annual ACM Symposium on Theory of Computing (STOC 2000)*, pages 28–37, 2000.
2. B. Bollobas. *Random Graphs*. Academic Press, London, 1985.
3. T. Bui, S. Chaudhuri, T. Leighton, and M. Sipser. Graph bisection algorithms with good average case behavior. *Combinatorica*, 7:171–191, 1987.
4. J. Díaz, M. D. Penrose, J. Petit, and M. Serna. Approximating layout problems on random geometric graphs. *Journal of Algorithms*, 39:78–116, 2001.
5. A. Frieze and M. R. Jerrum. Improved approximation algorithms for max k-cut and max bisection. *Algorithmica*, 18:61–67, 1997.
6. M. R. Garey and D. S. Johnson. *Computers and Intractability: A Guide to the Theory of NP-Completeness*. Freeman, San Francisco, 1979.
7. M. X. Goemans and D. P. Williamson. Improved approximation algorithms for maximum cut and satisfiability problems using semidefinite programming. *J. Assoc. Comput. Mach.*, 42(6):1115–1145, 1995.
8. S. Janson T. Łuczak and A. Ruciński. *Random graphs*. Wiley, New York, 2000.
9. M. Karpinski, M. Kowaluk and A. Lingas. Approximation algorithms for max-bisection on low degree regular graphs and planar graphs. Technical report, Department of Computer Science, University of Bonn, 2000.
10. A. V. Kostochka and L. S. Melnikov. On bounds of the bisection width of cubic graphs. In J. Nesetril and M. Fiedler, editors, *Fourth Czechoslovakian Symposium on Combinatorics, Graphs and Complexity*, pages 151–154. Elsevier Science Publishers, 1992.
11. R. W. Robinson and N. C. Wormald. Almost all cubic graphs are hamiltonian. *Random Structures and Algorithms*, 19:128–147, 2001.
12. N. C. Wormald. Differential equations for random processes and random graphs. *Annals of Applied Probability*, 5:1217–1235, 1995.
13. N. C. Wormald. The differential equation method for random graph processes and greedy algorithms. In M. Karoński and H. Prömel, editors, *Lectures on Approximation and Randomized Algorithms*, pages 73–155. PWN, Warsaw, 1999.
14. N. C. Wormald. Models of random regular graphs. In *Surveys in Combinatorics*, pages 239–298. Cambridge University Press, 1999.
15. N. C. Wormald. Analysis of greedy algorithms on graphs with bounded degrees. *Discrete Mathematics*, 2002. Submitted.
16. Y. Ye. A .699-approximation algorithm for Max-Bisection. *Math. Program.*, 90(1, Ser. A):101–111, 2001.

Small k-Dominating Sets of Regular Graphs

William Duckworth and Bernard Mans

Department of Computing
Macquarie University
Sydney NSW 2109 Australia
{billy,bmans}@ics.mq.edu.au

Abstract. A k-dominating set of a graph, G, is a set of vertices, $\mathcal{D} \subseteq V(G)$, such that for every vertex $v \in V(G)$, either $v \in \mathcal{D}$ or there exists a vertex $u \in \mathcal{D}$ that is at distance at most k from v in G. We are interested in finding k-dominating sets of small cardinality. In this paper we consider a simple, yet efficient, randomised greedy algorithm for finding a small k-dominating set of regular graphs. We analyse the average-case performance of this heuristic by analysing its performance on random regular graphs using differential equations. This, in turn, proves an upper bound on the size of a minimum k-dominating set of random regular graphs.

1 Introduction

A *dominating set* of a graph, G, is a set of vertices, $D \subseteq V(G)$, such that every vertex of G either belongs to D or is incident with a vertex in G. The problem of finding a *minimum* dominating set of a graph is one of the core, well-known, NP-hard optimisation problems in graph theory (see, for example, [5]). Johnson [7] showed that for general graphs on n vertices, this problem is approximable within $1+\log n$. Raz and Safra [10] showed that this problem is not approximable within $c \log n$ for some $c > 0$. When restricted to graphs of bounded degree $d \geq 3$, Papadimitriou and Yannakakis [11] showed that the minimum dominating set problem is APX-complete and is approximable within $\sum_{i=1}^{d+1} \frac{1}{i} - 1/2$.

A graph, G, is said to be d-regular if every vertex of G is incident with precisely d other vertices of G. When discussing any d-regular graph on n vertices, dn is assumed to be even. We consider graphs that are undirected, unweighted and contain no loops or multiple edges. Note that for d-regular graphs, the minimum dominating set problem is trivially approximable within $(d + 1)/2$.

Reed [12] showed that the size of a minimum dominating set of an n-vertex cubic (i.e. 3-regular) graph is at most $3n/8$ and gave an example of a cubic graph on eight vertices with no dominating set of size less than three, demonstrating the tightness of this bound.

As we consider *random* regular graphs, we need some notation. We use the notation **P** (probability), u.a.r. (uniformly at random) and **E** (expectation). We say that a property, $\mathcal{B} = \mathcal{B}_n$, of a random regular graph on n vertices holds a.a.s. (asymptotically almost surely) if $\lim_{n \to \infty} \mathbf{P}(\mathcal{B}_n)=1$.

J.D.P. Rolim and S. Vadhan (Eds.): RANDOM 2002, LNCS 2483, pp. 126–138, 2002.
© Springer-Verlag Berlin Heidelberg 2002

Molloy and Reed [9] showed that the size of a minimum dominating set, D, of a random cubic graph on n vertices a.a.s. satisfies $0.2636n \leq |D| \leq 0.3126n$. Duckworth and Wormald [3] improved these results by showing that for a random cubic graph on n vertices, the size of a minimum *independent* dominating set, I, a.a.s. satisfies $0.2641n \leq |I| \leq 0.27942n$. The upper bound was achieved by analysing the performance of a greedy heuristic on random cubic graphs using differential equations whereas the lower bound was achieved by means of a direct expectation argument. For $d > 3$, Zito [17] presented upper and lower bounds on the size of a minimum independent dominating set of random d-regular graphs.

A k-dominating set of a graph, G, is a set of vertices, $\mathcal{D} \subseteq V(G)$, such that for every vertex $v \in V(G)$, either $v \in \mathcal{D}$ or there exists a vertex $u \in \mathcal{D}$ that is at distance at most k from v in G. Chang and Nemhauser [1] showed that the problem of finding a *minimum* k-dominating set of a graph is NP-hard even when restricted to bipartite or chordal graphs of diameter $2k + 1$. The minimum k-dominating set problem has recently been studied in many contexts (see, for example, [4]). Slater [5, Chapter 12] gives a recent survey of results on the complexity and the approximability of this problem. In distributed environments, graphs representing network topologies often have bounded or even regular degree and small k-dominating sets are useful in bounding the compactness of routing tables (see, for example, [8]).

Note that for $k = 1$, the minimum k-dominating set problem is the more familiar dominating set problem. Throughout this paper we consider the size of a minimum k-dominating set of a random d-regular graph where d and k are constant, $d \geq 3$ and $k \geq 2$. For $k \geq 2$, as far as the authors are aware, no non-trivial approximation results were previously known for the minimum k-dominating set problem on regular graphs.

It is simple to verify that the size of a minimum k-dominating set, $\mathcal{D}$, of a random d-regular graph on n vertices a.a.s. satisfies

$$\frac{(d-2)n}{d(d-1)^k - 2} \leq |\mathcal{D}| \leq \frac{n}{2k+1}. \tag{1}$$

The lower bound is derived by considering the maximum number of vertices a vertex may dominate whilst the upper bound follows as a consequence of a result of Robinson and Wormald [13] that proves the a.a. sure Hamiltonicity of random regular graphs.

In this paper we consider a simple, yet efficient, randomised greedy algorithm for finding a small k-dominating set of regular graphs. We analyse the average-case performance of this heuristic by analysing its performance on random regular graphs using differential equations. This, in turn, proves an upper bound on the size of a minimum k-dominating set of random regular graphs. For a few small values of d and k, our results are summarised in Table 1. The columns ML give the a.a. sure upper bounds (scaled by n) on the size of a minimum k-dominating set of random d-regular graphs that are the main results of this paper. As a comparison, the columns LB give the corresponding trivial lower bounds (scaled by n) given by Equation (1).

Table 1. Bounding minimum k-dominating sets of a random d-regular graphs

	$d = 3$		$d = 4$		$d = 5$		$d = 6$		$d = 7$	
k	ML	LB	ML	LB	ML	LB	ML	LB	ML	LB
2	.1451	.1000	.1167	.0588	.0990	.0385	.0865	.0270	.0771	.0200
3	.0970	.0455	.0781	.0189	.0672	.0094	.0596	.0053	.0538	.0033
4	.0741	.0218	.0623	.0062	.0555	.0023	.0502	.0010	.0459	.0006

2   Random Graphs and Differential Equations

One method of analysing the performance of a randomised algorithm is to use a system of differential equations to express the expected changes in variables describing the state of the algorithm during its execution on random graphs. Wormald [16] gives an exposition of this method and Duckworth [2] applies this method to other graph-theoretic optimisation problems. In this section we describe the model we use to generate a regular graph u.a.r. (see, for example, [6]) and give an overview of the method of [16].

To generate an n-vertex, d-regular graph u.a.r., first take dn points in n buckets labelled $1 \ldots n$ with d points in each bucket and then choose u.a.r. a disjoint pairing of the dn points. The buckets represent the vertices of the randomly generated graph and each pair represents an edge whose end-points are given by the buckets of the points in the pair. For the generated graph to be regular and simple, no pair may contain two points from the same bucket and no two pairs may contain four points from just two buckets. With probability bounded below by a positive constant, loops and multiple edges do not occur (see, for example, Wormald [15, Section 2.2]).

Throughout the execution of our algorithm, edges are deleted and, as this happens, the degree of a vertex may change. In what follows, we denote the set of vertices of degree i in the graph, at time t, by $V_i = V_i(t)$ and let $Y_i = Y_i(t)$ denote $|V_i|$. (For such variables, in the remainder of the paper, the parameter t will be omitted where no ambiguity arises.) We can express the state of the graph at any point during the execution of the algorithm by considering the variables Y_i where $0 \leq i \leq d$. In order to analyse a randomised algorithm for finding a small k-dominating set, $\mathcal{D}$, of regular graphs, we calculate the expected change in this state over a predefined unit of time in relation to the expected change in the size of $\mathcal{D}$. Let $D = D(t)$ denote $|\mathcal{D}|$ at any stage of an algorithm (time t) and let $\mathbf{E}\Delta X$ denote the expected change in a random variable X conditional upon the history of the process. We then use equations representing $\mathbf{E}\Delta Y_i$ and $\mathbf{E}\Delta D$ to derive a system of differential equations. The solutions to these differential equations describe functions which represent the behaviour of the variables Y_i. Wormald [16] describes a general result which guarantees that the solutions of the differential equations almost surely approximate the variables Y_i and D with error $o(n)$. The expected size of $\mathcal{D}$ may be deduced from these results.

3 A Simple Heuristic

We say that our algorithm proceeds as a series of *operations*. Pseudo-code for the algorithm we present, k-greedy, is given in Figure 1; a more detailed description follows and this is represented pictorially in Figure 2.

Input: A d-regular n-vertex graph, G.
Output: A k-dominating set $\mathcal{D}$ for G.

$\mathcal{D} \leftarrow \emptyset$;
while ($|E(G)| > 0$) **do**
      Select u_1 u.a.r. from those vertices of minimum positive degree in G;
      Let S denote the neighbours of u_1;
      for ($q = 2 \ldots k{+}1$) **do**
          if ($S \neq \emptyset$)
              Select u_q u.a.r. from the vertices of maximum degree in S;
              Let S denote the neighbours of u_q;
              $v \leftarrow u_q$;
          else break;
          endif
      enddo
      Add v to $\mathcal{D}$;
      Delete all edges incident with vertices at distance at most k from v;
      Add any isolates that are created to $\mathcal{D}$;
enddo

Fig. 1. Algorithm k-greedy

For each operation, a vertex, u_1, is selected u.a.r. from those vertices of minimum positive degree. The choice of which vertex to add to $\mathcal{D}$ in an operation initially depends upon the degree(s) of the neighbour(s) of u_1. Select u_2 u.a.r. from those vertices of maximum degree amongst the neighbours of u_1 and delete all edges incident with u_1. Select u_3 u.a.r. from those vertices of maximum degree amongst the neighbours of u_2 and delete all edges incident with the other neighbours of u_2. Continue this process up to $k{+}1$ times. Each time, once u_j has been selected u.a.r. from the neighbours of u_{j-1}, delete all edges incident with vertices at distance at most $j - 3$ from the other neighbours of u_{j-1}. The final vertex of maximum degree chosen is the vertex, v, to be added to $\mathcal{D}$. Once v has been chosen, all vertices at distance at most k from v are dominated by v and all edges incident with these vertices are deleted. In Figure 2, a typical operation is represented. The shaded wavy lines represent the deletion of all edges incident with vertices at distance at most $j - 3$ from the neighbours of u_{j-1} (apart from u_j). The meaning of other notation in the figure will become apparent in the next section.

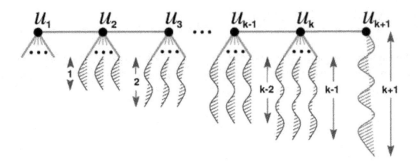

Fig. 2. Selecting a vertex to add to $\mathcal{D}$

The deletion of all the edges in an operation may cause the generation of non-dominated vertices of degree zero which we refer to as *isolates* (as they are isolated from the rest of the graph by the deletion of these edges). Such vertices will not be selected by subsequent operations of the algorithm, therefore, any isolates created in an operation are added to $\mathcal{D}$ as part of that operation along with v. This ensures that the set returned is k-dominating.

4 Algorithm Analysis

The analysis of our algorithm is carried out using differential equations and in this way we prove the following theorem.

Theorem 1. *For each $d \geq 3$ and $k \geq 2$, there exists a constant c such that the size of a minimum k-dominating set of a random d-regular graph on n vertices is asymptotically almost surely at most $cn + o(n)$, for d and k constant.*

Proof. The first operation of the algorithm is unique in the sense that when we select u_1 u.a.r. from those vertices of minimum positive degree, it is the only time that it is selected from V_d. This follows from the well known fact that random regular graphs are a.a.s. connected (see, for example, [15]).

We split the remainder of the analysis into $d - 1$ ordered *phases*, Phase 1, Phase 2, ..., Phase $d - 1$. Phase f, $1 \leq f \leq d - 2$ is informally defined as the period of time where the minimum positive degree of a vertex in the graph is either $d - f$ or $d - f - 1$ and any vertices in V_{d-f-1} that are generated are used up almost as fast as they are created so that Y_{d-f-1} remains small. Once the rate of generating vertices in V_{d-f-1} becomes larger than the rate they are used up, the algorithm moves into Phase $f + 1$. In Phase $d - 1$, the minimum positive degree of a vertex in the graph is 1 and each operation generates vertices in V_1 with positive probability. The precise distinction between phases will be clarified below.

Define a Type r operation to be an operation in which the vertex u_1 is selected u.a.r. from V_r. Also, define a *clutch* to be a series of operations in Phase

f from an operation of Type $d - f$ up to but not including the next operation of Type $d - f$. Initially one only needs to assume that after the initial operation of the algorithm, the process starts in Phase 1; in Phase f, $1 \leq f \leq d - 2$, all operations are of Type $d - f$ and $d - f - 1$; and in Phase $d - 1$, all operations are of Type 1 so that a clutch in Phase $d - 1$ consists of just one operation. We proceed with a general examination of the first $d - 2$ phases and then a general examination of the final phase. This allows us to give a formal definition of the distinction between the phases.

The initial operation of Phase f is of Type $d - f$ (at least a.a.s.). The next operation of the algorithm may be of Type $d - f$ or Type $d - f - 1$ depending on the size of the set V_{d-f-1}. In order to calculate the expected change in the variables Y_i for a clutch in Phase f, we first calculate the expected change in the variables Y_i for an operation of Type $w \in \{d - f, d - f - 1\}$ in Phase f, $1 \leq f \leq d - 2$.

For an operation of Type w in Phase f, a vertex, u_1, is chosen u.a.r. from V_w. Let s denote the sum of the degrees of all the vertices in the graph at a given stage (time t). Note that

$$s = s(t) = \sum_{i=1}^{d} i Y_i.$$

For our analysis it is convenient to assume that $s > \epsilon n$ for some arbitrarily small but fixed $\epsilon > 0$. Later we discuss the last operations of the algorithm where $s \leq \epsilon n$.

The probability that, when selecting a vertex of positive degree u.a.r. (at time t), the vertex is of degree j is P_j where

$$P_j = P_j(t) = \frac{j Y_j}{s}, \qquad 1 \leq j \leq d.$$

When an edge incident with a vertex of degree i is deleted, the expected change in Y_i due to changing the degree of the vertex from i to $i - 1$ (at time t) is $\rho_i + o(1)$ where

$$\rho_i = \rho_i(t) = P_{i+1} - P_i, \qquad 1 \leq i \leq d$$

and this equation is valid under the assumption that $P_{d+1} = 0$. To justify this, note that when the vertex of degree i was chosen, the sum of the degrees of all the vertices of degree i is $i Y_i$ and s is the sum of the degrees of all the vertices in G. In this case Y_i decreases; it increases if the vertex selected has degree $i + 1$. These two quantities are added because expectation is additive. The $o(1)$ term is due to the fact that the values of all these variables may change by a constant during the course of the operation being examined. Since $s > \epsilon n$ the error is in fact $O(1/n)$.

Once u_1 has been chosen u.a.r. from all the vertices in G of degree w, before deleting all of its incident edges, we investigate the degree(s) of its neighbour(s). We select a vertex, u_2, u.a.r. from those vertices of maximum degree amongst the neighbours of u_1. The probability that u_2 was in V_m, $d - f \leq m \leq d$, at the start of the operation (time t) is $\alpha + o(1)$ where

$$\alpha = \alpha(t) = (S_{d-f}^m)^w - (S_{d-f}^{m-1})^w$$

and

$$S_i^j = S_i^j(t) = \mathrm{P}_i + \mathrm{P}_{i+1} + \ldots + \mathrm{P}_j.$$

Note that m a.a.s. satisfies $d - f \leq m \leq d$ as this is Phase f.

The probability that there are r neighbours of u_1 in V_m, given that m is the maximum degree of all neighbours of u_1, is $\beta + o(1)$ where

$$\beta = \beta(t) = \frac{(\mathrm{P}_m)^r \binom{w}{r} (S_{d-f}^{m-1})^{w-r}}{\alpha}.$$

The expected number of neighbours of u_1 of degree j, $d - f \leq j \leq m - 1$, given that m was the maximum degree of all neighbours and there are r such vertices is $\gamma + o(1)$ where

$$\gamma = \gamma(t) = \frac{(\mathrm{P}_m)^r \binom{w}{r}}{\alpha} \sum_{a=0}^{w-r} a \binom{w-r}{a} (\mathrm{P}_j)^a \left(S_{d-f}^{m-1} - \mathrm{P}_j \right)^{w-r-a}$$

which simplifies to

$$\gamma = \frac{(\mathrm{P}_m)^r \binom{w}{r} (S_{d-f}^{m-1})^{w-r-1} (w-r)\mathrm{P}_j}{\alpha}.$$

The expected number of neighbours of degree i, $d - f \leq i \leq d$, incident with u_j, $j > 1$, is slightly different from that for u_1 as, if u_j had degree j' at the start of the operation, by the time all its incident edges are deleted, its degree is then $j' - 1$ as an edge between u_{j-1} and u_j is deleted. The analogous equations for u_j, $j > 1$, to those for α, β and γ are $\alpha' + o(1)$, $\beta' + o(1)$ and $\gamma' + o(1)$ where

$$\alpha' = \alpha'(t) = (S_{d-f}^m)^{w-1} - (S_{d-f}^{m-1})^{w-1},$$

$$\beta' = \beta'(t) = \frac{(\mathrm{P}_m)^r \binom{w-1}{r} (S_{d-f}^{m-1})^{w-r-1}}{\alpha} \quad \text{and}$$

$$\gamma' = \gamma'(t) = \frac{(\mathrm{P}_m)^r \binom{w-1}{r} (S_{d-f}^{m-1})^{w-r-2} (w-r-1)\mathrm{P}_j}{\alpha}.$$

When we select u_j, $j > 1$, of degree m', we select u_{j+1} and delete the remaining $m' - 1$ edges incident with u_j (one edge from u_j to u_{j-1} was already deleted). Apart from u_{j+1}, all other neighbours of u_j have all edges incident with vertices at distance at most $j - 3$ deleted. As random regular graphs a.a.s. contain few small cycles [6] the subgraph deleted from each of these vertices is a.a.s. a tree.

We describe the deletion of a tree of height h in *levels* where the root of the tree (a neighbour of u_j other than u_{j+1}) is at level 0 and all other vertices in the tree at distance ℓ from the root are at level ℓ. When an edge is deleted between a vertex at level ℓ in the tree and a vertex at level $\ell+1$ (at time t), the expected remaining degree of the vertex at level $\ell+1$ is $T+o(1)$ where

$$T = T(t) = \sum_{b=2}^{d}(b-1)\mathrm{P}_b.$$

For a tree with a root of degree j at the start of an operation, the expected number of vertices at level $\ell > 0$ in the tree (at time t), is $Q(\ell, j) + o(1)$ where

$$Q(\ell, j) = Q(\ell, j, t) = (j-1)T^{\ell-1}.$$

The expected number of vertices of degree i in the levels $1, \ldots, h-1$ when exposing a tree of height h from a root of degree j is $R(i, j, h) + o(1)$ where

$$R(i, j, h) = R(i, j, h, t) = \sum_{\ell=1}^{h-1}Q(\ell, j)\mathrm{P}_i.$$

Therefore, the expected change in Y_i when exposing a tree of height h from a neighbour of $u_{j'}$ of degree j (at time t) is $X(i, j, h) + o(1)$ where

$$X(i, j, h) = X(i, j, h, t) = -\delta_{ij} - R(i, j, h) + Q(h, j)\rho_i$$

in which δ_{ij} denotes the Kronecker delta function.

In Figure 2, the heights associated with the shaded wavy lines denote the heights of the trees deleted incident with their respective u_j.

We are now ready to compute the expected changes in the variables Y_i, for an operation of Type w (not a clutch yet) in Phase f. We note that, due to the intricate way each subsequent vertex of $\mathcal{D}$ is chosen, these equations are not of the simplest form and note that we have analysed several simpler heuristics, each of which give weaker results.

The expected change in Y_i when performing an operation of Type w in Phase f is $\phi(i, w, k) + o(1) = \phi(i, w, k, t) + o(1)$ where $\phi(i, w, 0)$ is given by

$$X(i, w, k+1),$$

$\phi(i, w, k)$ is given by

$$-\delta_{iw} + \sum_{m=d-f}^{d}\alpha[\phi(i, m, k-1) + \sum_{r=1}^{w}\beta(r-1)(\delta_{im-1} - \delta_{im}) + \sum_{j=d-f}^{m-1}\gamma(\delta_{ij-1} - \delta_{ij})]$$

and, for $0 < \kappa < k$, $\phi(i, w, \kappa)$ is given by

$$-\delta_{iw} + \sum_{m=d-f}^{d}\alpha'[\phi(i, m, \kappa-1) + \sum_{r=1}^{w-1}\beta'(r-1)X(i, m, k-\kappa) + \sum_{j=d-f}^{m-1}\gamma'X(i, j, k-\kappa)].$$

We define a *birth* to be the generation of a vertex in V_{d-f-1} by performing an operation in Phase f. The expected number of births from performing an operation of Type $w \in \{d-f, d-f-1\}$ (at time t) is $\nu(k, w) + o(1) = \nu(k, w, t) + o(1)$ where $\nu(0, w)$ is given by

$$Q(k+1, w)\mathrm{P}_{d-f},$$

$\nu(k, w)$ is given by

$$\sum_{m=d-f}^{d} \alpha[\nu(k-1, m) + \sum_{r=1}^{w} \beta(r-1)\delta_{md-f} + \sum_{j=d-f}^{m-1} \gamma\delta_{jd-f}]$$

and, for $0 < \kappa < k$, $\nu(\kappa, w)$ is given by

$$\sum_{m=d-f}^{d} \alpha'[\nu(\kappa-1, m) + \sum_{r=1}^{w-1} \beta'(r-1)Q(k-\kappa, m)\mathrm{P}_{d-f} + \sum_{j=d-f}^{m-1} \gamma'Q(k-\kappa, j)\mathrm{P}_{d-f}].$$

Here, we consider the probability that vertices of degree $d - f$ in the graph become vertices of degree $d - f - 1$ by deleting an edge incident with the vertex.

Consider the Type $d - f$ operation at the start of the clutch to be the first generation of a *birth-death* process in which the individuals are the vertices in V_{d-f-1}, each giving birth to a number of children (a.a.s. independent of the others) with expected number $\nu(k, d-f-1)$. Then, the expected number in the jth generation is $\nu(k, d-f)(\nu(k, d-f-1))^{j-1}$ and the expected total number of births in the clutch is

$$\frac{\nu(k, d-f)}{1 - \nu(k, d-f-1)}.$$

The equation giving the expected change in Y_i for a clutch in Phase f, $1 \le f \le d - 2$ is therefore given by

$$\mathbf{E}\Delta Y_i = \phi(i, d-f, k) + \frac{\nu(k, d-f)}{1 - \nu(k, d-f-1)}\phi(i, d-f-1, k) + o(1), \qquad 1 \le i \le d. \tag{2}$$

In the first $d - 2$ phases, since Y_{d-f-1} is very small and Y_{d-f} is at least a constant times n, the probability of an isolate being created is negligible in any one operation. Thus, the expected increase in the size of the k-dominating set is 1 for any single operation the first $d - 2$ phases. So the equation giving the expected increase in the size of the k-dominating set for a clutch in Phase f, $1 \le f \le d - 2$ is given by

$$\mathbf{E}\Delta D = 1 + \frac{\nu(k, d-f)}{1 - \nu(k, d-f-1)}. \tag{3}$$

For the final phase, all operations are of Type 1 and we must also consider the possibility of isolates being generated. The equation giving the expected change in Y_i for a clutch in Phase $d - 1$ is therefore given by

$$\mathbf{E}\Delta Y_i = \phi(i, 1, k), \qquad 1 \le i \le d. \tag{4}$$

The expected number of isolates generated for an operation of Type 1 in Phase $d-1$ is $\nu(k,1)$, therefore the equation giving the expected increase in the size of the k-dominating set for a clutch in Phase $d-1$ is

$$\mathbf{E}\Delta D = 1 + \nu(k,1). \tag{5}$$

Write $\phi(i,w,k,t) = n\bar{\phi}(i,w,k,t/n)$, $\mathrm{P}_i(t) = n\bar{\mathrm{P}}_i(t/n)$, $\alpha'(t) = n\bar{\alpha}'(t/n)$, $\gamma'(t) = n\bar{\gamma}'(t/n)$, $Q(\ell,j,t) = n\bar{Q}(\ell,j,t/n)$, $\beta(t) = n\bar{\beta}(t/n)$, $s(t) = n\xi(t/n)$, $X(i,j,h,t) = n\bar{X}(i,j,h,t/n)$, $\nu(k,w,t) = n\bar{\nu}(k,w,t/n)$, $\rho_i(t) = n\bar{\rho}_i(t/n)$, $\alpha(t) = n\bar{\alpha}(t/n)$, $S_i^j(t) = n\bar{S}_i^j(t/n)$, $\beta'(t) = n\bar{\beta}'(t/n)$, $\gamma(t) = n\bar{\gamma}(t/n)$, $T(t) = n\bar{T}(t/n)$, $Y_i(t) = nz_i(t/n)$ and $R(i,j,h,t) = n\bar{R}(i,j,h,t/n)$.

Equation (2) representing $\mathbf{E}\Delta Y_i$ for processing a clutch in Phase f forms the basis of a differential equation. The differential equation suggested is

$$\frac{dz_i}{dx} = \bar{\phi}(i,d-f,k) + \frac{\bar{\nu}(k,d-f)}{1-\bar{\nu}(k,d-f-1)}\bar{\phi}(i,d-f-1,k), \qquad 1 \le i \le d \tag{6}$$

where differentiation is with respect to x and xn represents the number, t, of clutches.

Equation (3) representing the expected increase in the size of $\mathcal{D}$ after processing a clutch in Phase f, $1 \le f \le d-2$, and writing $D(t) = nz(t/n)$ suggests the differential equation for z as

$$\frac{dz}{dx} = 1 + \frac{\bar{\nu}(k,d-f)}{1-\bar{\nu}(k,d-f-1)}. \tag{7}$$

For Phase $d-1$, equation (4) representing $\mathbf{E}\Delta Y_i$ for processing a clutch suggests the differential equation

$$\frac{dz_i}{dx} = \bar{\phi}(i,1,k), \qquad 1 \le i \le d. \tag{8}$$

Equation (5) representing the increase in the size of $\mathcal{D}$ after processing a vertex (and therefore a clutch) in Phase $d-1$, suggests the differential equation for z as

$$\frac{dz}{dx} = 1 + \bar{\nu}(k,1). \tag{9}$$

We use a result from [16] to show that during each phase, the functions representing the solutions to the differential equations almost surely approximate the variables Y_i/n and D/n with error o(1). For this we need some definitions.

Consider a probability space, S, whose elements are sequences $(q_0, q_1, \ldots)$ where each $q_t \in S$. We use h_t to denote $(q_0, q_1, \ldots, q_t)$, the *history* of the process up to time t, and H_t for its random counterpart. $S^{(n)+}$ denotes the set of all $h_t = (q_0, \ldots, q_t)$ where each $q_i \in S$, $t = 0, 1, \ldots$. All these things are indexed by n and we will consider asymptotics as $n \to \infty$.

For variables $Y_1, \ldots, Y_a$ defined on the components of the process and $W \subseteq \mathbb{R}^{a+1}$, define the *stopping time* $T_W = T_W(Y_1, \ldots, Y_a)$ to be the minimum t such that

$$(t/n, Y_1(t)/n, \ldots, Y_a(t)/n) \notin W.$$

The following is a restatement of [16, Theorem 6.1]. We refer the reader to that paper for explanations and to [14] for a similar result with virtually the same proof.

Theorem 2 ([16]). *Let $\widehat{W} = \widehat{W}(n) \subseteq \mathbb{R}^{a+1}$. For $1 \le l \le a$, where a is fixed, let $y_l : S^{(n)+} \to \mathbb{R}$ and $f_l : \mathbb{R}^{a+1} \to \mathbb{R}$, such that for some constant C_0 and all l, $|y_l(h_t)| < C_0 n$ for all $h_t \in S^{(n)+}$ for all n. Let $Y_l(t)$ denote the random counterpart of $y_l(h_t)$. Assume the following three conditions hold, where in (ii) and (iii) W is some bounded connected open set containing the closure of*

$$\{(0, z_1, \ldots, z_a) : \mathbf{P}(Y_l(0) = z_l n, 1 \le l \le a) \ne 0 \text{ for some } n\}.$$

(i) For some functions $\beta = \beta(n) \ge 1$ and $\gamma = \gamma(n)$, the probability that

$$\max_{1 \le l \le a} |Y_l(t+1) - Y_l(t)| \le \beta,$$

conditional upon H_t, is at least $1 - \gamma$ for $t < \min\{T_W, T_{\widehat{W}}\}$.
(ii) For some function $\lambda_1 = \lambda_1(n) = o(1)$, for all $l \le a$

$$|\mathbf{E}(Y_l(t+1) - Y_l(t) \mid H_t) - f_l(t/n, Y_1(t)/n, \ldots, Y_a(t)/n)| \le \lambda_1$$

for $t < \min\{T_W, T_{\widehat{W}}\}$.
(iii) Each function f_l is continuous and satisfies a Lipschitz condition, on

$$W \cap \{(t, z_1, \ldots, z_a) : t \ge 0\},$$

with the same Lipschitz constant for each l.

Then the following are true.

(a) For $(0, \hat{z}_1, \ldots, \hat{z}_a) \in W$ the system of differential equations

$$\frac{dz_l}{dx} = f_l(x, z_1, \ldots, z_a), \qquad l = 1, \ldots, a$$

has a unique solution in W for $z_l : \mathbb{R} \to \mathbb{R}$ passing through

$$z_l(0) = \hat{z}_l,$$

$1 \le l \le a$, and which extends to points arbitrarily close to the boundary of W;
(b) Let $\lambda > \lambda_1 + C_0 n \gamma$ with $\lambda = o(1)$. For a sufficiently large constant C, with probability $1 - O(n\gamma + \frac{\beta}{\lambda} \exp(-\frac{n\lambda^3}{\beta^3}))$,

$$Y_l(t) = n z_l(t/n) + O(\lambda n)$$

uniformly for $0 \le t \le \min\{\sigma n, T_{\widehat{W}}\}$ and for each l, where $z_l(x)$ is the solution in (a) with $\hat{z}_l = \frac{1}{n} Y_l(0)$, and $\sigma = \sigma(n)$ is the supremum of those x to which the solution can be extended before reaching within ℓ^∞-distance $C\lambda$ of the boundary of W.

For Phase f and for arbitrary small ϵ, define D_f to be the set of all (t, z_i, z) for which $t > -\epsilon$, $\xi > \epsilon$, $\bar{\nu}(k, d - f - 1) < 1 - \epsilon$, $z > -\epsilon$ and $z_i < 1 + \epsilon$ where $1 \leq i \leq d$. Then, D_f defines a domain for the process so that 2 may be applied within Phase f.

For part (i) of 2 we must ensure that $Y_i(t)$ does not change too quickly throughout the process. As long as the expected number of births in a clutch is bounded above, the probability of getting say n^ϵ births is $O(n^{-K})$ for any fixed K. As we assume d and k to be constant, this comes from a standard argument as in [16, page 141]. So part (i) of 2 holds with $\beta = n^\epsilon$ and $\gamma = n^{-K}$. Equations (2) and (3) verify part (ii) for a function λ_1 which goes to zero sufficiently slowly. Note in particular that since $\xi > \epsilon$ inside D_f, the assumption that $s > \epsilon n$ used in deriving these equations is justified. By the definition of the phase and the domain, D_f, it may be verified that the functions derived from equations (2) and (3) are continuous on D_f and its boundary. This implies that the functions are uniformly continuous. From this, the Lipschitz property of the functions required by 2 part (iii) may be deduced.

The conclusion of 2 therefore holds for each of the first $d - 2$ phases. This implies (by taking $\lambda = o(1)$ tending to zero sufficiently slowly) that with probability

$$1 - O(\lambda^{-1} \exp(-n\lambda^3)),$$

the random variables Y_i and D a.a.s. remain within $O(\lambda n)$ of the corresponding deterministic solutions to the differential equations (6) and (7) until a point arbitrarily close to where it leaves the domain D_f. Choosing, for example, $\lambda = n^{-1/4}$, makes this success probability $1 - o(n)$.

We compute the ratio dz_i/dz. Now, differentiation is with respect to z and all functions may be taken as functions of z. By (numerically) solving these differential equations, for each of the first $d - 2$ phases, we find that the solution hits a boundary of the domain at $\bar{\nu}(k, d - f - 1) = 1 - \epsilon$. At this point, we formally define Phase 1 as the period of time from time $t = 0$ to the time t_1 such that $z = t_1/n$ is the solution of $\bar{\nu}(k, d - 2) = 1 - \epsilon$ and Phase f, $2 \leq f \leq d - 2$, as the period of time from time t_{f-1} to time t_f where $z = t_{f'}/n$ is the solution of $\bar{\nu}(k, d - f' - 1) = 1 - \epsilon$ in Phase f'.

Some extra work is required to ensure that the phases proceed as defined informally and that Phase $f + 1$ follows Phase f without reverting back to previous phases. For reasons of brevity, we do not include the full details here. The work involves ensuring that around each change of phase, say from Phase f to Phase $f + 1$, the expected number of births from processing a vertex in V_{d-f-1} is at least a constant times n. For this, the main requirement is that $\nu(k, d - f - 1)$ is significantly larger than 1.

For Phase $d-1$, Theorem 2 applies as above except that here, a clutch consists of one Type 1 operation. Computing the ratio dz_i/dz and solving this, we see that the solution hits a boundary of D_{d-1} at $\xi = \epsilon$. From the point in Phase $d - 1$ after which Theorem 2 does not apply until the end of the algorithm, the change in each variable per operation is bounded by a constant, hence in $o(n)$ steps, the change in the variables is $o(n)$.

The equations were solved using a Runge-Kutta method with a step-size of 10^{-6}. The solution of $\xi = \epsilon$ in Phase $d - 1$ corresponds to the size of the k-dominating set (scaled by $1/n$) when no vertices remain. For $\epsilon = 0$ and for a few values of d and k, the constant, c, in Theorem 1 is given in Table 1 of Section 1.

This completes the proof of the theorem. $\square$

References

1. Chang, G.J. and Nemhauser, G.L.: The k-Domination and k-Stability Problems on Graphs. TR-540, School of Operations Research and Industrial Eng., Cornell University (1982)
2. Duckworth, W.: Greedy Algorithms and Cubic Graphs. PhD thesis, Department of Mathematics and Statistics, The University of Melbourne, Australia (2001)
3. Duckworth, W. and Wormald, N.C.: Minimum Independent Dominating Sets of Random Cubic Graphs. Random Structures and Algorithms. To Appear.
4. Favaron, O., Haynes, T.W. and Slater, P.J.: Distance-k Independent Domination Sequences. Journal of Combinatorial Mathematics and Combinatorial Computing (2000) **33** 225–237
5. Haynes, T.W., Hedetniemi, S.T. and Slater, P.J.: Domination in Graphs: Advanced topics. Marcel Dekker Inc. (1998) New York
6. Janson, S., Łuczak, T. and Rucinski, A.: Random Graphs. Wiley-Interscience (2000)
7. Johnson, D.S.: Approximation Algorithms for Combinatorial problems. In: Proceedings of the 5th Annual ACM STOC, Journal of Computer and System Sciences (1994) **9** 256–278
8. Kutten, S. and Peleg, D.: Fast Distributed Construction of Small k-dominating Sets and Applications. Journal of Algorithms (1998) **28**(1) 40–66
9. Molloy, M. and Reed, B.: The Dominating Number of a Random Cubic Graph. Random Structures and Algorithms (1995) **7**(3) 209–221
10. Raz, R. and Safra, S.: A Sub-Constant Error-Probability Low-Degree Test and a Sub-Constant Error-Probability PCP Characterization of NP. In: Proceedings of the 29th Annual ACM STOC (1999) 475–484 (electronic)
11. Papadimitriou, C.H. and Yannakakis, M.: Optimization, Approximation and Complexity Classes. Journal of Computer and System Sciences (1991) **43**(3) 425–440
12. Reed, B.: Paths, Stars and the Number Three. Combinatorics, Probability and Computing (1996) **5** 277–295
13. Robinson, R.W. and Wormald, N.C.: Almost All Regular Graphs are Hamiltonian. Random Structures and Algorithms (1994) 5(2) 363–374
14. Wormald, N.C.: Differential Equations for Random Processes and Random Graphs. Annals of Applied Probability (1995) **5** 1217–1235
15. Wormald, N.C.: Models of Random Regular Graphs. In: Surveys in Combinatorics (1999) 239–298 Cambridge University Press
16. Wormald, N.C.: The Differential Equation Method for Random Graph Processes and Greedy Algorithms. In: Lectures on Approximation and Randomized Algorithms (1999) 73–155, PWN, Warsaw
17. Zito, M.: Greedy Algorithms for Minimisation Problems in Random Regular Graphs. In: Proceedings of the 19th European Symposium on Algorithms. Lecture Notes in Computer Science (2001) **2161** 524–536, Springer-Verlag

Finding Sparse Induced Subgraphs of Semirandom Graphs

Amin Coja-Oghlan[*]

Humboldt-Universität zu Berlin, Institut für Informatik,
Unter den Linden 6, 10099 Berlin, Germany
`coja@informatik.hu-berlin.de`

Abstract. The aim of this paper is to present an SDP-based algorithm for finding a sparse induced subgraph of order $\Theta(n)$ hidden in a semirandom graph of order n. As an application we obtain an algorithm that requires only $O(n)$ random edges in order to k-color a semirandom k-colorable graph within polynomial expected time, thereby extending the results of Feige and Kilian [7] and of Subramanian [15].

1 Introduction

For several NP-hard optimization problems, such as the maximum independent set problem, graph coloring, and even 3-coloring, strong non-approximability results are known. In fact, both the size of a maximum independent set and the chromatic number of a graph cannot be approximated within $n^{1-\varepsilon}$, $\varepsilon > 0$, under reasonable complexity-theoretic assumptions [9], [6]. Also, it is hard to 4-color 3-colorable graphs [11]. Thus, since one cannot hope for algorithms that perform well in the worst case, it is natural to ask for algorithms that work acceptably in the "average case".

But what does "in the average case" mean? One may take random graphs for a first answer. Needless to say, the algorithmic theory of random graphs has become an important and fruitful field of investigation. However, algorithms for random graph problems tend to make extensive use of special properties of random graphs. But what if the "average case" instances we have in mind do not satisfy typical properties of random graphs?

Semirandom models may provide an answer. Instances of semirandom graph problems consist of a random share and a worst case part added by the adversary. The smaller the random share is, the harder the according algorithmic problem becomes. Because of the presence of an adversary, algorithms that rely on typical properties of random graphs will in general be useless in the semirandom case. Conversely, one may conclude that algorithms that perform well on semirandom instances are comparatively robust.

The first problem studied in this article is to find a sparse induced subgraph of a semirandom graph. This problem is a straightforward generalization of the

---

[*] Research supported by the Deutsche Forschungsgemeinschaft (grant DFG FOR 413/1-1)

J.D.P. Rolim and S. Vadhan (Eds.): RANDOM 2002, LNCS 2483, pp. 139–148, 2002.

independent set problem. An instance is made up as follows. Let $\alpha \in (0;1)$ be a constant, and let $V_1 = \{1, \ldots, \lfloor (1 - \alpha)n \rfloor\}$, $V_2 = \{\lfloor (1 - \alpha)n \rfloor + 1, \ldots, n\}$ and $V = V_1 \cup V_2$. First each vertex $v \in V_1$ chooses ω random neighbours in V_2, independently of all others. The result is a random bipartite graph G_0; this model of random bipartite graphs is also proposed in [7, Sec. 1.4]. Then the adversary adds V_1-V_2-edges and V_1-V_1-edges arbitrarily and inserts an arbitrary graph with maximum degree $\leq \Delta$ into V_2, where $\Delta \in \mathbb{N}_0 = \{0, 1, 2, 3, \ldots\}$ is some constant. Finally, the adversary permutes the vertices V according to some permutation $\sigma \in S_n$ in order to obtain the instance G. The algorithmic problem associated with the above model is to find a set $S \subset V$, $\#S \geq \alpha n$, such that the maximum degree of the subgraph $G[S]$ of G induced on S is at most Δ. Note that the decisions of the adversary are *not* random decisions. Hence, the only random object in the construction of G is the choice of the random bipartite graph G_0. Furthermore, the adversary is allowed to e.g. construct local maxima. Clearly, neither the bipartite graph G_0 nor the permutation σ are known to the algorithm.

We present an algorithm that on input (G, Δ) within polynomial expected time outputs a set $S \subset V$, $\#S \geq \alpha n$, such that $\Delta(G[S]) \leq \Delta$, provided $\omega \geq c(\alpha)\Delta$ is sufficiently large. Since ω is a constant w.r.t. n, the graph G_0 consists of only $O(n)$ random edges. As we shall see, by making ω large enough one can achieve that even the lth moment of the running time is bounded by a polynomial, for arbitrary l. Clearly, probability is taken over the choice of the random bipartite graph $G_0 = \mathcal{G}_{n,\alpha,\omega}$.

Related work has been done by Feige and Krauthgamer [8] and by Feige and Kilian [7]. In the semirandom model studied in [8] first a clique of size $\Omega(n^{1/2})$ is hidden in an otherwise random graph $G_{n,1/2}$. Then the adversary may remove edges that do not connect two vertices that belong to the hidden clique. It is shown that using SDP-techniques it is possible to recover the hidden clique with high probability (where probability is taken over the choice of the random graph $G_{n,1/2}$) and to certify the optimality of the solution produced by the algorithm. Note that in this model the adversary is not allowed to construct local maxima.

Instances $G = (V, E)$ of the semirandom model of the independent set problem studied in [7] are created in three steps. First, a set $S \subset V$ of αn vertices is chosen at random, where $n = \#V$ and $\alpha \in (0;1)$ is a constant. Then, each edge $\{v, w\}$, $v \in V \setminus S$, $w \in S$, is included into E with probability p independently of all other edges. Finally, the adversary may add further $(V \setminus S)$-V-edges. It is shown that in the case $p \geq (1 + \varepsilon) \ln(n)/(\alpha n)$, $\varepsilon > 0$, it is possible to recover the hidden set S with high probability (where probability is taken over the coin tosses of the process that created G). The problem becomes hard if $p \leq (1 - \varepsilon) \ln(n)/(\alpha n)$, because in this case there are many vertices in $V \setminus S$ that are not incident with any random edge.

Note that in the case $p \geq (1 + \varepsilon) \ln(n)/(\alpha n)$ every vertex of $V \setminus S$ is incident with $\Omega(\ln n)$ random edges with high probability, i.e. every vertex in $V \setminus S$ has chosen $\Omega(\ln n)$ random neighbours in S. Consequently, the "ω-neighbours model", where every vertex $v \in V \setminus S$ chooses ω random neighbours in S, in the case $p \geq (1 + \varepsilon) \ln(n)/(\alpha n)$ asymptotically coincides with the "binomial model", where each S-$V \setminus S$ edge is present with probability p. It is explicitly mentioned in [7] that the algorithm given for the binomial model also works for the ω-neighbours model, i.e. finds the hidden independent set with high probability. Thus, the new aspect in this current paper is that we find *sparse* hidden subgraphs (not just independent sets) within a polynomial average running time over *all* input instances of the ω-neighbours model. Moreover, we show that it is sufficient that ω grows with Δ linearly.

The second problem studied in this article is the k-coloring problem. Semirandom instances of the k-coloring problem are created as follows. First the adversary partitions the vertex set into k disjoint classes. No assumptions are made about the sizes of the k classes. Then, with respect to the classes produced by the adversary, a k-partite graph is chosen at random according to a random model analogous to the ω-neighbours model of random bipartite graphs above. Finally, the adversary may add edges that connect vertices in different classes. We shall see that there exists a constant $\omega(k)$ such that in order to k-color semirandom k-colorable graphs within expected polynomial time it suffices that each vertex chooses $\omega(k)$ random neighbours in each of the other color classes, thereby producing not more than $\Theta(n)$ random edges. Since by choosing $\omega(k)$ sufficiently large one can achieve that arbitrarily high moments of the running time are polynomials, our k-coloring algorithm seems to be quite robust.

Semirandom models for the k-coloring problem have been studied by Blum and Spencer [3], by Subramanian [15], and by Feige and Kilian [7]. In the semirandom model for k-coloring studied in [7], first the vertices $V = \{1, \ldots, n\}$ are partitioned into k disjoint sets $V_1, \ldots, V_k$ of size n/k each. Then, each edge $\{v, w\}$, $v \in V_i$, $w \in V_j$, $i \neq j$, is included into G with probability p, independently of all others. Finally, the adversary may add further edges $\{v, w\}$ as above, thereby obtainig the instance G. The algorithm given in [7] k-colors G with high probability, provided $p \geq (1+\varepsilon)k(\ln n)/n$. Although the "$\omega$-neighbours model" for k-colorable semirandom graphs is not explicitly mentioned in [7], the k-coloring algorithm presented there can easily be adapted to k-color instances of the ω-neighbours model (with color classes of arbitrary size) with high probability. Moreover, Feige and Kilian show that k-coloring is hard if $p \leq (1-\varepsilon)(\ln n)/n$. In [3] and [15] "binomial" semirandom models are studied, with $p = n^{\varepsilon-1}$ for $\varepsilon > 0$ constant. For the problem of coloring *random* graphs see e.g. Alon and Kahale [2], or Prömel and Steger [14].

Among the above, the only algorithm that solves the semirandom k-coloring problem within expected polynomial time is the one given in [15]; in fact, Subramanian's algorithm even finds an optimal coloring. In [5] the author has given algorithms for the binomial semirandom independent set problem and for the binomial semirandom k-coloring problem. In the case $p \gg \ln(n)/n$ these algorithms have a polynomial expected running time. Thus, concerning k-coloring, the new aspect in the current paper is that the algorithm k-colors semirandom graphs produced according to the ω-neighbours model. Indeed, the fact that we admit color classes of arbitrary sizes causes some new technical difficulties.

2   Models and Results

2.1   Sparse Induced Subgraphs

Let $\alpha \in (0; 1)$ be a constant, $V = \{1, \ldots, n\}$, $V_1 = \{1, \ldots, \lfloor(1 - \alpha)n\rfloor\}$, and $V_2 = V \setminus V_1$. Let $\omega \in \mathbb{N}$. By $\mathcal{G}_{n,\alpha,\omega}$ we denote the set of all (V_1, V_2)-bipartite graphs G_0 such that each vertex $v \in V_1$ in G_0 has degree ω. Hence $\#\mathcal{G}_{n,\alpha,\omega} = \binom{\lceil \alpha n \rceil}{\omega}^{\lfloor(1-\alpha)n\rfloor}$. We equip $\mathcal{G}_{n,\alpha,\omega}$ with the uniform distribution. Thus, the random bipartite graph $\mathcal{G}_{n,\alpha,\omega}$ is obtained by letting each vertex $v \in V_1$ choose precisely ω random neighbours in V_2, independently of all other vertices in V_1. Let $\Delta \in \mathbb{N}_0$ be a constant. Instances of the "hidden sparse subset" problem are produced as follows:

1. A bipartite graph $G_0 \in \mathcal{G}_{n,\alpha,\omega}$ is chosen at random.
2. The adversary adds some edges $\{v, w\}$, $v \in V_1$, $w \in V$, to G_0 in order to obtain G_1.
3. The adversary inserts a graph of maximum degree at most Δ into V_2, thereby obtaining G_2.
4. Finally, the adversary permutes the vertices of G_2, thereby completing the instance G.

For a bipartite graph $G_0 \in \mathcal{G}_{n,\alpha,\omega}$, we let $\mathcal{I}(G_0, \Delta)$ denote the set of all *instances over* G_0, i.e. the set of all graphs G that can be constructed from G_0 via 2.–4. above.

Note that the only random contribution to the construction of G is the choice of G_0. Thus, the graph G contains precisely $\lfloor(1 - \alpha n)\rfloor\omega$ random edges, and the adversary may add up to

$$\frac{1}{2}\Delta\lceil \alpha n \rceil + \lfloor(1 - \alpha)n\rfloor(\lceil \alpha n \rceil - \omega) + \binom{\lfloor(1 - \alpha)n\rfloor}{2} \tag{1}$$

edges. Consequently, (1) indicates that the algorithmic problem (on input (G, Δ), find an induced subgraph of order $\geq \alpha n$ and maximum degree $\leq \Delta$) becomes harder as ω decreases; for the smaller ω is, the more edges are under the rule of the adversary. Similarly, the problem becomes harder as Δ increases.

If $\mathcal{A}$ is an algorithm that on input $G = (V, E)$ and Δ outputs a set $S \subset V$ such that $\Delta(G[S]) \leq \Delta$ and $\#S \geq \alpha n$, provided G admits such a set S, then we say that $\mathcal{A}$ *finds* (α, Δ)-*sparse sets*. Let $R_{\mathcal{A}}(G)$ denote the running time of $\mathcal{A}$ on input G. We say that $\mathcal{A}$ *runs in polynomial expected time* if there exists a polynomial $f(n)$ such that the following condition is satisfied: For any map I that to each $G_0 \in \mathcal{G}_{n,\alpha,\omega}$ assigns an instance $I(G_0) \in \mathcal{I}(G_0, \Delta)$ the expectation of $R_{\mathcal{A}} \circ I$ is bounded by $f(n)$. This definition carries over to the case that $\mathcal{A}$ is randomized.

Theorem 1. *For each $\alpha \in (0; 1)$ there exists a number $\omega(\alpha) > 0$ such that for all $\Delta \in \mathbb{N}_0$ and all $\omega_0 \geq \omega(\alpha)(\Delta+1)$ there is a randomized algorithm that within polynomial expected time finds (α, Δ)-sparse sets in $\mathcal{I}(\mathcal{G}_{n,\alpha,\omega_0}, \Delta)$.*

In Section 3, we present an algorithm that has the properties stated in Theorem 1. Note that the theorem contains a result on the independent set problem as a special case ($\Delta = 0$); for this problem we can even construct a deterministic algorithm. We emphasize the fact that the number of random edges needed depends on Δ *linearly*. Though in Theorem 1 Δ is not allowed to grow with n, using the methods of this paper it is possible to design a randomized algorithm that finds (α, Δ)-sparse sets even in the case that $\Delta = \Delta(n)$ grows with n (details omitted).

2.2 k-Coloring

Let $k \geq 3$ be a fixed integer and let $V = \{1, \ldots, n\}$. Let $\chi : \{1, \ldots, n\} \rightarrow \{1, \ldots, k\}$ be a map such that for the sets $V_i = \chi^{-1}(i)$, $i = 1, \ldots, k$, we have

$$\#V_1 \leq \#V_2 \leq \cdots \leq \#V_k.$$

Let $\omega \in \mathbb{N}$. Then, one can obtain a random k-colorable graph G_0 by letting each vertex $v \in V_i$, $i = 1, \ldots, k - 1$, choose ω neighbours in V_j, $j = i + 1, \ldots, k$, independently of all other vertices. Clearly, χ is a proper k-coloring of G_0. By $\mathcal{G}_{n, \chi, \omega}$ we denote the set of all random k-colorable graphs that can be obtained in this manner. Our semirandom model of the k-coloring problem is as follows.

1. The adversary chooses an arbitrary map $\chi : V \rightarrow \{1, \ldots, k\}$, such that for the sets $V_i = \chi^{-1}(i)$, $i = 1, \ldots, k$, we have $\#V_1 \leq \cdots \leq \#V_k$.
2. Then, $G_0 \in \mathcal{G}_{n, \chi, \omega}$ is chosen randomly.
3. In order to obtain the instance G, the adversary adds to G_0 arbitrary edges $\{v, w\}$ such that $v \in V_i$ and $w \in V_j$, $i \neq j$.

For each pair (G_0, χ), by $\mathcal{I}(G_0, \chi)$ we denote the set of all *instances over* (G_0, χ), i.e. the set of all graphs G that can be constructed from G_0 and χ by 3. above. Let $\mathcal{A}$ be an algorithm that on input G outputs a k-coloring of G, provided $\chi(G) \leq k$. Let $R_{\mathcal{A}}(G)$ denote the running time of $\mathcal{A}$ on input G. We say that $\mathcal{A}$ k-colors (k, ω)-*semirandom graphs within expected polynomial time* if there exists a polynomial $f(n)$ such that for all $\chi : V \rightarrow \{1, \ldots, k\}$ the following condition is satisfied: For any map I that to each $G_0 \in \mathcal{G}_{n, \chi, \omega}$ assigns an instance $I(G_0) \in \mathcal{I}(G_0, \chi)$, the expectation of $R_{\mathcal{A}} \circ I$ is at most $f(n)$.

Theorem 2. *For each $k \in \{3, 4, 5, \ldots\}$ there exists $\omega(k) \in \{1, 2, \ldots\}$ such that the following holds. If $\omega_0 \geq \omega(k)$, then there is an algorithm $\mathcal{A}$ that k-colors (k, ω_0)-semirandom graphs within polynomial expected time. On the other hand, if $\omega_0 < \omega(k)$, then there is no such algorithm, unless $P = NP$.*

3 Finding Sparse Induced Subgraphs

The algorithm for the hidden sparse subset problem makes use of an SDP-based randomized procedure `Filter`. Similar methods have in [7] been applied to the semirandom independent set problem. The procedure `Filter` itself is based on

SDP-techniques developed in [1], [10], and [13]. Details can be found in my technical report [4]. As before, we let $\alpha \in (0;1)$, $\Delta \in \mathbb{N}_0$, $V = \{1, \ldots, n\}$, $V_1 = \{1, \ldots, \lfloor(1-\alpha)n\rfloor\}$, and $V_2 = V \setminus V_1$.

Proposition 3. *Let $\gamma > 0$. Assume that $G_0 \in \mathcal{G}_{n,\alpha,\omega(\Delta+1)}$ for some sufficiently large number ω depending only on α and γ, and that $G \in \mathcal{I}(G_0, \Delta)$. Let S be the sparse subset hidden in G. There exists a polynomial time randomized algorithm $\textit{Filter}(G, \Delta, \varepsilon)$ that with probability $\geq 1 - e^{-\gamma n}$, over both the coin tosses and the choice of G_0, outputs a set $\mathcal{M}$ of subsets of V that contains an element $I \in \mathcal{M}$ such that $\#I \cap S \geq (\alpha - 5\varepsilon)n$ and $\#I \setminus S \leq \varepsilon \#I$. Moreover, $\#\mathcal{M} = O(n^2)$.*

Thus, with extremely high probability $\texttt{Filter}(G, \Delta, \varepsilon)$ computes a set I that contains many vertices of the hidden set S but only few vertices of $V \setminus S$. (If we are just to find an independent set, $\texttt{Filter}$ can be derandomized [13].) The algorithm $\texttt{Find}$ shown in Figure 1 applies network flow techniques to I in order to compute a large subset $S' \subset S$. Finally, using flow techniques again, $\texttt{Find}$ tries to enlarge S' until S is recovered. For the flow techniques to work properly, it is essential that the random bipartite graph $\mathcal{G}_{n,\alpha,\omega}$ contained in G is an expanding graph. More precisely, we need the following estimate.

Lemma 4. *Let $\gamma > 0$ and $d \in \mathbb{N}$. There exists a number ω_0 that depends only on α and γ such that for all $\omega \geq \omega_0 \cdot d$ the following holds. Given l numbers $\eta_1, \ldots, \eta_l \in \{0, 1, 2, \ldots, n/2\}$, the probability that there exist pairwise disjoint sets $T_1, \ldots, T_l \subset V_1$ such that $\#T_i \leq \frac{\alpha n}{2d}$ and $\#N(T_i) \leq d\#T_i - \eta_i$ for all i, is*

$$\ll \left\{ \binom{n}{\eta_1} \cdots \binom{n}{\eta_l} \right\}^{-\gamma}. \tag{2}$$

Since the bound (2) is not exponentially small, an algorithm that runs within polynomial expected time has to take into account the possibility that the random bipartite graph $G_0 \in \mathcal{G}_{n,\alpha,\omega}$ contained in the input graph is not an expanding graph. To this end, $\texttt{Find}$ uses a variable η that measures how far from being an expanding graph G_0 is.

Proposition 6. *Let $\alpha \in (0;1)$, $\Delta \in \mathbb{N}_0$ and $k \in \mathbb{N}$ be fixed. Suppose that $\omega \geq \omega_0$ for some sufficiently large ω_0 depending only on α and k. Let $G \in \mathcal{I}(\mathcal{G}_{n,\alpha,\omega(\Delta+1)}, \Delta)$. Then $\textit{Find}(G, \Delta, \alpha)$ outputs a set $S' \subset V$ that satisfies $\#S' \geq \alpha n$ and $\Delta(G[S']) \leq \Delta$. Furthermore, the kth moment of the running time of $\textit{Find}$ is polynomial.*

For the rest of this section, we keep the notations and assumptions from Proposition 6. We denote the random bipartite graph that is contained in G by G_0. Moreover, S denotes the hidden sparse subset of G_0, $\#S = \lceil \alpha n \rceil$. We may assume that the set $\mathcal{M}$ computed by $\texttt{Filter}$ contains an element I such that $\#I \cap S \geq (\alpha - 5\varepsilon)n$ and $\#I \setminus S < \varepsilon \#I$, because the probability that $\texttt{Filter}$ fails is exponentially small.

In order to prove our assertion concerning the running time of $\texttt{Find}$, we shall show that $\texttt{Find}$ terminates with a correct output before η becomes too large. More precisely, let

Algorithm 5. Find(G, Δ, α)
Input: A graph $G = (V, E)$, numbers $\alpha \in (0; 1)$, $\Delta \in \mathbb{N}_0$.
Output: A set $S' \subset V$.

1. Let ω be a sufficiently large number depending only on α and Δ. (One can put $\omega = \omega_0$ where ω_0 is as in Proposition 6.) Let $\tilde{\Delta} = \max\{\Delta, 2\}$. Let $\mathcal{M} = \texttt{Filter}(G, \tilde{\Delta}, \varepsilon = \alpha^2/1000)$.
2. For $\eta = 0, \ldots, n/2$ do
3.   For all $I \in \mathcal{M}$ do
4a.   For all $D' \subset V$, $\#D' \leq \eta$ do
4b.   Construct the following network N:

- The vertices of N are s, t, s_v for $v \in I \setminus D'$, and t_w for $w \in V$.
- The arcs of N are (s, s_v) for $v \in I \setminus D'$, (t_w, t) for $w \in V$, and (s_v, t_w) for $\{v, w\} \in E$, $v \in I \setminus D'$.
- The capacity c is given by $c(s, s_v) = \lceil \frac{\alpha \tilde{\Delta}}{2\varepsilon} \rceil$ for $v \in I \setminus D'$, $c(t_w, t) = 2\tilde{\Delta} + 1$ for $w \in V$, and $c(s_v, t_w) = 1$ for $\{v, w\} \in E$, $v \in I \setminus D'$.

Compute a maximum integer flow f in N and put

$$L = \{v \in I \setminus D' | \ f(s, s_v) = c(s, s_v)\} \cup D'.$$

4c.   Let $S' = I \setminus L$.
5a.   If there is some set $Y \subset V \setminus S'$, $\#Y \leq \eta$, such that $\#S' \cup Y \geq \alpha n$ and $\Delta(G[S' \cup Y]) \leq \Delta$, then terminate with output $S' \cup Y$.
5b.   For any $D \subset V$, $\#D \leq \eta$, do
5c.   For $\tau = 0, 1, \ldots, \lceil \log_2(n) \rceil$ do
5d.   Compute $V' = \{v \in V | \ \#N(v) \cap I' \leq \Delta\} \setminus D$. Construct the following network N.

- The vertices of N are s, t, s_v for $v \in V' \setminus S'$, and t_w for $w \in V'$.
- The arcs of N are (s, s_v) for $v \in V' \setminus S'$, (t_w, t) for $w \in V'$, and (s_v, t_w) if $\{v, w\} \in E$, $v \in V' \setminus S'$, $w \in V'$.
- The capacity c of N is given by $c(s, s_v) = 3\tilde{\Delta}$ for all $v \in V' \setminus S'$, $c(t_w, t) = \tilde{\Delta} + 1$ for $w \in V'$, and $c(s_v, t_w) = 1$ for $\{v, w\} \in E$, $v \in V' \setminus S'$, $w \in V'$.

Compute a maximum integer flow f in N. Put $L = \{v \in V' \setminus I' | \ f(s, s_v) = c(s, s_v)\}$ and $S' = V' \setminus L$. If $\Delta(G[S']) \leq \Delta$ and $\#S' \geq \alpha n$, then terminate with output S'.

6. Find a set $S' \subset V$ that satisfies $\#S' \geq \alpha n$ and $\Delta(G[S']) \leq \Delta$ by enumerating all subsets of V.

Fig. 1. The Algorithm Find.

$$\xi^{**} = \max\{3\Delta\#U - \#N_{G_0}(U)|\ U \subset V \setminus S,\ \#U \le \alpha n/(6\Delta)\}.$$

Moreover, given numbers $\eta_1, \ldots, \eta_\Delta$, let $D_{\eta_1, \ldots, \eta_\Delta}$ denote the event that there exist pairwise disjoint sets $Z_1, \ldots, Z_\Delta \subset V \setminus S$, $\#Z_i \le \varepsilon n/\Delta$, such that $\eta_i = 0$ or $\#N_{G_0}(Z_i) \le \frac{\alpha\Delta}{2\varepsilon}\#Z_i - \eta_i$ for all i. Further, for $\xi \in \{0, \ldots, n/2\}$, let D_ξ be the event that there exist $\eta_1, \ldots, \eta_\Delta \ge 0$, $\sum_{i=1}^\Delta \eta_i = \xi$, such that $D_{\eta_1, \ldots, \eta_\Delta}$ occurs. Let ξ^* be the maximum ξ such that D_ξ occurs. We shall prove that Find terminates before η exceeds $\max\{0, \xi^*, \xi^{**}\}$. Thus, we may assume $\eta \ge \xi^*$, $\eta \ge \xi^{**}$.

Lemma 7. *There exists a set $D' \subset V$, $\#D' \le \eta$, such that at the end of 4c we have $S' \subset S$ and $\#S' \ge (\alpha - \frac{11\varepsilon}{\alpha})n$.*

Sketch of Proof. Decompose the set $I \setminus S$ into disjoint sets $Z_1, \ldots, Z_\Delta$ such that $\#Z_i \le \frac{\varepsilon n}{\Delta}$. Let M_i be a maximum $\lceil\frac{\alpha\Delta}{2\varepsilon}\rceil$-fold matching from Z_i to S. Let D' be the set of all unmatched vertices in Z_i, $i = 1, \ldots, \Delta$. Then $\#D' \le \eta$. Each matching M_i induces a flow h_i in N in the obvious way. Let $h = h_1 + \cdots + h_{\Delta+1}$. Because the sets Z_i are pairwise disjoint, h is an admissable flow. Consequently, any maximum integer flow f satisfies $f(s, s_v) = c(s, s_v)$ for all $v \in I \setminus (S \cup D')$. Thus $L \supset I \setminus (S \cup D')$. The choice of the capacity c implies $\#S' \ge (\alpha - \frac{11\varepsilon}{\alpha})n$. □

Lemma 8. *Suppose that step 5 is encountered with a set $S' \subset S$, $\#S' \ge (\alpha - \frac{11\varepsilon}{\alpha})n$. Then step 5 terminates and outputs a set $S'' \subset V$ such that $\#S'' \ge \alpha n$ and $\Delta(G[S]) \le \Delta$.*

Sketch of Proof. Observe that the property stated in Lemma 4 implies that either $\#S \setminus S' \le \eta$ or $\tilde{V} = \{v \in V|\ \#N(v) \cap S' \le \Delta\}$ satisfies $\#\tilde{V} \setminus S \le \frac{\#S \setminus S'}{\Delta}$. If $\#S \setminus S' \le \eta$, step 5a will terminate with a correct output. Thus asssume that $\#\tilde{V} \setminus S \le \frac{\#S \setminus S'}{\Delta}$. Then there exists a 3Δ-fold matching M^* from $\tilde{V} \setminus S$ with defect η. Let D be the set of all unmatched vertices. Then, similar arguments as in the proof of Lemma 7 prove the assertion. □

Because for a given η the time consumed by steps 3–5 of Find is $\left(\frac{n}{\eta}\right)^2$ times some polynomial, the estimate in Lemma 4 ensures that the kth moment of the running time is polynomial, provided $\omega_0 = \omega_0(\alpha, k)$ is sufficiently large. A careful analysis of Find shows that the techniques developed in this section enable us not only to eventually recover *some* sparse subset but even *the hidden set S*. This is the reason why similar methods can be applied to the k-coloring problem.

4   Coloring k-Colorable Graphs

In this section, we extend the methods of section 3 to obtain an algorithm that k-colors k-colorable semirandom graphs within expected polynomial time. Let $V = \{1, \ldots, n\}$ and let $\chi : V \to \{1, \ldots, k\}$ be a map such that for the sets $V_i = \chi^{-1}(i)$, $i = 1, \ldots, k$, we have $\#V_1 \le \#V_2 \le \cdots \le \#V_k$. Let $G =$

$(V,E) \in \mathcal{I}(G_0, \chi)$ for some k-colorable random graph $G_0 = \mathcal{G}_{n,\chi,\omega}$, where ω is a sufficiently large constant depending only on k. In order to k-color G, we proceed as follows. Because we know that G contains an independent set of size $\geq n/k$ (namely V_k), the same techniques as in the steps 1 and 3–5 of Find enable us to exhibit large independent sets of G (first, we let $\eta = 0$). Whenever a large independent set S_k of G has been found, we remove S_k from $G = G_k$, thereby obtaining G_{k-1}. Let $N_{k-1} = \#V(G_{k-1})$. In the case $S_k = V_k$, we know that $\alpha(G_{k-1}) \geq N_{k-1}/(k-1)$. Therefore, proceeding as in the steps 1 and 3–5 of Find again, we find large independent sets of G_{k-1}, and so on. Finally, it is easy to check whether G_2 is bipartite. In this case, the sets $S_3, \ldots, S_k$ together with a 2-coloring of G_2 yield the desired k-coloring of G. However, our algorithm will have to take care of the fact that the cardinalities $\#V_i$, $i = 1, \ldots, k$, are not a priori known. Moreover, though the outlined heuristic will be successful if the conditions stated in Lemma 4 (with $\eta = 0$) holds for all V_i, the algorithm is supposed to k-color G even if this is not the case. To this end, k-Color uses a variable T from which the value for η to be used in the steps 3–5 of Find can be computed; k-Color slowly increases T, thereby improving the probability of success. The algorithm k-Color makes use of the following subroutine.

Algorithm 9. FindSet(l, G_l, T, n)
Input: A graph $G_l = (V(G_l), E(G_l))$.
Output: An l-coloring of G_l or "fail".

1. If $l \leq 2$, then check within polynomial time if $\chi(G_l) \leq l$. If this is the case, output an l-coloring of G_l and terminate. If $l \leq 2$ but $\chi(G_l) > l$, terminate with output "fail".
2. Let $N_l = \#V(G_l)$. If $N_l \leq \log \max\{n, T\}$, then try to l-color G_l within time $O(N_l^3 (1 + 3^{1/3})^{N_l})$ using the algorithm proposed in [12]. In the case of success, return an l-coloring; otherwise, return "fail".
3a. For $n_l = N_l/l, 1 + N_l/l \ldots, N_l$ do
3b.   Let

$$\xi_l = \min\left\{ n/2, \max\left\{ \xi : \left(\frac{N_l}{\xi}\right)^\gamma \leq T \right\} \right\}$$

    for some sufficiently large constant γ.
    For $\eta_l = 0, \ldots \xi_l$ do
3c.   Let $\alpha_l = n_l/N_l$. Proceed as in the steps 1 and 3–5 of Find (with α replaced by α_l and η replaced by η_l, and $\Delta = 0$) in order to exhibit independent sets of G_l of size $\geq \alpha_l N_l = n_l$. Whenever a sufficiently large independent set S_l has been found, run

$$\text{FindSet}(l - 1, G_l - S_l, T, n).$$

    If FindSet$(l - 1, G_l - S_l, T, n)$ successfully $l - 1$-colors $G_l - S_l$, output the achieved l-coloring of G_l and terminate.
4. Terminate with output "fail".

Algorithm 10. k-Color(G)

Input: A k-colorable graph $G = (V = \{1, \ldots, n\}, E)$.

Output: A k-coloring of G.

1. For $T = 1, \ldots, 3^n$, run FindSet$(k, G_k = G, T, n)$; if FindSet succeeds, output a k-coloring of G and terminate.
2. k-color G using the algorithm described in [12].

The analysis of k-Color entails the following result.

Proposition 11. *Let $k \in \mathbb{N}$. Let G be a k-colorable graph. Then on input G the algorithm k-Color outputs a k-coloring of G. Furthermore, for each $l \in \mathbb{N}$ the algorithm k-Color k-colors (k, ω_l)-semirandom graphs such that the lth moment of the running time of $\mathcal{A}$ is polynomial, provided ω_l is sufficiently large.*

References

1. Alon, N., Kahale, N.: Approximating the independence number via the ϑ-function. Math. Programming **80** (1998) 253–264.
2. Alon, N., Kahale, N.: A spectral technique for coloring random 3-colorable graphs. SIAM J. Comput. **26** (1997) 1733–1748
3. Blum, A., Spencer, J.: Coloring random and semirandom k-colorable graphs. J. of Algorithms **19(2)** (1995) 203–234
4. Coja-Oghlan, A.: On NP-hard semi-random graph problems. Technical report **148**, Fachbereich Mathematik der Universität Hamburg (2002)
5. Coja-Oghlan, A.: Zum Färben k-färbbarer semizufälliger Graphen in erwarteter Polynomzeit mittels Semidefiniter Programmierung. Technical report **141**, Fachbereich Mathematik der Universität Hamburg (2002); an extended abstract version is to appear in Proc. 27th. Int. Symp. on Math. Found. of Comp. Sci. (2002)
6. Feige, U., Kilian, J.: Zero knowledge and the chromatic number, J. Comput. System Sci. **57** (1998), 187–199
7. Feige, U., Kilian, J.: Heuristics for semirandom graph problems. J. Comput. and System Sci. **63** (2001) 639–671
8. Feige, U., Krauthgamer, J.: Finding and certifying a large hidden clique in a semirandom graph. Random Structures & Algorithms **16** (2000) 195–208
9. Håstad, J.: Clique is hard to approximate within $n^{1-\varepsilon}$. Proc. 37th Annual Symp. on Foundations of Computer Science (1996) 627–636
10. Karger, D., Motwani, R., Sudan, M.: Approximate graph coloring by semidefinite programming. J. Assoc. Comput. Mach. **45** (1998) 246–265
11. Khanna, S., Linial, N., Safra, S.: On the hardness of approximating the chromatic number. Combinatorica **20** (2000) 393–415
12. Lawler, E.L.: A note on the complexity of the chromatic number problem. Information Processing Letters **5** (1976) 66–67
13. Mahajan, S., Ramesh, H.: Derandomizing semidefinite programming based approximation algorithms. Proc. 36th IEEE Symp. on Foundations of Computer Science (1995) 162–169
14. Prömel, H. J., Steger, A.: Coloring K_{l+1}-free graphs in linear expected time. Random Structures & Algorithms **3** (1992) 374-402
15. Subramanian, C.: Minimum coloring random and semirandom graphs in polynomial average time. J. of Algorithms **33** (1999) 112–123

Mixing in Time and Space for Lattice Spin Systems: A Combinatorial View

Martin Dyer[1][*], Alistair Sinclair[2][**], Eric Vigoda[3], and Dror Weitz[2]

[1] School of Computing, University of Leeds, Leeds LS2 9JT, United Kingdom
dyer@comp.leeds.ac.uk
[2] Computer Science Division, University of California at Berkeley, Berkeley, CA
94720-1776, USA {sinclair,dror}@cs.berkeley.edu
[3] Department of Computer Science, University of Chicago, Chicago, IL 60637, USA
vigoda@cs.uchicago.edu

Abstract. The paper considers spin systems on the d-dimensional integer lattice $\mathbb{Z}^d$ with nearest-neighbor interactions. A sharp equivalence is proved between exponential decay with distance of spin correlations (a *spatial* property of the equilibrium state) and "super-fast" mixing time of the Glauber dynamics (a *temporal* property of a Markov chain Monte Carlo algorithm). While such an equivalence is already known in various forms, the proofs in this paper are purely combinatorial and avoid the functional analysis machinery employed in previous proofs.

1 Introduction

Lattice spin systems are a class of models that originated in Statistical Physics, though interest in them has since expanded to many other areas, including Probability Theory, Statistics, Artificial Intelligence, and Theoretical Computer Science. A *(lattice) spin system* consists of a collection of sites which are the vertices of a regular lattice graph. A *configuration* of the spin system is an assignment of one of a finite set of *spins* to each site. The sites interact locally, according to potentials specified by the system, such that different combinations of spins on neighboring sites have different relative likelihoods. This interaction gives rise to a well-defined probability distribution over configurations of any finite subset (volume) of the sites, conditional on a fixed configuration of the sites on the boundary of this subset. Such a distribution is referred to as a *finite volume Gibbs distribution*, and is regarded as the equilibrium state of the given subset conditional on the given boundary configuration.

A *Glauber dynamics* is a Markov chain Monte Carlo algorithm used to sample from the Gibbs distribution. A step in this Markov chain is a random update of the spin of a single site (or of a finite set of sites), conditional on its neighboring

---

[*] Supported by EPSRC grant "Sharper Analysis of Randomised Algorithms: a Computational Approach" and by EC IST project RAND-APX.
[**] Supported in part by NSF grants CCR-9820951 and CCR-0121555, and by DARPA cooperative agreement F30602-00-2-060.

spins and in a manner which is reversible with respect to the Gibbs distribution. As a result, such a Markov chain converges to the corresponding Gibbs distribution. The Glauber dynamics plays a central role not just as an algorithm for sampling from the Gibbs distribution but also as a model for the physical process of reaching equilibrium.

A striking phenomenon in the field of spin systems, at least for "amenable" lattices such as the integer lattice $\mathbb{Z}^d$, is the equivalence of (*a priori* unrelated) notions of temporal and spatial mixing. By *temporal mixing* we mean that the Glauber dynamics converges "very fast" to its stationary Gibbs distribution, while by *spatial mixing* we mean that in the Gibbs distribution, correlations between the spins of different sites decay "very fast" with the (lattice) distance between them. This equivalence is interesting because it precisely relates the running time of an algorithm to purely spatial properties of the underlying model. In addition, a common heuristic in computer science is that local algorithms should work well (run fast) for local problems. The equivalence between temporal and spatial mixing is an example of the above heuristic in a restricted setting, where the relationship is formally proven and where there are precise interpretations for the terms "local algorithm", "local problem", and "run fast".

A number of references discuss the above equivalence. They give different proofs and use a variety of definitions of temporal and spatial mixing. Of these references we mention Stroock and Zegarlinski [11], who were the first to establish the above equivalence, Martinelli and Olivieri [9,10], who obtained sharper results by working with a weaker spatial mixing assumption, and Cesi [3], who recently simplified some of the proofs. See also [8] for a review of results in the field.

The references mentioned above make crucial use of functional analysis in their proofs, and usually discuss quantities such as the *spectral gap* and the *log Sobolev constant* of the dynamics as a measure of its temporal mixing. In this paper, we give purely combinatorial proofs of this equivalence, based on the elementary technique of coupling probability distributions. Although some of the ideas we use have appeared before, our main contribution lies in presenting a complete argument which is purely combinatorial, where the reader does not need to resort to functional analysis concepts.

We note that the result we present in the direction going from spatial mixing to temporal mixing (of the single site Glauber dynamics) is limited in the sense that it only applies to *monotone* systems. For general systems, however, we show that spatial mixing implies temporal mixing of a "finite-block" Glauber dynamics, in which a sufficiently large block of spins is updated at each step. The corresponding implication for the single site dynamics in the general case is known [3,8,10,11], but currently we do not have a combinatorial proof of it.

The remainder of the paper is organized as follows. Section 2 includes precise definitions and statements of results. In Sect. 3 we list a few basic tools we use in the proofs. In Sect. 4 we prove that temporal mixing implies spatial mixing while in Sects. 5 and 6 we prove that spatial mixing implies temporal mixing for monotone and general systems respectively.

2   Definitions and Statements of Results

2.1   Spin Systems

Consider the d-dimensional integer lattice as a graph with vertex set $V = \mathbb{Z}^d$ and edge set E, where $(v, u) \in E$, denoted $v \sim u$, if and only if $\sum_{i=1}^{d} |v_i - u_i| = 1$.[1] We use the statistical physics terminology and refer to the vertices as *sites*. For a finite subset $\Psi \subset V$, we define its *boundary* $\partial\Psi = \{v \notin \Psi : \exists u \in \Psi \text{ s.t. } v \sim u\}$. Each site is assigned a spin from the spin space $S = \{1, \ldots, q\}$, and the configuration space is denoted by $\Omega = \Omega_\Psi = S^\Psi$. Given a configuration $\sigma \in \Omega$, we write $\sigma[v]$ for the spin that σ assigns to v and abuse this notation with $\sigma[\Lambda]$ standing for the configuration of the subset Λ under σ.

We consider spin systems with nearest neighbor interactions (although everything we do can be generalized to finite range interactions). Namely, we have a (symmetric) pair potential[2] $U : S \times S \to \mathbb{R}$, and a self potential $W : S \to \mathbb{R}$. Then, for a finite subset Ψ and a boundary configuration $\tau \in \Omega_{\partial\Psi}$, the Hamiltonian $H_\Psi^\tau : \Omega_\Psi \to \mathbb{R}$ is defined as

$$H_\Psi^\tau(\sigma) = \sum_{v \in \partial\Psi, u \in \Psi, v \sim u} U(\tau[v], \sigma[u]) + \sum_{v, u \in \Psi, v \sim u} U(\sigma[v], \sigma[u]) + \sum_{v \in \Psi} W(\sigma[v]) \ .$$

The value this Hamiltonian assigns can be considered as the "energy" of σ when τ is the boundary configuration. The *finite volume Gibbs distribution* associated with the subset Ψ and the boundary configuration τ assigns probability to σ which is proportional to the inverse exponential of its energy. Formally,

$$\mu_\Psi^\tau(\sigma) = (Z_\Psi^\tau)^{-1} \exp(-H_\Psi^\tau(\sigma)) \ ,$$

where Z_Ψ^τ is the appropriate normalizing factor.

Example. Probably the best known spin system is the *ferromagnetic Ising model*. In this case, the spin space is $S = \{-1, +1\}$, while $U(s_1, s_2) = -\beta \cdot s_1 \cdot s_2$ and $W(s) = -\beta \cdot h \cdot s$, where $\beta \in \mathbb{R}^+$ is the inverse temperature and $h \in \mathbb{R}$ is an external field. Thus, the energy of a configuration is linear in the number of edges with disagreeing spins, as well as the number of spins with sign opposite to that of h. For example, if $h = 0$ and if we ignore the effect of the boundary configuration (the so-called "free-boundary condition") then the minimum energy (highest probability) configurations are the two constant configurations where all the spins have the same value (either $+1$ or -1).

---

[1] Most of our results hold — with suitable modifications — for any "amenable" lattice (see, e.g., [6] for a definition). For simplicity, in this paper we focus just on $\mathbb{Z}^d$.

[2] The given definition of the pair potential does not cover systems with hard constraints where U may be infinite. In general, the results in this paper apply to systems with hard constraints as well (see full version for details).

2.2   The Glauber Dynamics

We study the following simple Markov chain (X_t), known as the (heat-bath) *Glauber dynamics*, which is used to sample from μ_Ψ^τ. Given the current configuration $X_t \in \Omega_\Psi$, the transition $X_t \to X_{t+1}$ is defined as follows:

- Choose a vertex v uniformly at random from Ψ.
- Let $X_{t+1}[u] = X_t[u]$ for all $u \neq v$.
- Choose $X_{t+1}[v]$ from $\mu_{\{v\}}^{X_t'}$, where X_t' is the configuration of $\partial\{v\}$ defined by $X_t'[u] = X_t[u]$ for $u \in \Psi$ and $X_t'[u] = \tau[u]$ for $u \in \partial\Psi$.

It is not too difficult to verify that this Markov chain is reversible w.r.t. the Gibbs distribution μ_Ψ^τ and, in particular, that μ_Ψ^τ is the unique stationary distribution.

Remark. In the literature, a Glauber dynamics is usually any Markov chain that makes single site updates that are reversible w.r.t. to the single site Gibbs measure. Indeed, all the results below apply to any choice of Glauber dynamics. However, for definiteness we will assume the above definition throughout this paper.

We also discuss a generalization of the Glauber dynamics to a Markov chain where at each step a block of sites is updated rather than a single site. Let $Q_L = [1, \ldots, L]^d$ be the d-dimensional regular box of side length L. Consider all the translations of Q_L that intersect the subset Ψ and let

$$B(\Psi, L) = \left\{ \Lambda \neq \emptyset \mid \Lambda = (z + Q_L) \cap \Psi \text{ for some } z \in \mathbb{Z}^d \right\} \ .$$

We think of each $\Lambda \in B(\Psi, L)$ as a *block*. We then denote by $HB(L)$ the heat-bath block dynamics that makes updates to blocks from $B(\Psi, L)$. Given the current configuration X_t, the transition $X_t \to X_{t+1}$ is defined as follows:

- Choose a block Λ uniformly at random from $B(\Psi, L)$.
- Let $X_{t+1}[u] = X_t[u]$ for all $u \notin \Lambda$.
- Choose $X_{t+1}[\Lambda]$ from $\mu_\Lambda^{X_t'}$, where X_t' is the configuration of $\partial\Lambda$ defined by $X_t'[u] = X_t[u]$ for $u \in \Psi$ and $X_t'[u] = \tau[u]$ for $u \in \partial\Psi$.

2.3   Temporal and Spatial Mixing

The statements in this paper relate an appropriate notion of *temporal mixing* (convergence in time of the Glauber dynamics) with an appropriate notion of *spatial mixing* (decay of correlation with distance in the Gibbs distribution). The exact definitions are given below.

Let μ_1 and μ_2 be two arbitrary distributions on Ω_Ψ. We write $\|\mu_1 - \mu_2\| = \max_{A \subseteq \Omega_\Psi} |\mu_1(A) - \mu_2(A)|$ for the *total variation distance* between the two distributions, and $\|\mu_1 - \mu_2\|_\Lambda = \max_{A \subseteq \Omega_\Lambda} |\mu_1(A) - \mu_2(A)|$ for the distance when projecting the two distributions onto Ω_Λ for $\Lambda \subseteq \Psi$.

Definition 2.1. *We say that the Glauber dynamics has* optimal temporal mix-ing *if there exist constants b and $c > 0$ such that for any subset Ψ with any boundary configuration, the dynamics on Ψ has the following property. For any two instances (X_t) and (Y_t) of the chain and for any positive integer k, $\|X_{kn} - Y_{kn}\| \leq bn\exp(-c \cdot k)$, where $n = |\Psi|$ is the volume of Ψ.*

In particular, optimal temporal mixing means that the distance from the stationary measure $\|X_{kn} - \mu_\Lambda^\tau\| \leq bn\exp(-c \cdot k)$ for any instance (X_t). Before we move on to the definition of the spatial mixing notion, we pause to compare optimal temporal mixing as defined here with some of the other notions of tem-poral mixing found in the literature. The *mixing time* of a Markov chain (as a function of ϵ) is the time it takes to get within a variation distance of ϵ from the stationary measure. Notice that optimal temporal mixing is equivalent to a mixing time of $O(n\log(\frac{n}{\epsilon}))$. Optimal temporal mixing also implies that the spectral gap of the dynamics is at least $\frac{c}{n}$. While a spectral gap of $\Omega(\frac{1}{n})$ does not immediately imply optimal temporal mixing, it is not too difficult to see that if the log Sobolev constant associated with the dynamics is $\Omega(\frac{1}{n})$ then the dynamics has optimal temporal mixing. We notice that in fact, in the context of spin systems, all the above notions of temporal mixing are known to be equiv-alent when considered to hold uniformly in the subset Ψ and in the boundary configuration (since they are all equivalent to an appropriate notion of spatial mixing as below).

The corresponding spatial notion we consider states that changing the spin of a site on the boundary has an exponentially small effect on the configuration of sites far away from the changed site. The distance between two sites v and u is defined as the graph distance between them and denoted by $\mathrm{dist}(v, u)$.

Definition 2.2. *We say the system has* strong spatial mixing *if there exist con-stants β and $\alpha > 0$ such that for any two subsets Λ, Ψ where $\Lambda \subseteq \Psi$, any site $u \in \partial\Psi$, and any pair of boundary configurations τ and τ^u that differ only at u, $\|\mu_\Psi^\tau - \mu_\Psi^{\tau^u}\|_\Lambda \leq \beta|\Lambda|\exp(-\alpha \cdot \mathrm{dist}(u, \Lambda))$.*

Remark. In the literature, the definition of strong spatial mixing may vary, where the difference is in which class of subsets Ψ the assumption applies to (for ex-ample, Ψ may be restricted to be a regular box). We work with the strongest version by requiring it to apply to all subsets in order to simplify our arguments.

2.4   Monotone Systems

Some of the statements in this paper apply only to monotone systems. In a monotone system, each site v is associated with a linear ordering of the spin space, denoted by $\succeq_v$. Since the spin space is finite, each of the linear orderings has unique maximal and minimal elements, which we call the *plus* and *minus* elements respectively. The single site orderings give rise to a partial ordering $\succeq_\Psi$ of the configuration space. Specifically, $\sigma_1 \succeq_\Psi \sigma_2$ if and only if $\sigma_1[v] \succeq_v \sigma_2[v]$ for every $v \in \Psi$. The system is *monotone* with respect to the above partial order-ing if, for every subset Ψ and any two boundary configurations τ_1 and τ_2 such that $\tau_1 \succeq_{\partial\Psi} \tau_2$, the Gibbs measure $\mu_\Psi^{\tau_1}$ statistically dominates the Gibbs measure $\mu_\Psi^{\tau_2}$ with respect to $\succeq_\Psi$. Equivalently, the two distributions can be coupled such that with probability 1, $\sigma_1 \succeq_\Psi \sigma_2$, where σ_1 and σ_2 are a pair of coupled configurations chosen from $\mu_\Psi^{\tau_1}$ and $\mu_\Psi^{\tau_2}$ respectively. Notice that it is enough that the above property holds for all single sites to ensure that it holds for all subsets Ψ. Also, since the single site orderings are linear, the system is "realizably" monotone [5]. This means that, given a collection of boundary configurations $\tau_1, \tau_2, \ldots, \tau_k$, we can simultaneously couple the k corresponding Gibbs distributions such that if $\tau_i \succeq_{\partial\Psi} \tau_j$, the corresponding coupled configurations satisfy $\sigma_i \succeq_\Psi \sigma_j$ with probability 1 (simultaneously for each such pair i, j).

Many well known spin systems are monotone, including the Ising model and the hard-core model (independent sets).

2.5   Results

Several notions of temporal and spatial mixing for models on integer lattices are known to be equivalent to one another [3,8,9,11], though the proofs are often rather complex and cast in the language of functional analysis. In this paper we present combinatorial proofs of the following implications.

Theorem 2.3. *If the single site dynamics has optimal temporal mixing then the system has strong spatial mixing.*

For monotone systems we show the converse as well:

Theorem 2.4. *If a monotone system has strong spatial mixing then the single site dynamics has optimal temporal mixing.*

In the general case (without assuming monotonicity), we show that

Theorem 2.5. *If a system has strong spatial mixing then there exists a finite integer L for which the heat-bath block dynamics $HB(L)$ has optimal temporal mixing.*

The converse of Theorem 2.5 (optimal temporal mixing of $HB(L)$ implies strong spatial mixing) can be proved using the same ideas as in the proof of Theorem 2.3 (with the addition of a few minor technical details), so we skip it here.

Notice that strong spatial mixing implies optimal temporal mixing of the single site Glauber dynamics in the general case as well [3,8,10,11], but we have not yet been able to find a purely combinatorial proof of this implication.

3   Preliminaries

In this section we identify some of the common tools we use in our proofs.

3.1   Coupling and Mixing Time

A common tool for bounding the total variation distance between two distributions, and in particular for bounding the mixing time of Markov chains, is coupling. A *coupling* of μ_1 and μ_2 is any joint distribution whose marginals are μ_1 and μ_2 respectively. If σ_1 and σ_2 are a pair of random configurations chosen from a given coupling of μ_1 and μ_2 then $\Pr(\sigma_1 \neq \sigma_2)$ is an upper bound on the total variation distance between μ_1 and μ_2. Also, there is always an optimal coupling, i.e., a coupling such that $\Pr(\sigma_1 \neq \sigma_2) = \|\mu_1 - \mu_2\|$.

In the proofs we give in this paper we use the following coupling of the Glauber dynamics, which we call an *identity coupling*. An identity coupling is determined by specifying, for each site v, a simultaneous coupling of all the single site Gibbs distributions (ranging over all possible values for the configuration of the neighbors of v). Namely, we have a joint distribution γ_v whose marginals are $\mu_{\{v\}}^{\tau_1}, \ldots, \mu_{\{v\}}^{\tau_k}$, where the set $\{\tau_1, \ldots, \tau_k\} = \Omega_{\partial\{v\}}$. Given γ_v, we simultaneously couple a collection of instances of the Glauber dynamics $(X_t^1), (X_t^2), \ldots, (X_t^l)$ using a Markovian coupling (i.e., the joint distribution of $X_{t+1}^1, \ldots, X_{t+1}^l$ is a function only of the coupled configurations $X_t^1, \ldots, X_t^l$) where the coupled transition $(X_t^1, \ldots, X_t^l) \to (X_{t+1}^1, \ldots, X_{t+1}^l)$ is as follows:

- Choose a site v u.a.r. from Ψ (the same one for all chains).
- Choose a collection of spins $(s_1, \ldots, s_k)$ from the joint distribution γ_v.
- For every $1 \leq i \leq l$ set $X_{t+1}^i[v] = s_j$ if and only if $X_t^i[\partial\{v\}] = \tau_j$.

An important property of this coupling is that if $X_t^i[\partial\{v\}] = X_t^j[\partial\{v\}]$ then $X_{t+1}^i[v] = X_{t+1}^j[v]$ with probability 1. Notice that in a monotone system there exists a *monotone identity coupling*, i.e., a joint distribution γ_v such that whenever $\tau_i \succeq_{\partial\{v\}} \tau_j$, $s_i \succeq_v s_j$ with probability 1.

We say that an identity coupling has *optimal mixing* if for any two instances of the chain (X_t) and (Y_t), we have $\Pr(X_{kn} \neq Y_{kn}) \leq bn \exp(-c \cdot k)$, where the probability space is the coupling of X_{kn} and Y_{kn} resulting from the identity coupling of the two processes. Notice that optimal mixing of an identity coupling implies optimal temporal mixing of the dynamics. Finally, the *coupling time* of an identity coupling is the minimum T such that $\Pr(X_T \neq Y_T) \leq \frac{1}{e}$. As a result, $\Pr(X_{kT} \neq Y_{kT}) \leq e^{-k}$ for any positive integer k.

3.2   Bounding the Speed of Propagation of Information

A central idea in the analysis of the mixing time of the Glauber dynamics, in particular when using spatial mixing assumptions, is to bound the speed at which information propagates during the dynamical process. We cite a lemma of this sort following Van den Berg [1], where the bound comes from a *paths of disagreement* (or *disagreement percolation*) argument. Similar bounds can be found in [7,8]. The version we give here applies to the Glauber dynamics on any graph of bounded degree (as in [7]), rather than just for finite subsets of $\mathbb{Z}^d$.

Lemma 3.1. *Let* $G = (V, E)$ *be a graph of maximum degree* $\Delta > 1$, *and let* $n = |V|$. *Let* (X_t) *and* (Y_t) *be two copies of a Glauber dynamics on* G *such that the two initial configurations agree everywhere except on* $A \subseteq V$. *Let* $B \subseteq V$ *be another subset and let* $r = \text{dist}(A, B)$. *Then, for any positive integer* $k \leq \frac{r}{(\Delta-1)}$, *if we run the dynamics for* $T = kn$ *steps,* $\Pr(X_T[B] \neq Y_T[B]) \leq 4 \min\{|A|, |B|\} \left(\frac{(\Delta-1)k}{e \cdot r}\right)^r$, *where the probability space is the coupling of* X_T *and* Y_T *resulting from any identity coupling of* (X_t) *and* (Y_t). *In particular, if* $T = kn$ *and* $\text{dist}(A, B) \geq (\Delta - 1)k$, *then* $\Pr(X_T[B] \neq Y_T[B]) \leq 4 \min\{|A|, |B|\} e^{-\text{dist}(A,B)}$.

In words, Lemma 3.1 states that in kn steps, with high probability, information percolates a distance of at most $(\Delta - 1)k$.

Proof. Since we couple X_t and Y_t using an identity coupling, if at time zero v had the same spin in both chains and at time T the spins at v differ then it must be the case that at some time $t' \leq T$ the site chosen to be updated was v and immediately before the update of v at time t' the two chains had different spins at one of the neighbors of v. Carrying this argument inductively, if we assume that at time zero the only sites whose spins may differ are included in A then in order for a site v to have different spins at time T there must be a *path of disagreement* going from A to v. Specifically, there must be $v_0, v_1, \ldots, v_l = v$ and $0 < t_1 < t_2 < \ldots < t_l \leq T$ such that $v_0 \in A$ and for $1 \leq i \leq l$, $v_i \sim v_{i-1}$ and at time t_i the site chosen to be updated was v_i. Notice that for a given path $v_0, \ldots, v_l$ the probability of this event occurring is at most $\binom{T}{l}(\frac{1}{n})^l$. Now, if the two configurations at time T differ at some site in B, there must be a path of disagreement of length at least $r = \text{dist}(A, B)$ going from A to B. Since the number of (simple) paths of length l going from A to B is bounded from above by $\min\{|A|, |B|\} \Delta(\Delta - 1)^{l-1}$ we can conclude that the probability of a disagreement in B at time $T = kn$ is at most

$$\min\{|A|, |B|\} \cdot \frac{\Delta}{\Delta - 1} \cdot \sum_{l=r}^{\infty} (\Delta - 1)^l \binom{kn}{l} \left(\frac{1}{n}\right)^l \leq$$

$$\min\{|A|, |B|\} \cdot \frac{\Delta}{\Delta - 1} \cdot \sum_{l=r}^{\infty} \left(\frac{(\Delta - 1)k}{e \cdot l}\right)^l \leq 4 \min\{|A|, |B|\} \left(\frac{(\Delta - 1)k}{e \cdot r}\right)^r ,$$

where in the last inequality we used the fact that $r \geq (\Delta - 1)k$.   □

Remark. We will often use Lemma 3.1 in a setting where only a subset of the sites may be updated in the Markov chain (i.e., the spins on some sites - typically those on the boundary - are held fixed throughout the process). Notice that the proof above is still valid in this setting (regardless of whether or not the fixed spins disagree - i.e., are of sites in A). In fact, it is valid even if the two compared chains have different sets of fixed sites as long as the sites which are fixed in only one of the chains are all included in the subset A, i.e., we just assume that the spins of these sites disagree in the two chains. An important point to keep in mind in these scenarios is the meaning of the parameter n. Rather than the volume of the graph, n stands for the inverse of the probability that a given site is chosen to be updated (and it must be the same in both chains). Indeed, this is the only use we made of this parameter in the proof. The scenarios mentioned in this remark will become clearer when they arise in the proofs below.

4   From Temporal to Spatial Mixing

In this section we prove Theorem 2.3, which states that if the Glauber dynamics has optimal temporal mixing then strong spatial mixing holds. The first step in the proof is to derive a stronger notion of temporal mixing, given in Lemma 5.1 below, where we have better error bounds when *projecting* the two compared distributions on a smaller configuration space. This stronger notion, which we call *projected optimal mixing*, is not an immediate consequence of optimal temporal mixing and, in fact, the amenability of $\mathbb{Z}^d$ is crucial for this implication.

Lemma 4.1. *Optimal temporal mixing implies that there exist constants b' and $c' > 0$ such that, for any subset Ψ of volume n, any boundary configuration, any two instances (X_t) and (Y_t) of the chain on Ψ and any subset $\Lambda \subseteq \Psi$, we have $\|X_{kn} - Y_{kn}\|_\Lambda \leq b'|\Lambda| \exp(-c' \cdot k)$ for any positive integer k.*

Proof. The idea of the proof is one we use throughout this paper, which involves using Lemma 3.1 in order to localize the dynamics we consider. Namely, when we run the dynamics for kn steps, with high probability information from sites which are at distance at least $(2d - 1)k$ from Λ does not percolate into Λ. Therefore, if we take a subset Λ_k surrounding Λ and whose boundaries are at distance at least $(2d-1)k$ from Λ, we can assume that the sites on the boundary of Λ_k are fixed throughout the process. Thus, we can use the optimal temporal mixing bound for a dynamics on the local subset Λ_k, whose volume is smaller than that of Ψ. As shown below, the fact that the volume of Λ_k grows only sub-exponentially in k (here is where we use the amenability of $\mathbb{Z}^d$) gives the required bound. An additional point we need to make in order to carry out the above argument is that when running the dynamics on Ψ, with high probability, an appropriate portion of the time is spent in the subset Λ_k. This, however, is an easy consequence of the Chernoff bound.

We proceed with the formal proof. Consider the subset of all sites within distance $(2d - 1)k$ from Λ, and let Λ_k be the intersection of this subset with Ψ. Notice that $\text{dist}(\Lambda, \Psi \setminus \Lambda_k) \geq (2d - 1)k$ and that $|\Lambda_k| \leq [2(2d - 1)k]^d |\Lambda|$.

In addition to the chains (X_t) and (Y_t), we consider two additional chains, denoted by $(X_t^{\Lambda_k})$ and $(Y_t^{\Lambda_k})$, whose initial configurations inside Λ_k are the same as (X_t) and (Y_t) respectively. The configuration of $\Psi \setminus \Lambda_k$ is fixed to the same arbitrary configuration in both $(X_t^{\Lambda_k})$ and $(Y_t^{\Lambda_k})$ and remains fixed throughout the process, i.e., $(X_t^{\Lambda_k})$ and $(Y_t^{\Lambda_k})$ represent modified processes where, in a given step, if the chosen site to be updated is outside Λ_k then the spin of that site remains unchanged, while if it is in Λ_k then it is updated as usual. Notice that this modified process is the same as running the dynamics on Λ_k except that the probability of a site being chosen at a given step is $\frac{1}{|\Psi|}$ instead of $\frac{1}{|\Lambda_k|}$.

Using the triangle inequality, we have $\|X_{kn} - Y_{kn}\|_\Lambda \leq \|X_{kn} - X_{kn}^{\Lambda_k}\|_\Lambda + \|X_{kn}^{\Lambda_k} - Y_{kn}^{\Lambda_k}\|_\Lambda + \|Y_{kn}^{\Lambda_k} - Y_{kn}\|_\Lambda$. Lemma 3.1 (together with the remark following it) gives a bound on the first and third terms in the r.h.s. of the last inequality. To see this, couple (X_t) and $(X_t^{\Lambda_k})$ using a modified identity coupling, where an update of a site outside Λ_k in (X_t) is coupled with doing nothing in $(X_t^{\Lambda_k})$. Notice that at time zero the two chains agree on Λ_k. Disagreement may percolate from $\Psi \setminus \Lambda_k$ into the bulk of Λ_k as we run the chains, but since $\text{dist}(\Lambda, \Psi \setminus \Lambda_k) \geq (2d-1)k$, we can use Lemma 3.1 to deduce that $\|X_{kn} - X_{kn}^{\Lambda_k}\|_\Lambda \leq 4|\Lambda|e^{-(2d-1)k}$.

It remains to bound $\|X_{kn}^{\Lambda_k} - Y_{kn}^{\Lambda_k}\|_\Lambda$. Recall that both these chains have the same fixed configuration outside Λ_k so we can use the optimal temporal mixing assumption for a process on Λ_k. Notice that when running the chain $X_t^{\Lambda_k}$ for kn steps, on average $k|\Lambda_k|$ of the steps hit Λ_k. Using a Chernoff bound, with probability at least $1 - \exp(-\frac{k|\Lambda_k|}{8})$, the number of steps that hit Λ_k is at least $\frac{k|\Lambda_k|}{2}$. Thus, using the bound optimal temporal mixing gives for running a process on Λ_k for $\frac{k|\Lambda_k|}{2}$ steps, we have

$$\|X_{kn}^{\Lambda_k} - Y_{kn}^{\Lambda_k}\|_\Lambda \leq \|X_{kn}^{\Lambda_k} - Y_{kn}^{\Lambda_k}\|_{\Lambda_k} \leq b|\Lambda_k| \exp\left(-c \cdot \frac{k}{2}\right) + \exp\left(-\frac{k|\Lambda_k|}{8}\right) \leq$$

$$b[2(2d-1)k]^d |\Lambda| \exp\left(-c \cdot \frac{k}{2}\right) + \exp\left(-\frac{k}{8}\right) \leq b'|\Lambda| \exp(-c' \cdot k)$$

for appropriate constants b' and $c' > 0$. □

Proof of Theorem 2.3. Let Ψ be a subset of volume n, τ and τ^u be two boundary configurations that differ only at u, and let $\Lambda \subseteq \Psi$. Following Lemma 4.1, we assume the dynamics has projected optimal mixing and show that

$$\|\mu_\Psi^\tau - \mu_\Psi^{\tau^u}\|_\Lambda \leq b'|\Lambda| \exp\left(-\frac{c'}{2d-1} \cdot \text{dist}(u, \Lambda)\right) + 4|\Lambda|e^{-\text{dist}(u,\Lambda)} . \tag{1}$$

The idea of the proof is that when running the Glauber dynamics, the time needed in order for the projected distribution on Λ to be close to the stationary one is less than the time it takes for the disagreement at u to percolate into Λ. Formally, consider the following two instances of the Glauber dynamics on Ψ. The first, denoted by Z_t, is an instance with τ as the boundary configuration while the second, denoted by Z_t', is an instance with τ^u as the boundary configuration. The initial configuration of Ψ in both chains is chosen from $\mu_\Psi^{\tau^u}$. Notice that this is the stationary distribution of Z_t' and therefore $Z_t' = \mu_\Psi^{\tau^u}$ for all t.

Using the triangle inequality, we have $\|\mu_\Psi^\tau - \mu_\Psi^{\tau^u}\|_\Lambda = \|\mu_\Psi^\tau - Z_t'\|_\Lambda \leq \|\mu_\Psi^\tau - Z_t\|_\Lambda + \|Z_t - Z_t'\|_\Lambda$. By letting $t = \frac{\text{dist}(u,\Lambda)}{(2d-1)} \cdot n$ we can make sure both terms are small. We bound the first term using the temporal mixing assumption. Namely, for $t = \frac{\text{dist}(u,\Lambda)}{(2d-1)} \cdot n$ we have $\|\mu_\Psi^\tau - Z_t\|_\Lambda \leq b'|\Lambda| \exp(-c' \cdot \frac{\text{dist}(u,\Lambda)}{(2d-1)})$. We use Lemma 3.1 in order to bound the second term. Notice that Z_t and Z_t' have the same initial distribution on Ψ and thus they can be coupled such that at time zero they have the same configuration on Ψ with probability 1. We continue to couple the two processes using an identity coupling. Disagreement may percolate from u, but since $\mathrm{dist}(u, \Lambda) = (2d-1)\frac{t}{n}$ we have $\|Z_t - Z_t'\|_\Lambda \leq 4|\Lambda|e^{-\mathrm{dist}(u, \Lambda)}$. $\square$

Remark. Notice that the argument for showing that projected temporal mixing implies spatial mixing uses only Lemma 3.1 and can thus be carried out in models with any underlying finite degree graph. However, the proof of Lemma 4.1 uses the amenability of $\mathbb{Z}^d$ and breaks down for non-amenable graphs. Indeed, the Ising model on a tree at an appropriate temperature provides a counter-example to the claim of Lemma 4.1 in non-amenable graphs as can be concluded from [7].

5   From Spatial to Temporal Mixing: The Monotone Case

In this section we show that in monotone systems the strong spatial mixing assumption implies optimal temporal mixing of the single-site Glauber dynamics (Theorem 2.4). Actually, we state two theorems whose combination gives Theorem 2.4. The first theorem, whose proof uses ideas from the proof of Theorem 4.2 of [9], states that the strong spatial mixing assumption implies $O(n \log^2 n)$ coupling time of any monotone identity coupling, uniformly in the volume n and in the boundary configuration. The second theorem, which is based on Theorem 3.12 of [8], states (for general systems) that if there exists n_0 for which the coupling time of any identity coupling of the Glauber dynamics on subsets of volume n_0 is at most $\frac{c}{\log n_0} n_0^{1+1/d}$ for an appropriate constant c, uniformly in the boundary configuration, then this identity coupling has optimal mixing. In particular, any upper bound of $o(\frac{n^{1+1/d}}{\log n})$ on the asymptotic coupling time immediately implies that the identity coupling has optimal mixing.

Theorem 5.1. *Strong spatial mixing implies that the coupling time of any monotone identity coupling of the Glauber dynamics on any subset is at most $T(n) = cn(\log n)^2$ for some constant c, uniformly in n and in the boundary configuration.*

Proof. As in our earlier arguments, the idea of the proof is again to localize the dynamics, which allows us to use inductive bounds from smaller volume subsets. However, here we use strong spatial mixing to achieve the localization, rather than the bound on the speed of propagation of information from Lemma 3.1.

Fix a large enough n_0 (to be determined later). Clearly, by choosing an appropriate constant $c = c(n_0)$, the coupling time statement is true for all $n \leq n_0$. We now show the statement for $n > n_0$, by inductively assuming its validity for volumes $m \leq [2 \cdot \frac{2}{\alpha} \log(3e\beta n)]^d$, where α, β are the constants in the definition of strong spatial mixing (Definition 2.2).

Let Ψ be a subset of volume n with an arbitrary boundary configuration. Let (X_t) and (Y_t) be two instances of the chain with arbitrary initial configurations inside Ψ. We will show that after $T(n)$ steps, for every site $v \in \Psi$, the probability that the two spins at v differ is at most $\frac{1}{en}$, and therefore, the probability that two configurations (on Ψ) differ is at most $\frac{1}{e}$, as required.

Consider the regular box of radius $\frac{2}{\alpha}\log(3e\beta n)$ around v, and let Λ_v be the intersection of this box with Ψ. Let $m = |\Lambda_v|$ and notice that $m \leq [2 \cdot \frac{2}{\alpha}\log(3e\beta n)]^d$. We now introduce four additional chains that may only update sites in Λ_v. We will couple these chains along with (X_t) and (Y_t) such that, whenever the site chosen to be updated is outside Λ_v only X_t and Y_t are updated while the additional four chains remain unchanged. On the other hand, when the site to be updated belongs to Λ_v all six chains are updated simultaneously according to the monotone identity coupling. Below we describe the additional four chains and their initial configurations inside Ψ. Outside Ψ, all four chains have the same boundary configuration as (X_t) and (Y_t).

- $Q_t^{+,\Lambda_v}, Q_t^{-,\Lambda_v}$: the chains starting from the all $+$ $(-)$ configuration on Ψ.
- $Z_t^{+,\Lambda_v}, Z_t^{-,\Lambda_v}$: the chains starting from the all $+$ $(-)$ configuration outside Λ_v, while the initial configuration inside Λ_v is chosen from the (stationary) Gibbs measure on Λ_v with this boundary configuration.

Notice that we can simultaneously couple the six chains such that at time zero, with probability one, $Q_0^{+,\Lambda_v} \succeq X_0 \succeq Q_0^{-,\Lambda_v}$, $Q_0^{+,\Lambda_v} \succeq Y_0 \succeq Q_0^{-,\Lambda_v}$, and $Z_t^{+,\Lambda_v} \succeq Z_t^{-,\Lambda_v}$. Since we use a monotone identity coupling, we have by induction that these relations hold for all t. Thus, we have

$$\Pr(X_t[v] \neq Y_t[v]) \leq \Pr(Q_t^{+,\Lambda_v}[v] \neq Q_t^{-,\Lambda_v}[v]) \leq \Pr(Q_t^{+,\Lambda_v}[v] \neq Z_t^{+,\Lambda_v}[v]) +$$

$$\Pr(Z_t^{+,\Lambda_v}[v] \neq Z_t^{-,\Lambda_v}[v]) + \Pr(Z_t^{-,\Lambda_v}[v] \neq Q_t^{-,\Lambda_v}[v]) .$$

We use the strong spatial mixing assumption to bound the middle term of the last expression. Since Z_t^{+,Λ_v} and Z_t^{-,Λ_v} are the stationary Gibbs distributions on Λ_v with the corresponding boundary configurations, strong spatial mixing (together with the triangle inequality) gives $\|Z_t^{+,\Lambda_v} - Z_t^{-,\Lambda_v}\|_{\{v\}} \leq |\partial\Lambda_v \setminus \partial\Psi|\beta\exp(-\alpha \cdot \text{dist}(\partial\Lambda_v \setminus \partial\Psi, v))$. This does not necessarily guarantee the same bound on the probability of disagreement under the coupling because the coupling we use is not necessarily the optimal one. However, monotonicity guarantees that our coupling is within a factor of $q-1$ (recall that q is the size of the spin space) from the optimal coupling, as explained next. We embed the ordering associated with v in the linear ordering $1, 2, \ldots, q$ with integer arithmetic. Since the spins at v are coupled such that with probability one $Z_t^{+,\Lambda_v}[v] \geq Z_t^{-,\Lambda_v}[v]$,

$$\Pr(Z_t^{+,\Lambda_v}[v] \neq Z_t^{-,\Lambda_v}[v]) \leq \text{E}(Z_t^{+,\Lambda_v}[v] - Z_t^{-,\Lambda_v}[v]) =$$

$$\text{E}(Z_t^{+,\Lambda_v}[v]) - \text{E}(Z_t^{-,\Lambda_v}[v]) \leq (q-1)\|Z_t^{+,\Lambda_v} - Z_t^{-,\Lambda_v}\|_{\{v\}} \leq$$

$$(q-1)|\partial\Lambda_v \setminus \partial\Psi|\beta\exp(-\alpha \cdot \text{dist}(\partial\Lambda_v \setminus \partial\Psi, v)) \leq \frac{1}{3en}$$

for large enough n. Notice that in order to get the inequality in the middle line we used an optimal coupling of $Z_t^{+,\Lambda_v}[v]$ and $Z_t^{-,\Lambda_v}[v]$ together with the fact that the oscillation of any function whose range is $[1, q]$ is at most $q-1$.

We now show that $\Pr(Q_t^{+,\Lambda_v}[v] \neq Z_t^{+,\Lambda_v}[v]) \leq \frac{1}{3en}$ when $t = T(n)$ (by symmetry, the same holds for the minus chains), thus completing the proof.

Using a Chernoff bound, if we run the dynamics on Ψ for $cn(\log n)^2$ steps then with probability at least $1 - \frac{1}{6en}$ the number of steps in which Λ_v is hit is at least

$$\frac{1}{2}cm(\log n)^2 = (2\log n)cm\frac{\log n}{4} \geq (2\log n)cm(\log m)^2$$

for large enough n. If we assume that indeed Λ_v is hit this often then we can use the induction hypothesis to bound the probability that the spins at v differ because the two chains we are comparing have the same fixed boundaries outside Λ_v. Indeed, after $T(m) = cm(\log m)^2$ steps in Λ_v, the configurations (on the whole of Λ_v) disagree with probability at most $\frac{1}{e}$, and thus after $(2\log n)T(m)$ steps, they disagree with probability at most $\frac{1}{n^2}$. Hence, $\Pr(Q_{T(n)}^{+,\Lambda_v}[v] \neq Z_{T(n)}^{+,\Lambda_v}[v]) \leq \frac{1}{6en} + \frac{1}{n^2} \leq \frac{1}{3en}$ for large enough n, as required. $\square$

Theorem 5.2. *Suppose there exists an identity coupling such that for all subsets Λ of volume at most n_0, where n_0 is a sufficiently large constant, the coupling time of the given identity coupling on Λ is at most $\frac{1}{8(2d-1)}\frac{n_0^{1/d}}{\log n_0}|\Lambda|$ uniformly in the boundary configuration. Then for all n and for all subsets Ψ of volume n with any boundary configuration, $\Pr(X_{kn} \neq Y_{kn}) \leq |\Psi|\exp(-c \cdot k)$, where $c = 2(2d-1)n_0^{-1/d}$. Namely, this identity coupling has optimal mixing.*

Proof. Consider the Glauber dynamics on Ψ with an arbitrary boundary configuration. We will show that for any two instances of the chain (X_t) and (Y_t) and any $v \in \Psi$ we have $\Pr(X_{kn}[v] \neq Y_{kn}[v]) \leq \exp(-c \cdot k)$ under the given identity coupling. Using a union bound, this implies that $\Pr(X_{kn} \neq Y_{kn}) \leq |\Psi|\exp(-c \cdot k)$.

Let $l_0 = \lceil \frac{1}{c} \rceil = \lceil \frac{n_0^{1/d}}{2(2d-1)} \rceil$. As before, we will use Lemma 3.1 to localize the dynamics. Together with the hypothesis of the theorem, this will imply that after $l_0 n$ steps the spins at v agree with high probability. What we want, however, is that the probability of disagreement will continue to decay exponentially with the number of steps. Notice that such a result would follow if, once the spins at v agreed, they continued to agree through the rest of the process, but this is clearly not the case. However, using the amenability of $\mathbb{Z}^d$ and another localization argument, we can show that if all the spins within a large enough radius around v agree at a given time, then the spins at v will continue to agree for sufficiently many steps (depending on the radius of agreement). Bootstrapping from the sufficiently small probability of disagreement after $l_0 n$ steps, we get the required exponential decay.

We proceed with the formal proof. Let $\rho(k) = \max_{X_0, Y_0, v \in \Psi} \Pr(X_{kn}[v] \neq Y_{kn}[v])$. We have the following two claims.

Claim 1. *Under the hypothesis of the theorem,*

$$\rho(l_0) \leq \frac{1}{e2^d(n_0 + 1)} = \frac{1}{e2^d([2(2d-1)l_0]^d + 1)} .$$

Claim 2. *Without any assumptions, for any k_1 and k_2,*

$$\rho(k_1 + k_2) \leq [2(2d-1)k_2]^d\rho(k_1)\rho(k_2) + 4e^{-k_2} .$$

Theorem 5.2 follows from the combination of the above two claims. To see this, let $\phi(k) = 2^d([2(2d-1)k]^d + 1) \cdot \max\left\{\rho(k), 2e^{-\frac{k}{2}}\right\}$. Using Claim 2, we have by an explicit calculation that $\phi(2k) \le \phi(k)^2$. On the other hand, from Claim 1 we get that $\phi(l_0) \le \frac{1}{e}$ (where we used the fact that l_0 is large enough to handle the case of $\rho(l_0) < 2e^{-\frac{l_0}{2}}$). We then conclude that $\rho(k) \le \phi(k) \le \exp(-\frac{k}{l_0})$, as required.

The proofs of the above two claims use similar ideas to those used in the proof of Lemma 4.1, i.e., using Lemma 3.1 to localize the dynamics and appealing to the amenability of $\mathbb{Z}^d$. We refer the reader to the full version of this paper for the detailed proofs of both claims, and also to [8], where a proof of a claim similar to Claim 2 can be found.   □

As remarked at the beginning of this section, combining Theorems 5.1 and 5.2 immediately yields Theorem 2.4.

6   From Spatial to Temporal Mixing: The General Case

In this section we prove Theorem 2.5. Namely, we show that in general (without assuming monotonicity), strong spatial mixing implies that the heat-bath block dynamics has optimal temporal mixing if the blocks used are large enough. Using *path coupling* [2], the proof is reduced to showing that strong spatial mixing implies that a condition known as the *Dobrushin-Shlosman condition* holds. The last implication is known [4], but we include a simple proof of it here.

Proof of Theorem 2.5. Consider the heat-bath dynamics $HB(L)$ on a rectangle Ψ of volume n with an arbitrary boundary configuration. Notice that L is a large enough constant to be set later and will depend only on the dimension d and the constants from the definition of strong spatial mixing. In particular, L is uniform in n and the boundary configuration. Recall that the dynamics chooses a block to be updated from $B(\Psi, L)$, which is the set of translations of the box of side-length L that intersect Ψ. We denote the number of blocks by $m = |B(\Psi, L)|$ and notice that $n \le m \le L^d n$ (the lower bound is due to the fact that the number of translations that intersect Ψ is at least the volume of Ψ while the upper bound crudely uses the fact that each site is covered by L^d translations). Using the *path coupling* method [2], it is enough to show that there exists a constant $c > 0$ (independent of n and the boundary configuration) such that, for any site $u \in \Psi$ and any two configurations σ, σ^u that differ only at u, there exists a coupling of the two chains whose current configurations are σ and σ^u respectively such that after one step, the average *Hamming* distance between the two coupled configurations is at most $1 - \frac{c}{m}$, i.e, decreases by at least $\frac{c}{m}$.

We couple these two chains using a specific identity coupling. Namely, the block chosen to be updated is the same in both chains, and if the boundaries of that block are the same in both σ and σ^u then we couple the update of the block such that the configurations inside the block agree with probability one. If the boundaries are not the same (this can happen only if u is on the boundary of the chosen block), we use a coupling to be described below.

From the way we defined the heat-bath block dynamics, each site in Ψ is included in exactly L^d blocks. Since we use an identity coupling, if a block including the site u is chosen to be updated then the Hamming distance between the two configurations will be zero (i.e., decrease by one) since the boundaries of this block are the same in both σ and σ^u. The probability of choosing a block as above is $\frac{L^d}{m}$. Thus, it is enough to show that the contribution to the expected change in Hamming distance from choosing the other blocks is at most $\frac{L^d-c}{m}$.

As we already mentioned, the Hamming distance may increase only if the block chosen to be updated is one whose boundaries include u. Since there are at most $2dL^{d-1}$ such blocks, we will be done once we show that we can couple the update of each such block Λ such that the resulting average Hamming distance in Λ is strictly less then $\frac{L}{2d}$.

Let Λ be such that $u \in \partial\Lambda$. Let also $r = \frac{1}{2}(\frac{L}{4d})^{\frac{1}{d}}$, $\Lambda_r = \{v \in \Lambda \mid \text{dist}(v, u) \leq r\}$, and $\overline{\Lambda_r} = \Lambda \setminus \Lambda_r$. By the strong spatial mixing assumption, $\|\mu_\Lambda^\sigma - \mu_\Lambda^{\sigma^u}\|_{\overline{\Lambda_r}} \leq \beta|\overline{\Lambda_r}|\exp(-\alpha \cdot r) \leq L^{-d}$ for a large enough L. We can thus couple the update of Λ such that the two coupled configurations disagree over $\overline{\Lambda_r}$ with probability at most L^{-d}. A trivial upper bound on the resulting average Hamming distance in Λ in this coupling is then $|\Lambda_r| + L^{-d}|\overline{\Lambda_r}| \leq \frac{L}{4d} + 1$. $\qquad\square$

References

1. J. van den Berg, A uniqueness condition for Gibbs measures, with applications to the 2-dimensional Ising antiferromagnet, *Comm. Math. Phys.*, **152**, 161–166 (1993).
2. R. Bubley and M.E. Dyer, Path coupling: a technique for proving rapid mixing in markov chains, *Proc. 38th IEEE Symp. on Found. of Comp. Sc.*, 223–231 (1997).
3. F. Cesi, Quasi-factorization of the entropy and logarithmic Sobolev inequalities for Gibbs random fields, *Prob. Theory and Related Fields*, **120**, 569–584 (2001).
4. R.L. Dobrushin and S.B. Shlosman, Completely Analytical Gibbs Fields, in: J. Fritz, A. Jaffe, D. Szasz, *Statistical Mechanics and Dynamical Systems*, 371–403, Birkhauser, Boston (1985).
5. J.A. Fill and M. Machida, Stochastic monotonicity and realizable monotonicity, *Ann. of Prob.*, **29**, 938–978 (2001).
6. O. Häggström, J. Jonasson and R. Lyons, Explicit isoperimetric constants and phase transitions in the random-cluster model, *Ann. of Prob.*, **30**, 443–473 (2002).
7. C. Kenyon, E. Mossel and Y. Peres, Glauber dynamics on trees and hyperbolic graphs, *Proc. 42nd IEEE Symp. on Found. of Comp. Sc.*, 568–578 (2001).
8. F. Martinelli, Lectures on Glauber dynamics for discrete spin models, *Lectures on Probability Theory and Statistics (Saint-Flour, 1997)*, Lecture notes in Math. **1717**, 93–191, Springer, Berlin (1998).
9. F. Martinelli and E. Olivieri, Approach to equilibrium of Glauber dynamics in the one phase region I: The attractive case, *Comm. Math. Phys.* **161**, 447–486 (1994).
10. F. Martinelli and E. Olivieri, Approach to equilibrium of Glauber dynamics in the one phase region II: The general case, *Comm. Math. Phys.* **161**, 487–514 (1994).
11. D.W. Stroock and B. Zegarlinski, The logarithmic Sobolev inequality for discrete spin systems on a lattice, *Comm. Math. Phys.* **149**, 175–194 (1992).

Quantum Walks on the Hypercube

Cristopher Moore[1] and Alexander Russell[2]

[1] Computer Science Department, University of New Mexico, Albuquerque and the Santa Fe Institute `moore@cs.unm.edu`
[2] Department of Computer Science and Engineering, University of Connecticut, Storrs, Connecticut `acr@cse.uconn.edu`

Abstract. Recently, it has been shown that one-dimensional quantum walks can mix more quickly than classical random walks, suggesting that quantum Monte Carlo algorithms can outperform their classical counterparts. We study two quantum walks on the n-dimensional hypercube, one in discrete time and one in continuous time. In both cases we show that the instantaneous mixing time is $(\pi/4)n$ steps, faster than the $\Theta(n \log n)$ steps required by the classical walk. In the continuous-time case, the probability distribution is *exactly* uniform at this time. On the other hand, we show that the average mixing time as defined by Aharonov et al. [AAKV01] is $\Omega(n^{3/2})$ in the discrete-time case, slower than the classical walk, and nonexistent in the continuous-time case. This suggests that the instantaneous mixing time is a more relevant notion than the average mixing time for quantum walks on large, well-connected graphs. Our analysis treats interference between terms of different phase more carefully than is necessary for the walk on the cycle; previous general bounds predict an exponential average mixing time when applied to the hypercube.

1 Introduction

Random walks form one of the cornerstones of theoretical computer science. As algorithmic tools, they have been applied to a variety of central problems, such as estimating the volume of a convex body [DFK91,LK99], approximating the permanent [JS89,JSV00], and finding satisfying assignments for Boolean formulae [Sch99]. Furthermore, the basic technical phenomena appearing in the study of random walks (e.g., spectral decomposition, couplings, and Fourier analysis) also support several other important areas such as pseudorandomness and derandomization (see, e.g., [AS92, (§9,§15)]).

The development of efficient *quantum* algorithms for problems believed to be intractable for classical randomized computation, like integer factoring and discrete logarithm [Sho97], has prompted the investigation of *quantum walks*. This is a natural generalization of the traditional notion discussed above where, roughly, the process evolves in a unitary rather than stochastic fashion.

The notion of "mixing time," the first time when the distribution induced by a random walk is sufficiently close to the stationary distribution, plays a central role in the theory of classical random walks. For a given graph, then,

J.D.P. Rolim and S. Vadhan (Eds.): RANDOM 2002, LNCS 2483, pp. 164–178, 2002.
© Springer-Verlag Berlin Heidelberg 2002

it is natural to ask if a quantum walk can mix more quickly than its classical counterpart. (Since a unitary process cannot be mixing, we define a stochastic process from a quantum one by performing a measurement at a given time or a distribution of times.) Several recent articles [AAKV01,ABN+01,NV00] have answered this question in the affirmative, showing, for example, that a quantum walk on the n-cycle mixes in time $\mathcal{O}(n \log n)$, a substantial improvement over the classical random walk which requires $\Theta(n^2)$ steps to mix. Quantum walks were also defined in [Wat01], and used to show that undirected graph connectivity is contained in a version of quantum LOGSPACE. These articles raise the exciting possibility that quantum Monte Carlo algorithms could form a new family of quantum algorithms that work more quickly than their classical counterparts.

Two types of quantum walks exist in the literature. The first, introduced by [AAKV01,ABN+01,NV00], studies the behavior of a "directed particle" on the graph; we refer to these as *discrete-time* quantum walks. The second, introduced in [FG98,CFG01], defines the dynamics by treating the adjacency matrix of the graph as a Hamiltonian; we refer to these as *continuous-time* quantum walks. The landscape is further complicated by the existence of two distinct notions of mixing time. The "instantaneous" notion [ABN+01,NV00] focuses on particular times at which measurement induces a desired distribution, while the "average" notion [AAKV01], another natural way to convert a quantum process into a stochastic one, focuses on measurement times selected randomly from some interval. While the average mixing time is known to be at most polynomially faster than the classical mixing time [AAKV01], no such result is known for the instantaneous mixing time.

In this article, we analyze both the continuous-time and a discrete-time quantum walk on the hypercube. In both cases, the walk is shown to have an instantaneous mixing time at $(\pi/4)n$. Recall that the classical walk on the hypercube mixes in time $\Theta(n \log n)$, so the quantum walk is faster by a logarithmic factor. Moreover, since $\pi/4 < 1$ our results show that the quantum walk mixes in time less than the diameter of the graph, and in fact in the continuous-time case the probability distribution at $t = (\pi/4)n$ is *exactly* uniform. Both of these things happen due to a marvelous conspiracy of destructive interference between terms of different phase.

These walks show *i.)* a similarity between the two notions of quantum walks, and *ii.)* a disparity between the two notions of quantum mixing times. As mentioned above, both walks have an instantaneous mixing time at time $(\pi/4)n$. On the other hand, we show that the average mixing time of the discrete-time walk is $\Omega(n^{3/2})$, *slower* than the classical walk, and that for the continuous-time walk there is *no* time at which the time-averaged probability distribution is close to uniform in the sense of [AAKV01]. Our results suggest that for large graphs (large compared to their mixing time) the instantaneous notion of mixing time is more appropriate than the average one, since the probability distribution is close to uniform only in a narrow window of time.

The analysis of the hypercubic quantum walk exhibits a number of features markedly different from those appearing in previously studied walks. In particular, the dimension of the relevant Hilbert space is, for the hypercube, exponential in the length of the desired walk, while in the cycle these quantities are roughly equal. This requires that interference be handled in a more delicate way than is required for the walk on the cycle; in particular, the general bound of [AAKV01] yields an exponential upper bound on the mixing time for the discrete-time walk.

We begin by defining quantum walks and discussing various notions of mixing time. We then analyze the two quantum walks on the hypercube in Sections 2 and 3.

1.1   Quantum Walks and Mixing Times

Any graph $G = (V, E)$ gives rise to a familiar Markov chain by assigning probability $1/d$ to all edges leaving each vertex v of degree d. Let $P_u^t(v)$ be the probability of visiting a vertex v at step t of the random walk on G starting at u. If G is undirected, connected, and not bipartite, then $\lim_{t \to \infty} P_u^t$ exists[1] and is independent of u. A variety of well-developed techniques exist for establishing bounds on the rate at which P_u^t achieves this limit (e.g., [Vaz92]); if G happens to be the Cayley graph of a group — as are, for example, the cycle and the hypercube — then techniques from Fourier analysis can be applied [Dia88]. Below we will use some aspects of this approach, especially the Diaconis-Shahshahani bound on the total variation distance [DS81].

For simplicity, we restrict our discussion to quantum walks on Cayley graphs; more general treatments of quantum walks appear in [AAKV01,FG98].

Before describing the quantum walk models we set down some notation. For a group G and a set of generators Γ such that $\Gamma = \Gamma^{-1}$, let $X(G, \Gamma)$ denote the undirected Cayley graph of G with respect to Γ. For a finite set S, we let $L(S) = \{f : S \to \mathbb{C}\}$ denote the collection of $\mathbb{C}$-valued functions on S with $\sum_{s \in S} |f(s)|^2 = 1$. This is a Hilbert space under the natural inner product $\langle f|g \rangle = \sum_{s \in S} f(s) g(s)^*$. For a Hilbert space V, a linear operator $U : V \to V$ is *unitary* if for all $\boldsymbol{v}, \boldsymbol{w} \in V$, $\langle \boldsymbol{v}|\boldsymbol{w} \rangle = \langle U\boldsymbol{v}|U\boldsymbol{w} \rangle$; if U is represented as a matrix, this is equivalent to the condition that $U^\dagger = U^{-1}$ where $\dagger$ denotes the Hermitian conjugate.

There are two natural quantum walks that one can define for such graphs, which we now describe.

The discrete-time walk: This model, introduced by [AAKV01,ABN+01, NV00], augments the graph with a *direction space*, each basis vector of which corresponds one of the generators in Γ. (Note that this labels the neighborhood of each vertex of the Cayley graph in a consistent way.) A step of the walk then consists of the composition of two unitary transformations; a *shift* operator which leaves the direction unchanged while moving the particle in its current direction, and a *local transformation* which operates on the direction while leaving the position unchanged. To be precise, the quantum walk on $X(G, \Gamma)$ is defined on the space $L(G \times \Gamma) \cong L(G) \otimes L(\Gamma)$. Let $\{\delta_\gamma \mid \gamma \in \Gamma\}$ be the natural basis for $L(\Gamma)$, and $\{\delta_g | g \in G\}$ the natural basis for $L(G)$. Then the shift operator is

---

[1] In fact, this limit exists under more general circumstances; see e.g. [MR95].

$S : (\delta_g \otimes \delta_\gamma) \mapsto (\delta_{g\gamma} \otimes \delta_\gamma)$, and the local transformation is $\check{D} = \mathbf{1} \otimes D$ where D is defined on $L(\Gamma)$ alone and $\mathbf{1}$ is the identity on $L(G)$. Then one "step" of the walk corresponds to the operator $U = \check{D}V$. If we measure the position of the particle, but not its direction, at time t, we observe a vertex v with probability $P_t(v) = \sum_{\gamma \in \Gamma} |\langle U^t \psi_0 \mid \delta_v \otimes \delta_\gamma \rangle|^2$ where $\psi_0 \in L(G \times \Gamma)$ is the initial state.

The continuous-time walk: This model, introduced in [FG98], works directly on $L(G)$. The walk evolves by treating the adjacency matrix of the graph as a Hamiltonian and using the Schrödinger equation. Specifically, if H is the adjacency matrix of $X(G, \Gamma)$, the evolution of the system at time t is given by U^t, where $U^t \equiv e^{iHt}$ (here we use the matrix exponential, and U^t is unitary since H is real and symmetric). Then if we measure the position of the particle at time t, we observe a vertex v with probability $P_t(v) = |\langle U^t \psi_0 | \delta_v \rangle|^2$ where ψ_0 is the initial state. Note the analogy to classical Poisson processes: since $U^t = e^{iHt} = 1 + iHt + (iHt)^2/2 + \cdots$, the amplitude of making s steps is the coefficient $(it)^s/s!$ of H^s, which up to normalization is Poisson-distributed with mean t.

Remark. In [CFG01], the authors point out that defining quantum walks in continuous time allows unitarity without having to extend the graph with a direction space and a chosen local operation. On the other hand, it is harder to see how to carry out such a walk in a generically programmable way using only local information about the graph, for instance in a model where we query a graph to find out who our neighbors are. Instead, continuous-time walks might correspond to special-purpose analog computers, where we build in interactions corresponding to the desired Hamiltonian and allow the system to evolve in continuous time.

In both cases we start with an initial wave function concentrated at a single vertex corresponding to the identity u of the group. For the continuous-time walk, we have $\langle \psi_0 | \delta_v \rangle = 1$ if $v = u$ and 0 otherwise. For the discrete-time walk, we start with a uniform superposition over all possible directions,

$$\langle \psi_0 | \delta_v \otimes \delta_\gamma \rangle = \begin{array}{ll} 1/\sqrt{|\Gamma|} & \text{if } v = u, \\ 0 & \text{otherwise.} \end{array}$$

For the hypercube, $u = \mathbf{0} = (0, \ldots, 0)$.

In order to define a discrete-time quantum walk, one must select a local operator D on the direction space. In principle, this introduces some arbitrariness into the definition. However, if we wish D to respect the permutation symmetry of the n-cube, and if we wish to maximize the operator distance between D and the identity, it is easy to show [MR01] that we are forced to choose Grover's diffusion operator [Gro96] which we recall below. We call the resulting walk the "symmetric discrete-time quantum walk" on the n-cube. (Watrous [Wat01] also used Grover's operator to define quantum walks on undirected graphs.)

Since Grover's operator is close to the identity matrix for large n, one might imagine that it would take $\Omega(n^{1/2})$ steps to even change direction, giving the quantum walk a mixing time of $\approx n^{3/2}$, slower than the classical random walk. However, like many intuitions about quantum mechanics, this is simply wrong.

Since the evolution of the quantum walk is governed by a unitary operator rather than a stochastic one, there can be no "stationary distribution" $\lim_{t \to \infty} P_t$ unless P_t is constant for all t. In particular, for any $\epsilon > 0$, there are infinitely many positive times t for which $\|U^t - \mathbf{1}\| \leq \epsilon$ so that $\|U^t \psi_u - \psi_u\| \leq \epsilon$ and P_t is close to the initial distribution. However, there may be particular stopping times t which induce distributions close to, say, the uniform distribution, and we call these *instantaneous mixing times*. (For bipartite graphs such as the hypercube, the most we can hope for is that P_t and P_{t+1} are each close to uniform on the vertices of odd and even parity, say, respectively; to mix on the entire graph we can then flip a classical coin and stop at t or $t + 1$.)

Definition 1. *We say that t is an ϵ-instantaneous mixing time for a quantum walk if $\|P_t - \Pi\| \leq \epsilon$, where $\|A - B\| = \frac{1}{2} \sum_v |A(v) - B(v)|$ denotes total variation distance and Π denotes the uniform distribution, or, in the case of bipartite graphs, the uniform distribution on vertices of the appropriate parity.*

For these walks we show:

Theorem 1. *For the symmetric discrete-time quantum walk on the n-cube, $t = \lceil k(\pi/4)n \rceil$ is an ϵ-instantaneous mixing time with $\epsilon = \mathcal{O}(n^{-7/6})$ for all odd k.*

and, even more surprisingly,

Theorem 2. *For the continuous-time quantum walk on the n-cube, $t = k(\pi/4)n$ is a 0-instantaneous mixing time for all odd k.*

Thus in both cases the mixing time is $\Theta(n)$, as opposed to $\Theta(n \log n)$ as it is in the classical case.

Aharonov et al. [AAKV01] define another natural notion of mixing time for quantum walks, in which the stopping time t is selected uniformly from the set $\{0, \ldots, T - 1\}$. They show that the time-averaged distributions $\bar{P}_T = (1/T) \sum_{t=0}^{T-1} P_t$ do converge as $T \to \infty$ and study the rate at which this occurs. For a continuous-time random walk, we analogously define the distribution $\bar{P}_T(v) = (1/T) \int_0^T P_t(v) \, dt$. Then we call a time at which $\bar{P}_T$ is close to uniform an *average mixing time*:

Definition 2. *We say that T is an ϵ-average mixing time for a quantum walk if $\|\bar{P}_T - \Pi\| \leq \epsilon$, where Π denotes the uniform distribution.*

In this paper we also prove a lower bound on the ϵ-average mixing times for the hypercube. For the discrete-time walk, it is even longer than the mixing time of the classical random walk:

Theorem 3. *For the discrete-time quantum walk on the n-cube, the ϵ-average mixing time is $\Omega(n^{3/2}/\epsilon)$.*

This is surprising given that the instantaneous mixing time is only linear in n. However, the probability distribution is close to uniform only in narrow windows around the odd multiples of $(\pi/4)n$, so $\bar{P}_T$ is far from uniform for significantly longer times. We also observe that the general bound given in [AAKV01] yields an exponential upper bound on the average mixing time, showing that it is necessary to handle interference for walks on the hypercube more carefully than for those on the cycle.

For the continuous-time walk the situation is even worse: while it possesses 0-instantaneous mixing times at all odd multiples of $(\pi/4)n$, the limiting distribution $\lim_{T \to \infty} \bar{P}_T$ is not uniform, and we show the following:

Theorem 4. *For the continuous-time quantum walk on the n-cube, there exists $\epsilon > 0$ such that* no *time is an ϵ-average mixing time.*

Our results suggest that in both the discrete and continuous-time case, the instantaneous mixing time is a more relevant notion than the average mixing time for large, well-connected graphs.

2   The Symmetric Discrete-Time Walk

In this section we prove Theorem 1. We treat the n-cube as the Cayley graph of $\mathbb{Z}_2^n$ with the regular basis vectors $e_i = (0, \ldots, 1, \ldots, 0)$ with the 1 appearing in the ith place. Then the discrete-time walk takes place in the Hilbert space $L(\mathbb{Z}_2^n \times [n])$ where $[n] = \{1, \ldots, n\}$. Here the first component represents the position x of the particle in the hypercube, and the second component represents the particle's current "direction"; if this is i, the shift operator will flip the ith bit of x.

As in [AAKV01,NV00], we will not impose a group structure on the direction space, and will Fourier transform only over the position space. For this reason, we will express the wave function $\psi \in L(\mathbb{Z}_2^n \times [n])$ as a function $\Psi : \mathbb{Z}_2^n \to \mathbb{C}^n$, where the ith coordinate of $\Psi(x)$ is the projection of ψ into $\delta_x \otimes \delta_i$, i.e. the complex amplitude of the particle being at position x with direction i. The Fourier transform of such an element Ψ is $\tilde{\Psi} : \mathbb{Z}_2^n \to \mathbb{C}^n$, where $\tilde{\Psi}(k) = \sum_x (-1)^{k \cdot x} \Psi(x)$. Then the shift operator for the hypercube is $S : \Psi(x) \mapsto \sum_{i=1}^n \pi_i \Psi(x \oplus e_i)$, where e_i is the ith basis vector in the n-cube, and π_i is the projection operator for the ith direction. The reason for considering the Fourier transform above is that the shift operator is diagonal in the Fourier basis: specifically it maps $\tilde{\Psi}(k) \mapsto S_k \tilde{\Psi}(k)$ where

$$
S_k = \begin{pmatrix} (-1)^{k_1} & & & 0 \\ & (-1)^{k_2} & & \\ & & \ddots & \\ 0 & & & (-1)^{k_n} \end{pmatrix}
$$

For the local transformation, we use Grover's diffusion operator on n states, $D_{ij} = 2/n - \delta_{ij}$ where δ_{ij} is the identity matrix. The advantage of Grover's operator is that, like the n-cube itself, it is permutation symmetric. We use this symmetry to rearrange $U_k = S_k D$ to put the negated rows on the bottom,

$$U_{\boldsymbol{k}} = S_{\boldsymbol{k}}D = \begin{pmatrix} \begin{array}{cc|ccc} 2/n-1 & 2/n & & & \\ 2/n & 2/n-1 & & 2/n & \\ \vdots & & \ddots & & \\ \hline & & -2/n+1 & -2/n & \cdots \\ & -2/n & -2/n & -2/n+1 & \\ & & \vdots & & \ddots \end{array} \end{pmatrix}$$

where the top and bottom blocks have $n - k$ and k rows respectively; here k is the Hamming weight of $\boldsymbol{k}$.

The eigenvalues of $U_{\boldsymbol{k}}$ then depend only on k. Specifically, $U_{\boldsymbol{k}}$ has the eigenvalues $+1$ and -1 with multiplicity $k - 1$ and $n - k - 1$ respectively, plus the eigenvalues λ, λ^* where

$$\lambda = 1 - \frac{2k}{n} + \frac{2i}{n}\sqrt{k(n-k)} = e^{i\omega_k}$$

and $\omega_k \in [0, \pi]$ is described by

$$\cos\omega_k = 1 - \frac{2k}{n}, \quad \sin\omega_k = \frac{2}{n}\sqrt{k(n-k)}$$

Its eigenvectors with eigenvalue $+1$ span the $(k-1)$-dimensional subspace consisting of vectors with support on the k "flipped" directions that sum to zero, and similarly the eigenvectors with eigenvalue -1 span the $(n-k-1)$-dimensional subspace of vectors on the $n - k$ other directions that sum to zero. We call these the *trivial* eigenvectors. The eigenvectors with eigenvalue $\lambda, \lambda^* = e^{\pm i\omega_k}$ are

$$v_k, v_k^* = \frac{1}{\sqrt{2}}\Big(\underbrace{\frac{\mp i}{\sqrt{n-k}}}_{n-k}, \underbrace{\frac{1}{\sqrt{k}}}_{k} \Big).$$

We call these the *non-trivial* eigenvectors for a given $\boldsymbol{k}$. Over the space of positions and directions these eigenvectors are multiplied by the Fourier coefficient $(-1)^{\boldsymbol{k}\cdot\boldsymbol{x}}$, so as a function of $\boldsymbol{x}$ and direction $1 \le j \le n$ the two non-trivial eigenstates of the entire system, for a given $\boldsymbol{k}$, are

$$v_{\boldsymbol{k}}(\boldsymbol{x}, j) = (-1)^{\boldsymbol{k}\cdot\boldsymbol{x}} \frac{2^{-n/2}}{\sqrt{2}} \times \begin{cases} 1/\sqrt{k} & \text{if } \boldsymbol{k}_j = 1 \\ -i/\sqrt{n-k} & \text{if } \boldsymbol{k}_j = 0 \end{cases}$$

with eigenvalue $e^{i\omega_k}$, and its conjugate $v_{\boldsymbol{k}}^*$ with eigenvalue $e^{-i\omega_k}$.

We take for our initial wave function a particle at the origin $\boldsymbol{0} = (0, \dots, 0)$ in an equal superposition of directions. Since its position is a δ-function in real space it is uniform in Fourier space as well as over the direction space, giving $\tilde{\Psi}_0(\boldsymbol{k}) = (2^{-n/2}/\sqrt{n})(1, \dots, 1)$. This is perpendicular to all the trivial eigenvectors, so their amplitudes are all zero. The amplitude of its component along the non-trivial eigenvector $v_{\boldsymbol{k}}$ is

$$a_{\boldsymbol{k}} = \langle \Psi_0 | v_{\boldsymbol{k}} \rangle = \frac{2^{-n/2}}{\sqrt{2}}\left(\sqrt{\frac{k}{n}} - i\sqrt{1 - \frac{k}{n}} \right) \tag{1}$$

and the amplitude of $v_{\boldsymbol{k}}^*$ is $a_{\boldsymbol{k}}^*$. Note that $|a_{\boldsymbol{k}}|^2 = 2^{-n}/2$, so a particle is equally likely to appear in either non-trivial eigenstate with any given wave vector.

At this point, we note that there are an exponential number of eigenvectors in which the initial state has a non-zero amplitude. We observe below that for this reason the general bound of Aharonov et al. [AAKV01] yields an exponential (upper) bound on the mixing time. In general, this bound performs poorly whenever the number of important eigenvalues is greater than the mixing time.

Instead, we will use the Diaconis-Shahshahani bound on the total variation distance in terms of the Fourier coefficients of the probability [Dia88]. If $P_t(\boldsymbol{x})$ is the probability of the particle being observed at position $\boldsymbol{x}$ at time t, and Π is the uniform distribution, then the total variation distance is bounded by

$$\|P_t - \Pi\|^2 \leq \frac{1}{4} \sum_{\substack{\boldsymbol{k} \neq (0,\ldots,0) \\ \boldsymbol{k} \neq (1,\ldots,1)}} \left|\tilde{P}_t(\boldsymbol{k})\right|^2 = \frac{1}{4} \sum_{k=1}^{n-1} \binom{n}{k} \left|\tilde{P}_t(k)\right|^2. \tag{2}$$

Here we exclude both the constant term and the parity term $\boldsymbol{k} = (1,\ldots,1)$, since as stated above we only ask that P_t approaches $\Pi = 2^{n-1}$, i.e. becomes uniform on the vertices of appropriate parity.

To find $\tilde{P}_t(\boldsymbol{k})$, we first need $\tilde{\Psi}_t(\boldsymbol{k})$. As Nayak and Vishwanath [NV00] did for the walk on the line, we start by calculating the tth matrix power of $U_{\boldsymbol{k}}$. This is

$$U_{\boldsymbol{k}}^t = \left(\begin{array}{ccc|ccc} a+(-1)^t & a & \cdots & & & \\ a & a+(-1)^t & & & c & \\ \vdots & & \ddots & & & \\ \hline & & & b-(-1)^t & b & \cdots \\ & -c & & b & b-(-1)^t & \\ & & & \vdots & & \ddots \end{array}\right)$$

where

$$a = \frac{\cos\omega_k t - (-1)^t}{n-k}, \quad b = \frac{\cos\omega_k t + (-1)^t}{k}, \quad \text{and} \quad c = \frac{\sin\omega_k t}{\sqrt{k(n-k)}}$$

Starting with the uniform initial state, the wave function after t steps is

$$\tilde{\Psi}_t(\boldsymbol{k}) = \frac{2^{-n/2}}{\sqrt{n}} \Big(\underbrace{\cos\omega_k t + \sqrt{\frac{k}{n-k}}\sin\omega_k t}_{n-k}, \underbrace{\cos\omega_k t - \sqrt{\frac{n-k}{k}}\sin\omega_k t}_{k}\Big) \tag{3}$$

In the next two sections we will use this diagonalization to bound the average and instantaneous mixing times as $\Omega(n^{3/2})$ and $\Theta(n)$ respectively.

2.1   Bounds on the Average Mixing Time of the Discrete-Time Walk

In this section, we prove Theorem 3. To do this, it is sufficient to calculate the amplitude at the origin. Fourier transforming Equation 3 back to real space at $\boldsymbol{x} = (0,\ldots,0)$ gives

$$\psi_t(\mathbf{0}) = 2^{-n/2} \sum_k \tilde{\Psi}_t(\mathbf{k}) = \frac{2^{-n}}{\sqrt{n}} \sum_{k=0}^{n} \binom{n}{k} \underbrace{(\cos \omega_k t, \ldots, \cos \omega_k t)}_{n}$$

The probability the particle is observed at the origin after t steps is then

$$P_t(\mathbf{0}) = |\psi_t(\mathbf{0})|^2 = \left(2^{-n} \sum_{k=0}^{n} \binom{n}{k} \cos \omega_k t \right)^2.$$

Let $k = (1-x)(n/2)$. For small x, k is near the peak of the binomial distribution, and $\omega_k = \cos^{-1} x = (\pi/2) - x + \mathcal{O}(x^3)$ so the angle θ between ω_k for successive values of k is roughly constant, $\theta = (2/n) + \mathcal{O}(x^2)$ leading to constructive interference if $\theta t \approx 2\pi$. Specifically, let t_m be the even integer closest to $\pi m n$ for integer m. Then $\cos \omega_k t_m = \pm \cos(2\pi k m + \mathcal{O}(x^3 m n)) = \pm(1 - \mathcal{O}(x^6 m^2 n^2))$. By a standard Chernoff bound, $2^{-n} \sum_{k \le (1-x)n/2} \binom{n}{k} = o(1)$ so long as $x = \omega(n^{-1/2})$. Let $x = \nu(n) n^{-1/2}$ where $\nu(n)$ is a function that goes to infinity slowly as a function of n. We then write

$$P_t(\mathbf{0}) = \left(o(1) + 2^{-n} \sum_{k=(1-x)n/2}^{(1+x)n/2} \binom{n}{k} (1 - \mathcal{O}(x^6 m^2 n^2)) \right)^2 = 1 - \mathcal{O}\left(\frac{\nu(n)^6 m^2}{n} \right)$$

which is $1 - o(1)$ as long as $m = o(n^{1/2} \nu(n)^{-3})$, in which case $t_m = o(n^{3/2} \nu(n)^{-3})$.

For a function $\psi : \mathbb{Z}_2^n \times [n] \to \mathbb{C}$ with $\|\psi\|_2 = 1$ and a set $S \subset \mathbb{Z}_2^n$, we say that ψ *is c-supported on* S if the probability $\mathbf{x} \in S$ is at least c, i.e. $\sum_{\mathbf{x} \in S, d \in [n]} |\psi(\mathbf{x}, d)|^2 \ge c$. The discussion above shows that ψ_{t_m} is $(1 - o(1))$-supported on $\{\mathbf{0}\}$ for appropriate $t_m \approx \pi m n$. Note that if ψ is c-supported on $\{\mathbf{0}\}$ then, since U is local, $U^k \psi$ must be $(c - (1 - c) = 2c - 1)$-supported on the set W_k of vertices of weight k. (The factor of $-(1 - c)$ is due to potential cancellation with portions of the wave function outside $\{\mathbf{0}\}$.) In particular, at times $t_m + k$, for $|k| \le n/2 - n$, ψ_{t_m+k} is $(1 - o(1))$-supported on $W_{n/2-x}$. If $x = x(n) = \omega(\sqrt{n})$, then $|W_{(1/2-\delta)n}|/2^n = o(1)$ and, evidently, the average $(1/T) \sum_i^T P_i$ has total variation distance $1 - o(1)$ from the uniform distribution if $T = o(n^{3/2})$. Therefore, for any $\epsilon > 0$ the ϵ-average mixing time is $\Omega(n^{3/2})$.

Thus we see that in the sense of [AAKV01], the discrete-time quantum walk is actually *slower* than the classical walk. In the next section, however, we show that its instantaneous mixing time is only linear in n.

We now observe that the general bound of [AAKV01] predicts an average mixing time for the n-cube which is exponential in n. In that article it is shown that the variation distance between $\bar{P}_T$ and the uniform distribution (or more generally, the limiting distribution $\lim_{T \to \infty} \bar{P}_T$) is bounded by a sum over distinct pairs of eigenvalues,

$$\|\bar{P}_T - \Pi\| \le \frac{2}{T} \sum_{i,j \text{ s.t. } \lambda_i \ne \lambda_j} \frac{|a_i|^2}{|\lambda_i - \lambda_j|} \tag{4}$$

where $a_i = \langle \psi_0 | v_i \rangle$ is the component of the initial state along the eigenvector v_i. (Since this bound includes eigenvalues λ_j for which $a_j = 0$, we note using the same reasoning as in [AAKV01] that it also holds when we replace $|a_i|^2$ with $|a_i a_j^*|$.)

For the quantum walk on the cycle of length n, this bound gives an average mixing time of $\mathcal{O}(n \log n)$. For the n-cube, however, there are exponentially many pairs of eigenvectors with distinct eigenvalues, all of which have a non-zero component in the initial state. Specifically, for each Hamming weight k there are $\binom{n}{k}$ non-trivial eigenvectors each with eigenvalue $e^{i\omega_k}$ and $e^{-i\omega_k}$. These complex conjugates are distinct from each other for $0 < k < n$, and eigenvalues with distinct k are also distinct. The number of distinct pairs is then

$$\sum_{k=1}^{n-1} \binom{n}{k}^2 + 4 \sum_{k,k'=0}^{n} \binom{n}{k}\binom{n}{k'} = \Omega(4^n)$$

Taking $|a_k| = 2^{-n/2}/\sqrt{2}$ from Equation 1 and the fact that $|\lambda_i - \lambda_j| \leq 2$ since the λ_i lie on the unit circle, we see that Equation 4 gives an upper bound on the ϵ-average mixing time of size $\Omega(2^n/\epsilon)$. In general, this bound will give a mixing time of $\mathcal{O}(M/\epsilon)$ whenever the initial state is distributed roughly equally over M eigenvectors, and when these are roughly equally distributed over $\omega(1)$ distinct eigenvalues.

2.2   The Instantaneous Mixing Time of the Discrete-Time Walk

To prove Theorem 1 we could calculate $\Psi_t(x)$ by Fourier transforming $\tilde{P}_t(k)$ back to real space for all x. However, this calculation turns out to be significantly more awkward than calculating the Fourier transform of the probability distribution, $\tilde{P}_t(k)$, which we need to apply the Diaconis-Shahshahani bound. Since $P_t(x) = \Psi_t(x)\Psi_t(x)^*$, and since multiplications in real space are convolutions in Fourier space, we perform a convolution over $\mathbb{Z}_2^n$:

$$\tilde{P}_t(k) = \sum_{k'} \tilde{\Psi}_t(k') \cdot \tilde{\Psi}_t(k \oplus k')$$

where the inner product is defined on the direction space, $u \cdot v = \sum_{i=1}^n u_i v_i^*$. We write this as a sum over j, the number of bits of overlap between k' and k, and l, the number of bits of k' outside the bits of k (and so overlapping with $k \oplus k'$). Thus k' has weight $j + l$, and $k \oplus k'$ has weight $k - j + l$.

Calculating the dot product $\tilde{\Psi}_t(k') \cdot \tilde{\Psi}_t(k \oplus k')$ explicitly from Equation 3 as a function of these weights and overlaps gives

$$\tilde{P}_t(k) = \frac{1}{2^n} \sum_{j=0}^{k} \sum_{l=0}^{n-k} \binom{k}{j}\binom{n-k}{l} \left[\cos\omega_{j+l}t \, \cos\omega_{k-j+l}t + A \, \sin\omega_{j+l}t \, \sin\omega_{k-j+l}t \right]$$

$$(5)$$

where

$$A = \frac{\cos\omega_k - \cos\omega_{j+l}\,\cos\omega_{k-j+l}}{\sin\omega_{j+l}\,\sin\omega_{k-j+l}}$$

The reader can check that this gives $\tilde{P}_t(0) = 1$ for the trivial Fourier component where $k = 0$, and $\tilde{P}_t(n) = (-1)^t$ for the parity term where $k = n$.

Using the identities $\cos a \cos b = (1/2)(\cos(a-b)+\cos(a+b))$ and $\sin a \sin b = (1/2)(\cos(a-b)-\cos(a+b))$ we can re write Equation 5 as

$$
\tilde{P}_t(k) = \frac{1}{2^n} \sum_{j=0}^{k} \sum_{l=0}^{n-k} \binom{k}{j}\binom{n-k}{l} \left[\left(\frac{1-A}{2}\right) \cos \omega_+ t + \left(\frac{1+A}{2}\right) \cos \omega_- t \right]
$$

$$
\equiv \frac{1}{2^n} \sum_{j=0}^{k} \sum_{l=0}^{n-k} \binom{k}{j}\binom{n-k}{l} Y \tag{6}
$$

where $\omega_\pm = \omega_{j+l} \pm \omega_{k-j+l}$.

The terms $\cos \omega_\pm t$ in Y are rapidly oscillating with a frequency that increases with t. Thus, unlike the walk on the cycle, the phase is rapidly oscillating everywhere, as a function of either l or j. This will make the dominant contribution to $\tilde{P}_t(k)$ exponentially small when $t/n = \pi/4$, giving us a small variation distance when we sum over all $\boldsymbol{k}$.

To give some intuition for the remainder of the proof, we pause here to note that if Equation 6 were an integral rather than a sum, we could immediately approximate the rate of oscillation of Y to first order at the peaks of the binomials, where $j = k/2$ and $l = (n-k)/2$. One can check that $d\omega_k/dk \geq 2/n$ and hence $d\omega_+/dl = d\omega_-/dj \geq 4/n$. Since $|A| \leq 1$, we would then write

$$
\tilde{P}_t(k) = \mathcal{O} \left(\frac{1}{2^n} \sum_{j=0}^{k} \sum_{l=0}^{n-k} \binom{k}{j}\binom{n-k}{l} \left(e^{4ijt/n} + e^{4ilt/n} \right) \right)
$$

which, using the binomial theorem, would give

$$
\left| \tilde{P}_t(k) \right| = \mathcal{O} \left(\left| \frac{1 + e^{4it/n}}{2} \right|^k + \left| \frac{1 + e^{4it/n}}{2} \right|^{n-k} = \cos^k \frac{2t}{n} + \cos^{n-k} \frac{2t}{n} \right) \tag{7}
$$

In this case the Diaconis-Shahshahani bound and the binomial theorem give

$$
\|P_t - \Pi\|^2 \leq \frac{1}{4} \sum_{0 < k < n} \binom{n}{k} \left(\cos^k \frac{2t}{n} + \cos^{n-k} \frac{2t}{n} \right)^2
$$

$$
\leq \frac{1}{2} \left[\left(2\cos^2 \frac{2t}{n} \right)^n + \left(1 + \cos^2 \frac{2t}{n} \right)^n - 1 \right]
$$

If we could take t to be the non-integer value $(\pi/4)n$, these cosines would be zero and P_t would be exactly uniform.

This will turn out to be essentially the right answer. But since Equation 6 is a sum, not an integral, we have to be wary of *resonances* where the oscillations are such that the phase changes by a multiple of 2π between adjacent terms, in which case these terms will interfere constructively rather than destructively. Thus to show that the first-order oscillation indeed dominates, we have a significant amount of work left to do. The details of managing these resonances can be found in [MR01]. The process can be summarized as follows: *i.)* we compute the Fourier transform of the quantity Y in Equation 6, since the sum of Equation 6 can be calculated for a single Fourier basis function using the binomial theorem; *ii.)* we bound the Fourier transform of Y using the method of stationary phase for asymptotic integrals. The dominant stationary point corresponds to the first-order oscillation, but there are also lower-order stationary points corresponding to faster oscillations; so *iii.)* we use an entropy bound to show that the contribution of the other stationary points is exponentially small.

To illustrate our result, we have calculated the probability distribution, and the total variation distance from the uniform distribution (up to parity), as a function of time for hypercubes of dimension 50, 100, and 200. In order to do this exactly, we use the walk's permutation symmetry to collapse its dynamics to a function only of Hamming distance. In Figure 1(a) we see that the total variation distance becomes small when $t = (\pi/4)n$, even though this is *less than the diameter n of the graph*. Since the vast majority of the vertices are a Hamming distance $n/2$ away from the origin, it is possible for the total variation distance to be small (even though it is not the case that P_t differs by a $\Theta(1)$ multiplicative factor between any two vertices, a stronger notion of mixing). What is more surprising, and what the "conspiracy of interference" gives us, is that the probability distribution is close to uniform on a plateau across the hypercube's equator as shown in Figure 1(b).

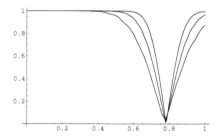
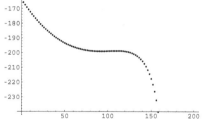

(a) Variation distance at time t as a function of t/n.

(b) Log$_2$ Probability as a function of Hamming weight.

Fig. 1. Graph (a) shows an exact calculation of the total variation distance after t steps of the quantum walk for hypercubes of dimension 50, 100, and 200, as a function of t/n. At $t/n = \pi/4$ the variation distance is small even though the walk has not had time to cross the entire graph. This happens because the distribution is roughly uniform across the equator of the n-cube where the vast majority of the vertices are located. Note that the window in which the variation distance is small gets narrower as n increases. Graph (b) shows the log$_2$ probability distribution on the 200-dimensional hypercube as a function of Hamming distance from the starting point after $157 \approx (\pi/4)n$ steps. The probability distribution has a plateau of 2^{-199} at the equator, matching the uniform distribution up to parity.

3   The Continuous-Time Walk

In the case of the hypercube, the continuous-time walk turns out to be particularly easy to analyze. The adjacency matrix, normalized by the degree, is $H(\boldsymbol{x}, \boldsymbol{y}) = 1/n$ if $\boldsymbol{x}$ and $\boldsymbol{y}$ are adjacent, and 0 otherwise. Interpreting H as the Hamiltonian treats it as the energy operator, and of course increasing the energy makes the system run faster; we normalize by the degree n in order to keep the maximum energy of the system, and so the rate at which transitions occur, constant as a function of n.

The eigenvectors of H and U^t are simply the Fourier basis functions: if $v_{\boldsymbol{k}}(\boldsymbol{x}) = (-1)^{\boldsymbol{k} \cdot \boldsymbol{x}}$ then $H v_{\boldsymbol{k}} = (1 - 2k/n) \, v_{\boldsymbol{k}}$ and $U^t v_{\boldsymbol{k}} = e^{it(1-2k/n)} \, v_{\boldsymbol{k}}$ where we again use k to denote the Hamming weight of $\boldsymbol{k}$. If our initial wave vector has a particle at $\boldsymbol{0}$, then its initial Fourier spectrum is uniform, and at time t we have $\tilde{\Psi}_t(\boldsymbol{k}) = 2^{-n/2} \, e^{it\left(1 - \frac{2k}{n}\right)}$. Again writing the probability P as the convolution of Ψ with Ψ^* in Fourier space, we have

$$\tilde{P}_t(\boldsymbol{k}) = \sum_{\boldsymbol{k}'} \tilde{\Psi}_t(\boldsymbol{k}') \, \tilde{\Psi}_t^*(\boldsymbol{k} \oplus \boldsymbol{k}') = \frac{1}{2^n} \sum_{\boldsymbol{k}'} e^{2it\left(|\boldsymbol{k} \oplus \boldsymbol{k}'| - k'\right)/n}$$

We write this as a sum over all possible overlaps j between $\boldsymbol{k}'$ and $\boldsymbol{k}$, and overlaps l between $\boldsymbol{k}'$ and $\boldsymbol{k} \oplus \boldsymbol{k}'$. Since $k' = j + l$ and $|\boldsymbol{k} \oplus \boldsymbol{k}'| = k - j + l$, this gives

$$\tilde{P}_t(k) = \frac{1}{2^n} \sum_{j=0}^{k} \sum_{l=0}^{n-k} e^{2it(k-2j)/n} = \cos^k \frac{2t}{n} \tag{8}$$

We now prove Theorem 2. Using Equation 8, the Diaconis-Shahshahani bound on the total variation distance is

$$\|P_t - \Pi\|^2 \leq \frac{1}{4} \sum_{k=1}^{n} \binom{n}{k} \left|\tilde{P}_t(k)\right|^2 = \left(1 + \cos^2 \frac{2t}{n}\right)^n - 1$$

At $t = (\pi/4)n$ and its odd multiples, this gives a zero total variation distance, showing that if we sample at these times the probability distribution is *exactly* uniform. Note that this is possible even when $t < n$ since the continuous-time walk has some probability for taking more than t steps (and, in fact, paths with different numbers of steps interfere with each other). Thus the continuous-time walk has exactly the same instantaneous mixing time as the discrete-time one, but with such a beautiful conspiracy of interference that every position has an identical probability.

In fact, the continuous-time walk can be viewed as n noninteracting qubits using the hypercube's structure as a weak product graph [MR01]. However, it is interesting that the constant in front of the mixing time is exactly the same as that for the symmetric discrete-time walk.

Finally, we prove Theorem 4 by considering the Fourier transform of $\bar{P}_T = (1/T) \sum_{t=0}^{T-1} P_t$. For the discrete-time walk, Equation 6 shows that for $k > 0$, the Fourier coefficient of $\bar{P}_T$ consists of a sum of oscillating terms proportional to $\cos \omega_\pm t$. As $T \to \infty$, these oscillations cancel, so we are left with just the constant term $k = 0$ and $\bar{P}_T$ indeed approaches the uniform distribution. For the continuous-time walk, on the other hand, Equation 8 shows that the kth Fourier coefficient of $\bar{P}_T$ approaches the average over t of $\cos^k 2t/n$. In fact, it is equal to this average whenever T is a multiple of $(\pi/2)n$. For k odd this average is zero, but for k even it is

$$\frac{1}{\pi} \int_0^\pi \mathrm{d}x \, \cos^k x = \frac{2^k \pi}{\Gamma\left(\frac{1}{2} - \frac{k}{2}\right)^2 k!}$$

Since these Fourier coefficients do not vanish, $\bar{P}_T$ does not approach the uniform distribution even in the limit $T \to \infty$. In particular, the Fourier coefficient of $\bar{P}_T$ for $k = 2$ is

$$\widetilde{P}_T(2) = \frac{1}{T} \int_0^T \mathrm{d}t \, \cos^2 \frac{2t}{n} = \frac{1}{2} + \frac{\sin 4T/n}{8T/n} \tag{9}$$

This integral is minimized when $T = 1.12335\,n$, at which point $\widetilde{P}_T(2) = 0.39138+$. Since $\widetilde{P}_T(2)$ is bounded below by this, it is easy to show that the total variation distance $\|\bar{P}_T - \Pi\|$ is bounded away from zero as a result. Thus there exists $\epsilon > 0$ such that no ϵ-average mixing time exists.

Acknowledgments. We are grateful to Dorit Aharonov, Julia Kempe, Mark Newman, Tony O'Connor, Leonard Schulman, and Umesh Vazirani for helpful conversations, and to McGill University and the Bellairs Research Institute for hosting a conference at which a significant part of this work was done. This work was supported by NSF grant PHY-0071139 and NSF CAREER award CCR-93065.

References

[AAKV01] Dorit Aharonov, Andris Ambainis, Julia Kempe, and Umesh Vazirani. Quantum walks on graphs. In ACM [ACM01].

[ABN$^+$01] Andris Ambainis, Eric Bach, Ashwin Nayak, Ashvin Vishwanath, and John Watrous. One-dimensional quantum walks. In ACM [ACM01].

[ACM01] *Proc. 33rd Annual ACM Symposium on Theory of Computing* (STOC) 2001.

[AS92] Noga Alon and Joel H. Spencer. *The Probabilistic Method.* Wiley & Sons, 1992.

[BH75] Norman Bleistein and Richard Handelsman. *Asymptotic expansions of integrals.* Holt, Rinehart and Winston, 1975.

[CFG01] Andrew Childs, Edward Farhi, and Sam Gutmann. An example of the difference between quantum and classical random walks. Preprint, `quant-ph/0103020`.

[DFK91] Martin Dyer, Alan Frieze, and Ravi Kannan. A random polynomial-time algorithm for approximating the volume of convex bodies. *Journal of the ACM,* 38(1):1–17, January 1991.

[Dia88] Persi Diaconis. *Group Representations in Probability and Statistics.* Lecture notes–Monograph series. Institute of Mathematical Statistics, 1988.

[DS81] Persi Diaconis and Mehrdad Shahshahani. Generating a random permutation with random transpositions. *Z. Wahrscheinlichkeitstheorie Verw. Gebiete*, 57:159–179, 1981.

[FG98] Edward Farhi and Sam Gutmann. Quantum computation and decision trees. *Phys. Rev. A*, 58:915–, 1998.

[GG81] Ofer Gabber and Zvi Galil. Explicit constructions of linear-sized superconcentrators. *Journal of Computer and System Sciences*, 22(3):407–420, June 1981.

[Gro96] Lov K. Grover. A fast quantum mechanical algorithm for database search. *Proc. 28th Annual ACM Symposium on the Theory of Computing* (STOC) 1996.

[JS89] Mark Jerrum and Alistair Sinclair. Approximating the permanent. *SIAM Journal on Computing*, 18(6):1149–1178, December 1989.

[JSV00] Mark Jerrum, Alistair Sinclair, and Eric Vigoda. A polynomial-time approximation algorithm for the permanent of a matrix with non-negative entries. Technical Report TR00-079, Electronic Colloquium on Computational Complexity, 2000.

[LK99] László Lovász and Ravi Kannan. Faster mixing via average conductance. *Proc. 31st Annual ACM Symposium on Theory of Computing* (STOC) 1999.

[Lub94] Alexander Lubotzky. *Discrete Groups, Expanding Graphs, and Invariant Measures*, volume 125 of *Progress in Mathematics*. Birkhäuser Verlag, 1994.

[MR01] Cristopher Moore and Alexander Russell. Quantum Walks on the Hypercube. Preprint, `quant-ph/0104137`.

[MR95] Rajeev Motwani and Prabhakar Raghavan. *Randomized Algorithms*. Cambridge University Press, 1995.

[Nis90] Noam Nisan. Pseudorandom generators for space-bounded computation. *Proc. 22nd Annual ACM Symposium on Theory of Computing* (STOC) 1990.

[NN93] Joseph Naor and Moni Naor. Small-bias probability spaces: Efficient constructions and applications. *SIAM Journal on Computing*, 22(4):838–856, August 1993.

[NV00] Ashwin Nayak and Ashvin Vishawanath. Quantum walk on the line. Los Alamos preprint archive, `quant-ph/0010117`, 2000.

[Per92] Rene Peralta. On the distribution of quadratic residues and nonresidues modulo a prime number. *Mathematics of Computation*, 58(197):433–440, 1992.

[Sch99] Uwe Schöning. A probabilistic algorithm for k-SAT and constraint satisfaction problems. In *40th Annual Symposium on Foundations of Computer Science*, pages 17–19. IEEE, 1999.

[Sho97] Peter W. Shor. Polynomial-time algorithms for prime factorization and discrete logarithms on a quantum computer. *SIAM Journal on Computing*, 26(5):1484–1509, October 1997.

[Vaz92] Umesh Vazirani. Rapidly mixing markov chains. In Béla Bollobás, editor, *Probabilistic Combinatorics and Its Applications*, volume 44 of *Proceedings of Symposia in Applied Mathematics*. American Mathematical Society, 1992.

[Wat01] John Watrous. Quantum simulations of classical random walks and undirected graph connectivity. *Journal of Computer and System Sciences*, 62:376–391, 2001.

Randomness-Optimal Characterization of Two NP Proof Systems

Alfredo De Santis[1], Giovanni Di Crescenzo[2], and Giuseppe Persiano[1]

[1] Dipartimento di Informatica ed Applicazioni,
Università di Salerno, 84081 Baronissi (SA), Italy
{ads,giuper}@dia.unisa.it
[2] Telcordia Technologies Inc., Morristown, NJ, 07960, USA
(Part of this work done while visiting Università di Salerno.)
giovanni@research.telcordia.com

Abstract. We investigate quantitative aspects of randomness in two types of proof systems for NP: two-round public-coin witness-indistinguishable proof systems and non-interactive zero-knowledge proof systems. Our main results are the following:

- if NP has a 2-round public-coin witness-indistinguishable proof system then it has one using $\Theta(n^\epsilon + \log(1/s))$ random bits,
- if NP has a non-interactive zero-knowledge proof system then it has one using $\Theta(n^\epsilon + \log(1/s))$ random bits,

where s is the soundness error, n the length of the input, and ϵ can be any constant > 0. These results only assume that NP $\neq$ average-BPP. As a consequence, assuming the existence of one-way functions, both classes of proof systems are characterized by the same randomness complexity as BPP algorithms.

In order to achieve these results, we formulate and investigate the problem of randomness-efficient error reduction for two-round public-coin witness-indistinguishable proofs and improve some of our previous results in [13] on randomness-efficient non-interactive zero-knowledge proofs.

1 Introduction

The class of languages NP is often defined in terms of a proof system between a computationally unbounded prover and a polynomial-time bounded verifier. On input string x, a candidate for membership to a language L, the prover computes a witness w of size polynomial in $|x|$. The verifier, given w, can efficiently verify that $x \in L$. Several probabilistic variants of such proof system for NP have been extensively studied in the literature: interactive proofs [24], interactive zero-knowledge proofs [24], interactive witness-indistinguishable proofs [18], non-interactive zero-knowledge proofs [6,5], probabilistically checkable proofs [16], multi-prover interactive proofs [3] etc. The use of random bits either by one or both parties is a crucial ingredient of all these proof systems.

In this paper we concentrate on two types of probabilistic proof systems for NP: non-interactive zero-knowledge (ZK) proof systems (first obtained by

J.D.P. Rolim and S. Vadhan (Eds.): RANDOM 2002, LNCS 2483, pp. 179–193, 2002.
© Springer-Verlag Berlin Heidelberg 2002

[6]), and two-round public-coin witness-indistinguishable (WI) proof systems (first obtained by [15]). In both proof systems, as for the definitional NP proof system, the prover only sends a single message sent by the prover to the verifier. In the first type, the additional ingredient is a public random string which helps in computing and verifying the proof; furthermore, the prover convinces the verifier that $x \in L$ with high probability and the verifier does not obtain any additional information he could not compute before (this is the zero-knowledge property). In the second type, an additional step is a random string sent by the verifier to the prover before the prover's message, which also helps in computing and verifying the proof; furthermore, the prover convinces the verifier that $x \in L$ with high probability and the verifier cannot tell which witness was being used by the prover (this is the witness-indistinguishability property). Since randomness seems essential in order to achieve both the zero-knowledge and the witness-indistinguishability property, the two mentioned proof systems seem to work in a minimal enough setting for extending the NP proof system in order to achieve these properties. Then the question of how much randomness is essential to this purpose comes out naturally.

In this paper we characterize both types of proof systems by showing that any such proof system can be transformed into one of the same type which only uses $\Theta(n^\epsilon + \log(1/s))$ random bits, where n is the size of the instance, s is the soundness error, and ϵ can be any constant > 0. In the case of non-interactive ZK, this performance is optimal (up to a constant) for typical values of s (e.g., $s = 2^{-n}$) unless NP=RP; the same is true in the case of 2-round WI, unless there exist 1-round WI proof systems, which is currently a very interesting open question. Our transformations require the only additional assumption that NP $\neq$ average-BPP and apply to any language in NP. We also note that the randomness complexity of the obtained proof system does not depend on the polynomial describing the size of the verifying predicate for the language.

Previously, non-interactive ZK proof systems for NP-complete languages have been proposed in [5,6,17,27,29,8,12,13] and all had much larger randomness complexity, depending on the polynomial describing the size of the verifying predicate for the language, with the only exception of [13], who presented a proof system using $\Theta(n^\epsilon + \log(1/s))$ random bits, based on a specific intractability assumption (hardness of deciding quadratic residuosity). We improve the latter proof system on two counts: by reducing the assumption to just NP $\neq$ average-BPP, and by allowing our proof system to enjoy the same randomness complexity even when satisfying adaptive soundness (which was not achieved by [13]).

The only previous 2-round WI proof systems have been proposed in [15], based on non-interactive ZK proof systems or approximate verifiable pseudo-random generators. By combining it with results from [13], their best proposed proof system can be shown to achieve randomness complexity $\Theta(n^\epsilon \cdot \log(1/s))$, assuming the hardness of deciding quadratic residuosity. We improve this proof system by reducing both the randomness and the strength of the complexity assumption.

Our main tools are randomness-efficient error reduction techniques, previously studied for applications to BPP algorithms (in particular, we use [9,1]), to Arthur-Merlin proofs [2], and to non-interactive ZK proofs [13]. In particular, we study the problem of randomness-efficient error reduction for non-interactive WI proof systems and 2-round WI proof systems. We also present a new randomness-efficient technique for achieving adaptive soundness in non-interactive zero-knowledge proof systems.

Several formal definitions and proofs are omitted for lack of space.

2 Definitions

We recall notions and definitions useful for the rest of the paper, including formal definitions for combinatorial hitting set generators, two-round public-coin witness-indistinguishable proof systems, and non-interactive zero-knowledge proof systems.

Randomness-efficient error reduction. The problem of error reduction for BPP algorithms consists of reducing the probability that the algorithm errs in deciding whether an input x is in the given language L or not, from a constant (e.g., $2/3$) to an arbitrarily small value (e.g., 2^{-k} for any k). A simple way to achieve this is to perform repeated executions of the original algorithm using uniformly and independently distributed random bits each time and then taking the majority of the outcomes. The problem of randomness-efficient error reduction for BPP algorithms consists of reducing the error by saving on the amount of random bits used with respect to the mentioned simple approach. Several interesting randomness-efficient error reduction techniques have been given in the literature [28,35,9,33,1,10,26,20,37], culminating in the unconditional result that any BPP algorithm can be modified into one that, on input an n-bit instance, can achieve error δ, using only $\Theta(m + \log(1/\delta))$ random bit (here, m is the number of random bits used by the original BPP algorithm obtaining constant error and can be reduced to n^ϵ, for any $\epsilon > 0$, by assuming the existence of pseudo-random generators or one-way functions). Most of the mentioned techniques are based on some generators which we abstract under the name of "combinatorial hitting set generators" (for lack of a better name), and that we formally define below.

Definition 1. Let a, b, k be positive integers and $\delta \in \{0, 1\}$. We say that a function $G : \{0, 1\}^a \rightarrow \{0, 1\}^{kb}$ is an (a, b, k, δ)-combinatorial hitting set generator if and only if for any set $S \subseteq \{0, 1\}^b$, $|S| \geq 2^{b-1}$, it holds that

$$\text{Prob}[\, s \leftarrow \{0, 1\}^a; (y_1, \ldots, y_k) \leftarrow G(s) \, : \, y_i \notin S \text{ for } i = 1, \ldots, k\,] \leq \delta.$$

where $|y_i| = b$, for $i = 1, \ldots, k$.

We will use the following two combinatorial hitting set generators.

The first, given in [9], uses pairwise independent functions, and achieves parameters $(2b, b, \Theta(1/\delta), \delta)$, for any positive integer b and any $\delta > 0$.

The second, given in [1] and further studied in [10,26,20], uses random walks on explicitly constructed expanders, and achieves parameters $(b + \Theta(\log(1/\delta)), b, \Theta(\log(1/\delta)), \delta)$, for any positive integer b and any $\delta > 0$. We note that explicitly constructed expanders are available from the literature (see, e.g., [19,30,34]).

NP, Interactive and non-interactive proof systems. The class NP is often defined in terms of a proof system: on input an instance x to a language L, a prover computes a witness w certifying that $x \in L$ and sends w to a verifier who can verify in polynomial time that $x \in L$.

Interactive proof systems were originally introduced in the seminal paper [24]. In an interactive proof system the prover and the verifier are modeled as interactive and probabilistic Turing machines, the latter running in polynomial time. An execution of the proof system consists of polynomially many rounds of interaction, during which the prover tries to convince the verifier that their common input x belongs to L; at the end of the interaction the verifier decides whether to accept or reject. An interactive proof system satisfies two requirements: completeness and soundness. Completeness states that if x belongs to L, at the end of the interaction with the prover, the verifier accepts with high probability. Soundness states that for any x not in L, the probability that any prover, can interact with the verifier and make the verifier accept, is very small. An interactive proof system is called *public-coin* if the verifier's messages only consist of random bits. A *2-round* interactive proof system consists of a single message sent by the verifier, followed by a single message sent by the prover.

Non-interactive proof systems were originally introduced and studied by [6, 5], in the so-called "public random string model". In this model all parties have access to a short public random string, and the proof consists of a single message sent by the prover to the verifier. A non-interactive proof system for a certain language L is a pair of algorithms, a prover and a verifier, the latter running in polynomial time, such that the prover, on input string x and the public random string, can compute a proof that convinces the verifier that the statement '$x \in L$' is true. Such proof systems satisfy two requirements: completeness and soundness. Completeness states that if the input x is in L, with high probability, the proof computed by the prover makes the verifier accept. Soundness states that for any x not in L, the probability that any prover, given the reference string, can compute a proof that makes the verifier accept, is very small.

Zero-Knowledge and Witness-Indistinguishability. In both interactive and non-interactive proof systems, at the end of the interaction, the verifier might be able to compute additional information other than the fact that the theorem is true. The notion of zero-knowledge [24] formalizes the property for a proof system of not revealing any additional information to a verifier other than the only fact that the theorem is true. The notion of witness-indistinguishability [18] formalizes the property for a proof system for a language in NP of not revealing which witness the prover is using while convincing the verifier that the theorem is true. We will use in the sequel parameterized versions of both notions, as follows. A non-interactive ZK proof system with parameters $(1, s, r(s, n))$ is a non-interactive ZK proof system where the completeness error is 0, the soundness error is s, and the length of the reference string is $r(s, n)$, where n is the length of the common input. Similarly, a 2-round public-coin WI proof system with parameters $(1, s, r(s, n))$ is a 2-round public-coin WI proof system where the completeness error is 0, the soundness error is s, and the length of the random string sent by the verifier is $r(s, n)$, where n is the length of the common input. We also define zaps [15], which are 2-round public-coin WI proof systems with special soundness and witness-indistinguishable properties (specifically, the soundness is adaptive in the sense that a cheating prover is allowed to choose a false theorem after seeing the reference string, and the witness-indistinguishability holds for any string sent by the possibly cheating verifier) .

Pseudo-Random Generators. Pseudo-random generators were introduced in [7]. Since then, they have been very often applied for minimizing randomness in algorithms and as atomic tools in proof systems and cryptographic protocols. Informally, a pseudo-random generator is a deterministic function that, given as input a short seed of random bits, returns an output polynomially longer than the seed, and such that the output is polynomially indistinguishable from a truly random string of the same length. In the sequel, we will denote as (a, b)-*pseudo-random generator* a pseudo-random generator with domain $\{0, 1\}^a$ and codomain $\{0, 1\}^b$, where $a < b$ and $b = \text{poly}(a)$.

Commitment Schemes. A commitment scheme [4] is a very popular cryptographic protocol, having several applications in the construction of proof systems and of more elaborated cryptographic protocols. We will use commitment schemes that are non-interactive and are executed in the public random string model. Informally, a non-interactive commitment scheme (C,R) in the public-random-string model is a two-phase protocol between two probabilistic polynomial time parties C and R, called the committer and the receiver, respectively, such that the following is true. In the first phase (the commitment phase), given the public random string σ, C commits to bit b by computing a pair of keys (com, dec) and sending com (the commitment key) to R. Given just σ and the commitment key, the polynomial-time receiver R cannot guess the bit with probability significantly better than $1/2$ (this is the hiding property). In the second phase (the decommitment phase) C reveals the bit b and the key dec (the decommitment key) to R. Now R checks whether the decommitment key is valid; if not, R outputs a special string $\bot$, meaning that he rejects the decommitment from C; otherwise, R can efficiently compute the bit b revealed by C and is convinced that b was indeed chosen by C in the first phase (this is the binding property).

The scheme in [31] can be made non-interactive in the public random string model. C and D use an $(r, 3r)$-pseudo-random generator G, for some security parameter r, and a public random string R of length $3r$. In order to commit to a bit b, computes $com = G(seed)$ if $b = 0$ or $com = G(seed) \oplus R$ if $b = 1$, and $dec = seed$, where $seed \in \{0, 1\}^r$ is uniformly chosen. Note that if $|R| = |com| = 4r$, then this scheme can use the same R to compute commitments to polynomially many bits.

3   Randomness-Efficient Error Reduction Preserving WI, ZK

In this section we discuss the conditions under which randomness-efficient error reduction techniques can preserve the properties of witness-indistinguishability and zero-knowledge, both for non-interactive and for 2-round proof systems.

With respect to witness-indistinguishability, we obtain that all combinatorial hitting set generators satisfying a certain condition preserve this property for non-interactive proof systems and they can be modified so that they preserve this property for 2-round proof systems. The condition, which we formally define later, is satisfied by many generators in the literature.

With respect to zero-knowledge, we know from [13] that no combinatorial hitting set generator in the literature is known to preserve this property but all of them can be combined with known techniques based on the existence of one-way functions so that they preserve this property for non-interactive proof systems. Here we combine the result in [13] with the technique in [9] to obtain low-randomness constant-error proof systems for proving a (fixed) polynomial number of theorems. These will be useful for later constructions of low-randomness non-interactive ZK proof systems for NP.

We define a property of combinatorial hitting set generators in Section 3.1, and then study randomness-efficient error reduction techniques for non-interactive WI proof systems in Section 3.2, 2-round WI proof systems in Section 3.3, and non-interactive ZK proof systems in Section 3.4.

3.1   A Property of Combinatorial Hitting Set Generators

Informally, the property we require in the following definition is that, given a random value z, it is possible to efficiently reconstruct a k-tuple that is a valid output of the combinatorial hitting set generator, and that contains z as any component of the k-tuple output.

Definition 2. An (a, b, k, δ)-combinatorial hitting set generator G, is *reconstructible* if, there exists an efficient probabilistic algorithm R such that, for any $i = 1, \ldots, k$, the two distributions D_0^i, D_1 are equal, where

$$D_0^i = \{z \leftarrow \{0,1\}^b; (s', y_1, \ldots, y_k) \leftarrow R(1^k, z, i) : (s'; y_1, \ldots, y_{i-1}, z, y_{i+1}, \ldots, y_k)\},$$
$$D_1 = \{s \leftarrow \{0,1\}^a; (y_1, \ldots, y_k) \leftarrow G(s) : (s; y_1, \ldots, y_k)\}.$$

We can show the following

Lemma 1. The combinatorial hitting set generators in [9,1] are reconstructible.

Proof. The generator G in [9] is based on pairwise independent functions over field $\mathrm{GF}[2^b]$, for some positive integer b, and is a combinatorial hitting set generator with parameters $(2b, b, \Theta(1/\delta), \delta)$, for any positive integer b and any $\delta > 0$. Specifically, G is defined as is $G(1^k, c, d) = (y_1, \ldots, y_k)$, where $y_i = c \cdot i + d$ for all $i = 1, \ldots, k$, $k = \Theta(1/\delta)$, c, d are random field elements and the arithmetic is over GF$[2^b]$. In order to show that G is a reconstructible generator, we define algorithm R as follows. On input $1^k, z, i$, R randomly chooses c, d such that $c \cdot i + d = z$ and then sets $y_j = c \cdot j + d$, for $j = 1, \ldots, k$. It is easy to show that the distribution of the output of R is the same as the distribution of the input-output pair of G. The generator G in [1] is based on random walks on explicitly constructible expanders. Specifically, G uniformly chooses a starting node and then returns all nodes (including the starting one) visited while performing a k-step random walk on the expander. In order to show that G is a reconstructible generator, we define algorithm R as follows. Algorithm R uniformly chooses a starting node z, sets $y_i = z$, and returns all nodes visited while performing an $(i - 1)$-step backward random walk starting from z and a $(k - i)$-step forward random walk starting from z. Now we need to show that the output of R and the input-output pair of G are equally distributed. This can be proved by using the fact that an expander is a regular graph, and the fact that the distribution of a node obtained by performing a random step backward or forward from a uniformly chosen node is again uniform.   □

3.2   Error Reduction for Non-interactive WI Proofs

We show that reconstructible combinatorial hitting set generators are sufficient to give a randomness-efficient error-reduction for non-interactive WI proof systems. In particular, we obtain the following

Theorem 1. *If there exists a non-interactive WI proof system for a language L with parameters $(1, 1/2, r)$, then there exist two non-interactive WI proof systems for L with the following parameters:*
1. *$(1, 1/k, 2r)$, for any positive integer k;*
2. *$(1, \delta, r + \Theta(\log(1/\delta)))$, for any $\delta > 0$.*

Proof. Let $CHSG : \{0,1\}^a \to \{0,1\}^{kb}$ be a combinatorial hitting set generator with parameters (a, b, k, δ) (to be set later). Also, denote by (A,B) the assumed non-interactive WI proof system for L with parameters $(1, 1/2, r)$, and consider the following non-interactive WI proof system (P,V) for L with parameters $(1, \delta, r')$ (to be set later).

The Proof System (P,V):
1. P and V share common input x and a reference string σ of r' bit
2. P has as additional input a witness w for x
3. P and V compute $(y_1, \ldots, y_k) = CHSG(\sigma)$ for some k (to be set later)
4. For $i = 1, \ldots, k$, P runs algorithm A on input x and witness w and using y_i as a reference string thus obtaining proof π_i
5. P sends $(\pi_1, \ldots, \pi_k)$ to V
6. V uses algorithm B to verify that proofs $\pi_1, \ldots, \pi_k$ are accepting on reference strings $y_1, \ldots, y_k$, respectively; if so, P accepts; otherwise P rejects.

We can show that (P,V) is a non-interactive WI proof system for L with parameters $(1, 1/k, 2r)$ for any k, if $CHSG$ is the generator from [9], or with parameters $(1, \delta, r + \Theta(\log(1/\delta)))$, for any $\delta > 0$ if $CHSG$ is the generator from [1].   □

3.3   Error Reduction for 2-Round WI Proofs

If in the protocols given by Theorem 1, the random string is sent by the verifier to the prover (rather than being publicly available), we obtain analogue claims on randomness-efficient error reduction for zaps.

By combining combinatorial hitting set generators with a technique in [15], we can prove a stronger theorem, by assuming that the original proof system is only a non-interactive WI proof system. We obtain the following

Theorem 2. If there exists a non-interactive WI proof system for language L with parameters $(1, 1/2, r)$, then there exist a zap for L with parameters $(1, \delta, \Theta(r + \log(1/\delta)))$, for any $\delta > 0$.

Proof. We use a non-interactive WI proof system for L (A,B) with parameters $(1, 1/2, r)$, and the combinatorial hitting set generator from [1], denoted as $CHSG : \{0, 1\}^a \to \{0, 1\}^{kb}$, with parameters slightly different than before; namely, $(\Theta(r + \log(1/\delta)), r, \Theta(r + \log(1/\delta)), \delta \cdot \Theta(2^{-r}))$, for any $\delta > 0$ (we note that we will not use the fact that this generator is reconstructible). We define the following 2-round public-coin WI proof system (P,V) for L with parameters $(1, \delta, r')$ (to be set later).

The Proof System (P,V):
1. Let x be the input such that P wants to convince V that '$x \in L$' and let w be P's witness for x
2. V sends to P a random string σ of length $r' = \Theta(r + \log(1/\delta))$
3. P and V compute $(y_1, \ldots, y_k) = CHSG(\sigma)$ for $k = \Theta(r + \log(1/\delta))$
4. P uniformly chooses a string $v \in \{0, 1\}^r$
5. For $i = 1, \ldots, k$, P runs algorithm A on input x and witness w and using $v \oplus y_i$ as B's message, thus obtaining proof π_i
6. P sends $(\pi_1, \ldots, \pi_k)$ to V
7. V uses algorithm B to verify that proofs $\pi_1, \ldots, \pi_k$ are accepting on reference strings $y_1, \ldots, y_k$, respectively; if so, P accepts; otherwise P rejects.

We can show that (P,V) is a zap for L with parameters $(1, \delta, \Theta(r + \log(1/\delta)))$, for any $\delta > 0$.   □

3.4   Error Reduction for Non-interactive ZK Proofs

We first recall a theorem from [13] and then present one corollary for proving a (fixed) polynomial number of theorems with a constant-error proof system, that will be used later in our constructions.

Theorem 3. [13] Let s be a function, let $\epsilon \in \{0, 1\}$, and let L be an NP-complete language and assume the existence of one-way functions. If L has a non-interactive ZK proof system (A,B) with parameters $(1, 1/2, r(1/2, n))$ and A runs in time polynomial in n given a polynomial-size witness, then L has a non-interactive ZK proof system (P,V) with parameters $(1, s, r(1/2, p(n)) + \Theta(\log(1/s)))$, for some polynomial p.

The above theorem was obtained in [13] by using a $(n^\epsilon + \Theta(\log(1/s)), n^\epsilon, \Theta(\log(1/s)), s)$-combinatorial hitting set generator. By replacing this generator with one with parameters $(2n^\epsilon, n^\epsilon, \Theta(1/s), s)$ (such as the one in [9]), we obtain the following

Corollary 1. Let k be a polynomial, let L be an NP-complete language and assume the existence of one-way functions. If L has a non-interactive ZK proof system with parameters $(1, 1/2, r(1/2, n))$ and A runs in time polynomial in n given a polynomial-size witness, then L has a non-interactive ZK proof system with parameters $(1, 1/2, 2r(1/2, p(n)))$, for proving $k(n)$ theorems, for some polynomial p, where n is the size of the theorems proved.

4   Low Randomness Proof Systems for NP

In this section we present our main constructions of two low-randomness proof systems for any language in NP. We present a preliminary protocol in Section 4.2, that is 'almost' a non-interactive ZK proof system with constant error soundness (it satisfies completeness and zero-knowledge but only a weak version of soundness). Then, by reducing the soundness error of this construction with the randomness-efficient error-reduction techniques in Section 3, we obtain our two main results. We also discuss the optimality of the number of random bits used by both proof systems.

4.1   A Preliminary Protocol

As a preliminary protocol to be later combined with the randomness-efficient techniques in Section 3, we could design a non-interactive ZK proof system for an NP-complete language with parameters $(1, 1/2, \Theta(n^\epsilon))$, for any $\epsilon > 0$, and only assuming the existence of a one-way function, and then directly apply these techniques. Instead, we design a preliminary protocol which only uses $\Theta(n^\epsilon)$ randomness, satisfies completeness and zero-knowledge, but only satisfies a weak version of soundness (roughly speaking, soundness holds only with respect to provers who choose certain commitments honestly). The application of the techniques in Section 3 will take care of a randomness-efficient reduction of the soundness error with respect to any prover. With this different approach we obtain better randomness complexity (improved by a large constant), better communication complexity (improved by a linear factor) and greater simplicity of the overall construction.

We now present a transformation from an arbitrary non-interactive WI proof system for an NP-complete language L into a zero-knowledge protocol using very small randomness. Fix a constant $\epsilon > 0$; then our transformation uses the following tools:

- a non-interactive WI proof system (A,B) for L with parameters $(1, s, r(s, n))$;
- a combinatorial hitting set generator $CHSG$ with parameters $(2n^\epsilon, n^\epsilon, \Theta(1/s), s)$ (this can be obtained by using the generator in [9]);

- an $(n^{\epsilon_0}, 4n^{\epsilon_0})$-pseudo-random generator PRG, where $\epsilon_0 > 0$ can be set to an arbitrary constant. (This can be based on any one-way function using the result in [25].)
- a non-interactive commitment scheme (C,R) using n^{ϵ_0} random bits to commit to any polynomial number of bits. (This can be based on any pseudo-random generator using the construction in [31], reviewed in Section 2, and therefore on any one-way function using the result of [25].)
- a standard polynomial time reduction from NP-complete languages ([21]).

An informal description. In order to avoid using randomness proportional to the polynomial describing the size of the predicate which on input instance x, and witness w, verifies that $x \in L$, we reduce proving the original statement to proving polynomially many smaller statements whose size only depends on the security parameter (which can be set equal to n^{ϵ}, for any $\epsilon > 0$). Assume for a moment that we can do this reduction; then, proving each of these statements can be done very efficiently, but we are left with the problem of proving *all* of these statements efficiently. This problem is solved by using the hitting set generator $CHSG$ and its returned strings as reference strings for proving all these statements with constant soundness error (note that the zero-knowledge property is preserved because we use the techniques underlying Theorem 3 and Corollary 1). Therefore, it only remains to describe how proving the statement $x \in L$ is efficiently reduced to proving polynomially many smaller statements. First, prover and verifier reduce the statement $(x \in L) \lor$ (a portion of σ is pseudorandom) to a 3SAT instance ϕ using known witness-preserving polynomial-time reductions (here, we use the techniques of [14,17] which also plays an important role in the application of Theorem 3 and Corollary 1). Then the prover computes a commitment to the truth values of all literals in ϕ (this step requires only n^{ϵ} random bits to commit to polynomially many bits if we use, for instance, the scheme in [31]). Finally the prover proves that the committed bits are consistent (namely, all pairs of commitments to $(l_i, \neg l_i)$ hide bits $(b_i, 1 - b_i)$, for some $b_i \in \{0,1\}$), and that the committed bits satisfy the formula (namely, each triple of commitments to literals in each clause hides at least one bit equal to 1). Proving these statements is reduced to proving equivalent statements for the NP-complete language L and using the original proof system (A,B). Note that now the size of each of these statements only depends on the security parameter of the commitment scheme and not on the polynomial describing the size of the verifying predicate for L.

Formal description. We denote by (P,V) the proof system for L obtained using the following transformation.

Input to P: witness w for instance x;

The proof system (P,V): On input an n-bit instance x and a reference string $\sigma = \sigma_1 \circ \sigma_2 \circ \sigma_3$, where $|\sigma_1| = 2n^{\epsilon}$, $|\sigma_2| = 3n^{\epsilon}$, and $|\sigma_3| = 2r(1/2, p(n))$, for some p specified later, P and V do the following.

    P and V: Let $T_{\sigma_1} = $ '$\exists r_1 \in \{0,1\}^{n^{\epsilon}}$ such that $\sigma_1 = PRG(r_1)$'; reduce statement '$(x \in L) \lor T_{\sigma_1}$' to statement '$\phi \in 3SAT$'; let p_1 be such that $|\phi| \le p_1(n)$ and let t be the assignment satisfying ϕ thus obtained.

P and V: Let $(\sigma_{31}, \ldots, \sigma_{3k}) = CHSG(\sigma_3)$, for $k \leq p_1(n)^3 + p_1(n)$;

P.1: Compute commitment/decommitment keys (com_l, dec_l) for bit $t(l)$, for each literal l in ϕ, using scheme (C,R) on a single reference string σ_2. Send com_l to V, for each literal l in ϕ.

P.2: For each variable x_i in ϕ, let $VT_i =$ '$\exists \ dec_{x_i}, dec_{1-x_i}$ s.t. $R(\sigma_2, com_{x_i}, dec_{x_i}) = 1, R(\sigma_2, com_{x_i}, dec_{x_i}) = 1, dec_{x_i} \neq dec_{1-x_i}$', and reduce statement VT_i to '$vz_i \in L$'.

P.3: For each clause $c_j = (l_1 \vee l_2 \vee l_3)$ in ϕ, let $CT_j =$ '$\exists \ dec_{l_1}, dec_{l_2}, dec_{l_3}$ s.t. $R(\sigma_2, com_{l_1}, dec_{l_1}) = 1, R(\sigma_2, com_{l_2}, dec_{l_2}) = 1, R(\sigma_2, com_{l_3}, dec_{l_3}) = 1$, and $(dec_{l_1} = 1) \vee (dec_{l_2} = 1) \vee (dec_{l_3} = 1)$'; reduce statement '$T_{\sigma_1} \vee CT_j$' to statement '$cz_j \in L$'; let $p_2(n)$ be such that $|vz_i| \leq p_2(n)$ and $|cz_j| \leq p_2(n)$ for all i, j.

P and V: In the following steps use proof system (A,B) with parameters $(1, 1/2, r(1/2, p(n)))$, where $p = p_2$ is a polynomial such that $|vz_i| \leq p(n)$ and $|cz_j| \leq p(n)$ for all i, j.

P.4: For all statements vz_i, compute proofs $\pi(vz)_i$ using algorithm A and σ_{3i} as a reference string and send $\pi(vz)_i$ to V.

P.5: For all statements cz_j, compute proofs $\pi(cz)_j$ using algorithm A and σ_{3j} as a reference string; send $\pi(cz)_j$ to V.

V.1: Given commitment keys com_l for each literal l in ϕ, perform the same reductions done by P in steps P.2 and P.3 above, thus obtaining statements $vz_i \in L$ and $cz_j \in L$.

V.2: Verify that proofs $\pi(vz)_i, \pi(cz)_j$ are accepting by using algorithm B and σ_{3i}, σ_{3j} as reference strings, respectively, for $i = 1, \ldots, p_1(n)$ and $j = p_1(n) + 1, \ldots, p_1(n) + p_1(n)^3$. If all verifications are satisfied then accept else reject.

First of all we consider the randomness complexity of (P,V). By setting polynomial p as done in (P,V), and ϵ_0 such that $r(1/2, p(n)) = \Theta(n^\epsilon)$ (note that p depends on ϵ_0), we see that the length of the reference string used by (P,V) is $\Theta(n^\epsilon)$. We now note that the proof system (P,V) satisfies the requirements of completeness, (weak) soundness and zero-knowledge. The completeness follows immediately from the completeness of (A,B). For the (weak) soundness, assume that $x \notin L$ and consider the class of provers that compute the commitments com_l honestly. Then we can apply Corollary 1 with $k(n) = p_1(n)^3 + p_1(n)$ and obtain that the error probability of (P,V) is at most $1/2$. For the zero-knowledge, the proof requires no further technical complication than the proof in [13]; in particular, note that since we are using the technique of [17], a simulator can generate a pseudo-random σ_1 and have an assignment for formula ϕ, and then carry out the rest of the simulation by actually running the prover's algorithm.

4.2   Randomness Optimal Non-interactive ZK for NP

By properly composing the protocol in Section 4.1, with the randomness-efficient error reduction technique in Theorem 1, we obtain the following

Theorem 4. Let L be a language in NP. If there exists a non-interactive ZK proof system for L with parameters $(1, s, r(s, n))$ and there exist one-way functions then for any $\epsilon > 0$ there exists a non-interactive ZK proof system for L with parameters $(1, s, \Theta(n^\epsilon + \log(1/s)))$.

Proof. It is enough to prove the theorem for an NP-complete language L', since, after a standard polynomial-time reduction, the size of an instance to an arbitrary NP language and the size of an instance to the NP-complete language are polynomially related, and therefore we can obtain analogue randomness complexity (since the dependency of the randomness complexity as a function of the instance size n is $\Theta(n^\epsilon)$ for any $\epsilon > 0$).

The composition. We consider protocol (P,V) in Section 4.2 for L' and apply the error reduction procedure of Theorem 1 to the proofs computed in steps P.4 and P.5, by using the reconstructible combinatorial hitting set generator of [1] with parameters $(\Theta(n^\epsilon + \log(1/\delta)), \Theta(n^\epsilon), \Theta(n^\epsilon + \log(1/\delta)), \delta)$. Note that we have increased the number of strings output by this generator by a factor $\Theta(n^\epsilon)$.

Properties. We are adding a random string of length $\Theta(n^\epsilon + \log(1/\delta))$ to the length of the random string used by (P,V), and therefore the total randomness used is $\Theta(n^\epsilon + \log(1/\delta))$. Completeness follows from the completeness of (P,V). Zero-knowledge follows from the zero-knowledge of (P,V), and the result in Theorem 1 saying that witness-indistinguishability is preserved by the reconstructible generator in [1]. Now we see that the soundness error is δ, with respect to any prover. First of all we note that for each of the $p_1(n)^3 + p_1(n)$ proofs computed by P, a prover has at most $2^{9n^{\epsilon_0}}$ choices of the statement vt_i or ct_j, since each of them contains at most 3 commitments, each of length at most $n^{3\epsilon_0}$ random bits. On the other hand, we are using a generator which decreases the error from $1/2$ to $2^{-\Theta(n^\epsilon + \log(1/\delta))}$; therefore, by proper choices of constants, the overall soundness error is at most $(p_1(n)^3 + p_1(n)) \cdot 2^{9n^{\epsilon_0}} \cdot 2^{-\Theta(n^\epsilon + \log(1/\delta))} \leq \delta$.

On the assumptions of Theorem 4. We note that our transformation is based on the assumption of the existence of any one-way function. However, we can apply results from [32] to further weaken this assumption. Specifically, [32] prove that any zero-knowledge proof system for a language not in average-BPP implies the existence of a one-way function (their proof is, in fact, particularly simplified in the case of non-interactive zero-knowledge proof systems). Then we can strengthen Theorem 4 by replacing the assumption of existence of one-way function with the assumption that NP $\neq$ average-BPP. Alternatively, by using another result from [32], we can replace the assumption of the existence of a one-way function with that of the existence of a language in NP that is hard (i.e., not in BPP) for all sufficiently large instances.

Optimality. We note that a result of [22] can be rephrased by saying that only RP languages have non-interactive ZK proof systems having soundness error $s = 0$. If $s > 0$, it is easy to prove a lower bound of $\log(1/s)$ on the number of random bits of the reference string of any such proof system (or otherwise the soundess property can be contradicted). Therefore, if $s \leq 2^{-n^\epsilon}$ (which includes the typical case $s = 2^{-n}$), the number of random bits used by our proof system is optimal (up to a constant) unless NP is in RP.

4.3   Randomness-Optimal 2-Round WI for NP

Similarly as for non-interactive ZK, we properly compose the protocol in Section 4.1 with a randomness-efficient error reduction technique (for a stronger result, only assuming the existence of a non-interactive ZK proof system, we can combine it with the technique from Theorem 2). We obtain the following

Theorem 5. Let L be a language in NP. If there exists a zap for L with parameters $(1, s, r(s, n))$ then for any $\epsilon > 0$ there exists a zap for L with parameters $(1, s, \Theta(n^\epsilon + \log(1/s)))$.

The proof of this theorem proceeds analogously to that of Theorem 4.

Optimality. If the soundness error s of a zap is > 0, then it is easy to prove a lower bound of $\log(1/s)$ on the number of random bits sent by the verifier in any such proof system (or otherwise the soundess property can be contradicted). Therefore, whenever $s \leq 2^{-n^\epsilon}$ (which includes the typical case $s = 2^{-n}$), the number of random bits used by our proof system is optimal (up to a constant) among 2-round proof systems unless NP is in RP. We note however that given the current state of the art, it is unknown whether 1-round WI proof systems with soundness error $s = 0$ exist. We consider settling this question an interesting open problem.

5   Efficient Adaptive Soundness and Other Applications

A technique used in our results can also be used to efficiently achieve adaptive soundness. Let us first recall this notion (introduced and called 'strong soundness' in [5]). The property of adaptive soundness in proof systems in the public random string model consists of requiring that the proof system is sound even against cheating prover who choose false statements based on the reference string. More formally, an *adaptively-sound* non-interactive ZK proof system in the public random string model is defined as a non-interactive ZK proof system, where the soundness requirement is replaced by the following:

2': *Adaptive Soundness.* For all algorithms P', it holds that

$$\text{Prob}\left[\sigma \leftarrow \{0,1\}^r; (x, Proof) \leftarrow P' : x \notin L \text{ and } V(\sigma, x, Proof) = 1\right] \leq s.$$

We note that the only known way to achieve adaptive soundness is to start with a proof system (A,B) satisfying ordinary soundness with parameters $(1, 1/2, r)$, and obtain a proof system (C,D) by running $n + 1$ independent executions of the original proof system, where n is the size of the input. The soundness error is then 2^{-n-1} for a fixed choice of the input, and $2^n \cdot 2^{-n-1} = 1/2$ for any possible choice. The proof system of (C,D) has then parameters $(1, 1/2, nr)$.

A better technique would be to apply Theorem 3 to proof system (A,B). As a consequence, the resulting protocol would have parameters $(1, 1/2, \Theta(n + r))$.

We abstract our technique used for the proof system (P,V) in Section 4.1. First, we reduce an n-bit statement to polynomially-many n^ϵ-bit statements and then we prove all these statements on reference strings obtained using the hitting set generators of [9] and [1]. Using this technique, the resulting protocol would have parameters $(1, 1/2, \Theta(n^{\epsilon} + r))$.

Other than adaptive soundness, additional requirements to the definition of non-interactive ZK proof systems have been presented in the literature, motivated by cryptographic scenarios, such as *adaptive zero-knowledge* [17], *non-malleability* [36], and *robustness* [11]. We note that our techniques immediately extend to satisfy all of these requirements with only constant overhead in the randomness complexity, by combining them with the techniques in [17,36,11].

We also note that we only focused on reducing the public randomness for non-interactive ZK proof systems or the randomness of the verifier for 2-round WI proof systems, ignoring the randomness of the prover, since the latter can always be reduced by the prover using pseudo-random bits instead of random ones.

References

1. M. Ajtai, J. Komlos, and E. Szemeredi, *Deterministic Simulation in Logspace*, Proc. of STOC 87.
2. M. Bellare, O. Goldreich, and S. Goldwasser, *Randomness in Interactive Proof Systems*, in Proc. of FOCS 90, pp. 563–572.
3. M. Ben-Or, S. Goldwasser, J. Kilian and A. Wigderson, *Multi-Prover Interactive Proofs: How to Remove Intractability Assumptions*, Proc. of STOC 88.
4. M. Blum, *Coin Flipping by Telephone*, Proc. IEEE Spring COMPCOM (1982), 133–137.
5. M. Blum, A. De Santis, S. Micali, and G. Persiano, *Non-Interactive Zero-Knowledge*, SIAM Jou. on Computing, vol. 20, no. 6, Dec 1991, pp. 1084–1118.
6. M. Blum, P. Feldman, and S. Micali, *Non-Interactive Zero-Knowledge and Applications*, Proc. of STOC 88.
7. M. Blum and S. Micali, *How to Generate Cryptographically Strong Sequence of Pseudo-Random Bits*, SIAM J. on Computing, vol. 13, no. 4, 1984, pp. 850–864.
8. J. Boyar and R. Peralta, *Short Discreet Proofs*, Proc. of EUROCRYPT 96.
9. B. Chor and O. Goldreich, *On the Power of Two-Point Based Sampling*, Journal of Complexity, vol. 5, pp. 96–106, 1989.
10. A. Cohen and A. Wigderson, *Dispersers, Deterministic Amplification and Weak Random Sources*, Proc. of FOCS 89.
11. A. De Santis, G. Di Crescenzo, R. Ostrovsky, G. Persiano, and A. Sahai, *Robust Non-Interactive Zero Knowledge*, in Proc. of CRYPTO 2001.
12. A. De Santis, G. Di Crescenzo, and G. Persiano, *Randomness-Efficient Non-Interactive Zero-Knowledge*, Proc. of ICALP 97.
13. A. De Santis, G. Di Crescenzo, and G. Persiano, *Non-Interactive Zero-Knowledge: A Low-Randomness Characterization of NP*, Proc. of ICALP 99.
14. A. De Santis and M. Yung, *Cryptographic applications of the meta-proof and the many-prover systems*, Proc. of CRYPTO 90.
15. C. Dwork and M. Naor, *Zaps and Their Applications*, Proc. of FOCS 2000.
16. U. Feige, S. Goldwasser, L. Lovasz, S. Safra and M. Szegedy, *Approximating Clique is Almost NP-complete*, Proc. of FOCS 91.

17. U. Feige, D. Lapidot, and A. Shamir, *Multiple Non-Interactive Zero-Knowledge Proofs Under General Assumptions*, SIAM Jou. on Computing, 29(1), 1999, p. 1–28.
18. U. Feige and A. Shamir, *Witness-Indistinguishable and Witness-Hiding Protocols*, Proc. of STOC 90.
19. O. Gabber and Z. Galil, *Explicit Constructions of Linear Sized Superconcentrators*, Journal of Computer and System Sciences, vol. 22, pp. 407–420, 1981.
20. Gillman, *A Chernoff Bound for Random Walks on Expanders*, Proc. of STOC 93.
21. M. Garey e D. Johnson, *Computers and Intractability: a Guide to the Theory of NP-Completeness,* W. H. Freeman & Co., New York, 1979.
22. O. Goldreich and Y. Oren, *Definitions and Properties of Zero-Knowledge Proof Systems*, Journal of Cryptology, vol. 7, 1994, pp. 1–32.
23. S. Goldwasser, and S. Micali, *Probabilistic Encryption*, in Journal of Computer and System Sciences, vol. 28, n. 2, 1984, pp. 270–299.
24. S. Goldwasser, S. Micali, and C. Rackoff, *The Knowledge Complexity of Interactive Proof-Systems*, SIAM J. on Computing, vol. 18, n. 1, 1989.
25. J. Hastad, R. Impagliazzo, L. Levin and M. Luby, *Construction of A Pseudo-Random Generator from Any One-Way Function*, SIAM Jou. on Computing, vol. 28, n. 4, pp. 1364–1396.
26. R. Impagliazzo and D. Zuckerman, *How to Recycle Random Bits*, Proc. of FOCS 89.
27. J. Kilian, *On the complexity of bounded-interaction and non-interactive zero-knowledge proofs*, Proc. of FOCS 94.
28. R. Karp, N. Pippenger, and M. Sipser, *Expanders, Randomness, or Time vs. Space*, in Proc. of 1st Structures of Complexity Theory, 1986.
29. J. Kilian, and E. Petrank, *An efficient zero-knowledge proof system for NP under general assumptions*, Journal of Cryptology, vol. 11, n. 1, pp. 1–28.
30. A. Lubotzky, R. Phillips, and P. Sarnak, *Explicit Expanders and the Ramanujan Conjectures*, Proc. of STOC 86.
31. M. Naor, *Bit Commitment from Pseudo-Randomness*, Proc. of CRYPTO 89.
32. R. Ostrovsky and A. Wigderson, *One-way Functions are Essential for Non-Trivial Zero-knowledge*, in Proc. of the 2nd Israel Symposium on Theory of Computing and Systems (ISTCS-93).
33. N. Nisan, *Pseudorandom bits for constant depth circuits*, Combinatorica, 11, pp. 63–70, 1991.
34. O. Reingold, S. Vadhan and A. Wigderson, *Entropy Waves, The Zig-Zag Graph Product, and New Constant-Degree Expanders and Extractors*, in Proc. of FOCS 2000.
35. M. Sipser, *A Complexity-Theoretic Aproach to Randomness*, in Proc. of STOC 1983.
36. A. Sahai, *Non-Malleable Non-Interactive Zero Knowledge and Adaptive Chosen-Ciphertext Security*, in Proc. of FOCS 1999.
37. D. Zuckerman, *Randomness-Optimal Oblivious Sampling*, in Proc. of STOC 97.

A Probabilistic-Time Hierarchy Theorem for "Slightly Non-uniform" Algorithms

Boaz Barak

Department of Computer Science, Weizmann Institute of Science, Rehovot, ISRAEL.
`boaz@wisdom.weizmann.ac.il`

Abstract. Unlike other complexity measures such as deterministic and nondeterministic time and space, and non-uniform size, it is not known whether probabilistic time has a strict hierarchy. For example, as far as we know it may be that **BPP** is contained in the class **BPtime**(n). In fact, it may even be that the class **BPtime**$(n^{\log n})$ is contained in the class **BPtime**(n).

In this work we prove that a hierarchy theorem does hold for "slightly non-uniform" probabilistic machines. Namely, we prove that for every function $a:\mathbb{N} \to \mathbb{N}$ where $\log \log n \leq a(n) \leq \log n$, and for every constant $d \geq 1$,

$$\mathbf{BPtime}(n^d)_{/a(n)} \subsetneq \mathbf{BPP}_{/a(n)}$$

here **BPtime**$(t(n))_{/a(n)}$ is defined to be the class of languages that are accepted by probabilistic Turing machines of running time $t(n)$ and description size $a(n)$. We actually obtain the stronger result that the class **BPP**$_{/\log \log n}$ is *not* contained in the class **BPtime**$(n^d)_{/\log n}$ for every constant $d \geq 1$.

We also discuss conditions under which a hierarchy theorem can be proven for *fully uniform* Turing machines. In particular we observe that such a theorem does hold if **BPP** has a complete problem.

1   Introduction

Complexity theory studies the problems that can be solved using a bounded amount of computational resources. These resources include time, space, non-determinism, randomness and non-uniform advice. In general the relation between the power of *different* resources is unknown (e.g., it is not known whether or not a polynomial amount of non-determinism can be exchanged for less than exponential amount of deterministic time, as can be done by trivial exhaustive search). However, it is usually the case that one can prove that if we are given more of the same resource then we *can* solve new problems. Such theorems are called *hierarchy theorems* and exist, for example, in the case of deterministic and non-deterministic time and space, and non-uniform advice. In all these cases one can prove that for "nice" functions $f : N \to \mathbb{N}$, there are problems that can be solved using $f(n)$ amount of the resource and cannot be solved using $\frac{f(n)}{(\log f(n))^2}$ amount. In particular the following inequalities are known for

J.D.P. Rolim and S. Vadhan (Eds.): RANDOM 2002, LNCS 2483, pp. 194–208, 2002.
© Springer-Verlag Berlin Heidelberg 2002

every constant $d \geq 1$: $\mathbf{Dtime}(n^d) \subsetneqq \mathbf{P}$, $\mathbf{Ntime}(n^d) \subsetneqq \mathbf{NP}$, $\mathbf{Size}(n^d) \subsetneqq \mathbf{P}_{/\mathbf{poly}}$, $\mathbf{Dspace}(n^d) \subsetneqq \mathbf{PSPACE}$.

A notable exception is the case of *probabilistic time*. For probabilistic time no such theorem is known. In particular as far as we know it may be the case that $\mathbf{BPtime}(n) = \mathbf{BPP}$ and it may even be the case that $\mathbf{BPtime}(n)$ is equal to classes with probabilistic *super-polynomial* time (e.g. it may hold that $\mathbf{BPtime}(n) = \mathbf{BPtime}(n^{\log n})$). In fact, not much is known beyond the trivial observation that $\mathbf{BPtime}(n) \subsetneqq \mathbf{Dtime}(f(n)) \subseteq \mathbf{BPtime}(f(n))$ for every (time-constructible) function $f(\cdot)$ such that $f(n) = 2^{\omega(n)}$. The best unconditional bound on $\mathbf{BPtime}(n)$ is that $\mathbf{BPtime}(n) \subsetneqq \mathbf{BPtime}(f(n))$ for any (time-constructible) function $f(\cdot)$ such that $f^{(c)}(n) = 2^{\omega(n)}$ for some constant c (where $f^{(c)}$ denotes the composition of f with itself c times) [1].[1]

What is different with probabilistic time? The main difference seems to be that a probabilistic Turing machine that decides a $\mathbf{BPtime}(t)$ language must possess a very special property. It must hold that for any $x \in \{0,1\}^*$ either $\Pr[M(x) = 1] > 2/3$ or $\Pr[M(x) = 1] < 1/3$. It is undecidable to test whether a machine M satisfies this property and it is also unknown whether one can test whether a machine M satisfies this property for every $x \in \{0,1\}^n$ using less than 2^n steps. Because of this difference, the standard diagonalization techniques that allow us to obtain hierarchy theorems for $\mathbf{P}$,$\mathbf{NP}$ and $\mathbf{PSPACE}$ do not work for $\mathbf{BPP}$.[2]

Note that it is believed that $\mathbf{BPtime}$ *does* have a strict hierarchy. Indeed, Impagliazzo and Wigderson [4] show that if there is a problem in $\mathbf{E} = \mathbf{Dtime}(2^{O(n)})$ that requires exponential-sized circuits then for every constant d there exists a constant c such that $\mathbf{BPtime}(n^d) \subseteq \mathbf{Dtime}(n^c)$ and thus $\mathbf{BPtime}(n^d) \subsetneqq \mathbf{Dtime}(n^{c+1}) \subseteq \mathbf{BPP}$.

1.1   Our Results

Our main result is that a hierarchy theorem does hold for probabilistic machines that are "slightly non-uniform". That is, we define the class $\mathbf{BPtime}(t(n))_{/a(n)}$ to contain all languages that are accepted by a $t(n)$-time probabilistic Turing machine with description size $a(n)$ (or almost equivalently, all languages that are accepted by a $t(n)$-time Turing machine that gets $a(n)$ bits of advice). Our main theorem is the following:

$$\mathbf{BPP}_{/\log \log n} \not\subseteq \mathbf{BPtime}(n^d)$$

for every constant $d \geq 1$. Using a padding argument, we obtain as a corollary that for every constant $d \geq 1$ and every function $a : \mathbb{N} \to \mathbb{N}$ that satisfies

1. $\omega(\log \log n) \leq a(n) \leq \log n$
2. $a(\cdot)$ is "nice" in the sense that $a(n^{O(1)}) = \Theta(a(n))$.[3]

---

[1] Allender *et al* strengthen this and show that for such functions $f(\cdot)$ there exists a language $L \in \mathbf{BPtime}(f(n))$ such that every $\mathbf{BPtime}(n)$ machine will fail to decide L almost everywhere [2]. See also [3,1] for *conditional* hierarchy theorems.

[2] The hierarchy theorem for non-uniform circuits is proven using a counting argument.

[3] Note that this bound on the rate of growth is satisfied by the function $\log n$.

It holds that

$$\mathbf{BPtime}(n^d)_{/\gamma a(n)} \subsetneq \mathbf{BPtime}(n^{d+1})_{/\gamma a(n)}$$

for some constant γ.

We also explore conditions under which hierarchy theorems hold for fully uniform probabilistic time. We show (using the standard diagonalization techniques) that if **BPP** has a complete problem (under a specific notion of completeness defined below), then $\mathbf{BPtime}(n^d) \subsetneq \mathbf{BPtime}(n^{d+1})$ for every constant d. By applying the Sipser-Lautenman Theorem [5,6], we obtain that if **BPP** $\supseteq$ **NP** then **BPP** does have a complete problem (under this notion) and so in this case $\mathbf{BPtime}(n^d) \subsetneq \mathbf{BPtime}(n^{d+1})$. As a corollary we obtain that $\mathbf{BPtime}(n) \neq \mathbf{NP}$.

1.2   Techniques

Our main technique is the use of an *instance checker* [7] for an **EXP**-complete language in order to establish a hierarchy theorem. The existence of instance checkers for such languages is implied by the existence of **PCP** proof systems for **EXP** [8,9]. The main observation we use is that languages with an instance checker have an *optimal* algorithm in some specific sense.[4] We use an instance checker to decide whether $x \in L$ for an **EXP**-complete language L, when given a set of oracles such that one of these oracles decides L. Instance checkers were used before in a similar setting by Trevisan and Vadhan [12].

Organization. Section 2 contains some basic notations and definitions. It also contains some basic upward scaling results that are proven using a padding argument. Section 3 contains the observation that the existence of a **BPP**-complete language implies a hierarchy theorem for **BPtime**, along with some corollaries. Section 4 contains our main result, which is a hierarchy theorem for "slightly non-uniform" probabilistic Turing machines.

2   Preliminaries

We identify a language $L \subseteq \{0,1\}^*$ with its characteristic function. That is, we say that $L(x)$ is equal to 1 if $x \in L$, and to 0 otherwise. For a probabilistic Turing machine M and an input x, we define $t_M(x)$ to be the maximum number of steps M takes when given x as input, where this maximum is over all possible coin-tosses of M. Let $t : \mathbb{N} \to \mathbb{N}$ be some function, we say that $L \in \mathbf{BPtime}(t)$ if there exists a probabilistic Turing machine M such that for any $x \in \{0,1\}^*$, $t_M(x) = O(t(|x|))$ and $\Pr[M(x) = L(x)] > \frac{2}{3}$. The class **BPP** is defined to be $\cup_{c \in \mathbb{N}} \mathbf{BPtime}(n^c)$. We say that $L \in \mathbf{Dtime}(t)$ if there exists a *deterministic* Turing machine M such that $M(x) = L(x)$ for every $x \in \{0,1\}^*$. The class **EXP** is defined to be $\cup_{c \in \mathbb{N}} \mathbf{Dtime}(2^{n^c})$.

---

[4] This is similar to the optimal algorithm for **NP**, that uses the self-reducibility of **NP**-complete languages [10,11].

We will sometimes identify a Turing machine M with its *description* as a string in $\{0,1\}^*$. The *description size* of M, denoted $|M|$, is the length of this string. For convenience, we assume that the description of Turing machines allows padding. That is, we assume that if M is a Turing machine with description of size k, then for every $n > k$, there exists a Turing machine M' with description size n with the same output distribution and running time as M. We will use a *universal Turing machine* $\mathcal{U}$ such that on input a Turing machine M and $x \in \{0,1\}^*$, the machine $\mathcal{U}$ runs for $poly(|M|t_M(x))$ steps, and $\mathcal{U}(M,x)$ is distributed identically to $M(x)$. For a Turing machine M, and a number m, we denote by M^m the machine M, when restricted to run for at most m steps by some time-out mechanism.

2.1   Scale-up Lemmas

Time-constructible functions. We define time-constructible functions as follows (we add the extra requirement that $n \leq f(n) \leq 2^n$ since we are only interested in functions of this form):

Definition 2.1. *Let $f:\mathbb{N} \to \mathbb{N}$ be a function. We say that f is* time-constructible *if $n \leq f(n) \leq 2^n$ and there is a deterministic Turing machine M such that on input n , M runs for at most $f(n)$ steps and outputs $f(n)$.*

It is a basic fact in complexity theory that inclusions for complexity classes "scale-up" (and so by the contrapositive argument separations "scale-down"). This fact, which is proved using a padding argument, will be useful to us. We will now state (without proofs) some specific lemmas of this form that we will use.

Lemma 2.2. *Let $d \geq 1$ be some constant. If $\mathbf{BPtime}(n^d) = \mathbf{BPtime}(n^{d+1})$ then $\mathbf{BPtime}(n^{d+1}) = \mathbf{BPtime}(n^{(d+1)\cdot\frac{d+1}{d}})$.*

By repeatedly applying Lemma 2.2, we obtain

Corollary 2.3. *Let $d \geq 1$ be some constant. Suppose that $\mathbf{BPtime}(n^d) = \mathbf{BPtime}(n^{d+1})$. Then, $\mathbf{BPtime}(n^d) = \mathbf{BPP}$.*

Another scaling up result is the following:

Lemma 2.4. *Let $d \geq 1$ be some constant. If $\mathbf{BPtime}(n^d) = \mathbf{BPP}$ then $\mathbf{BPtime}(t(n)) = \mathbf{BPtime}(t(n)^c)$ for every constant $c \geq 1$ and time-constructible function $t:\mathbb{N} \to \mathbb{N}$ that satisfies $t(n) \geq n^d$.*

Together, Corollary 2.3 and Lemma 2.4 imply that

Corollary 2.5. *For every $d \geq 1$, if there exists a time-constructible function $t : \mathbb{N} \to \mathbb{N}$ and a constant $c > 1$ such that $t(n) \geq n^d$ and $\mathbf{BPtime}(t(n)) \subsetneqq \mathbf{BPtime}(t(n)^c)$ then $\mathbf{BPtime}(n^d) \subsetneqq \mathbf{BPtime}(n^{d+1})$*

3   A Hierarchy Theorem Using a Complete Problem

In this section we will define a restricted type of reductions and prove that if **BPP** has a complete problem under reductions of this type then probabilistic time has a strict hierarchy. Actually, a weaker form of this section's main result holds also for the standard notion of a Karp reduction (see Remark 3.6). The observations of this section are not hard and may be known, but we still think their inclusion here is helpful.

3.1   Fixed-Polynomial Karp Completeness

We define a notion of hardness of languages that is somewhat different from the standard notion of hardness. The first difference is that we want to define **BPtime**-hardness rather than **BPP**-hardness. This means that we will say that a language L is **BPtime**-hard if for any **BPtime**(t) language L' there is a reduction between L' and L. Naturally, the complexity of the reduction itself will have to depend on the complexity of L' (i.e., on t), and a natural choice is to require that the complexity of the reduction should be at most polynomial in t. Indeed, we make this requirement and we strengthen it by requiring that the complexity of the reduction should be at most t^c where c is some fixed constant *independent* of L' and t. This may seem like a strict requirement, but we note that hard languages for other classes also satisfy a similar requirement. For example, there is a $O(t^3)$ time reduction from any **Ntime**(t) language and the language $3SAT$. The formal definition follows:

Definition 3.1. *Let L be a language. We say that L is **BPtime**-hard if there exists a constant c such that for any time-constructible function $t:\mathbb{N} \to \mathbb{N}$ and any language $L' \in$ **BPtime**(t) there exists a deterministic $t(|x|)^c$-time computable function $f : \{0,1\}^* \to \{0,1\}^*$ such that for any $x \in \{0,1\}^*$ it holds that $x \in L' \iff f(x) \in L$.*

*We say that L is **BPP**-complete if L is **BPtime**-hard and $L \in$ **BPP**.*

3.2   A Detour: Promise Problems

We will now present an example for a family of languages such that all languages in this family are **BPtime**-hard by Definition 3.1. It is conjectured that there exists a language in this family that is in **BPP** and so is **BPP**-complete. To define this family we need to recall the notion of *promise problem*.

Definition 3.2. *A promise problem Π is a pair of sets (Π_Y, Π_N) where $\Pi_Y, \Pi_N \subseteq \{0,1\}^*$ and Π_Y , Π_N are disjoint (i.e., $\Pi_Y \cap \Pi_N = \emptyset$).*

Definition 3.3. *Let $t:\mathbb{N} \to \mathbb{N}$ be some function. We say that $\Pi = (\Pi_Y, \Pi_N)$ is in **PromiseBPtime**(t) if there exists a probabilistic $t(n)$-time machine M such that $x \in \Pi_Y \implies \Pr[M(x) = 1] > \frac{2}{3}$ and $x \in \Pi_N \implies \Pr[M(x) = 1] < \frac{1}{3}$. We define **PromiseBPP** $\stackrel{def}{=} \cup_{c \in \mathbb{N}}$**PromiseBPtime**($n^c$).*

We note that unlike the case of **BPtime**, there is a hierarchy theorem for **PromiseBPtime** and in particular it is known that **PromiseBPtime**$(n^d) \subsetneq$ **PromiseBPtime**(n^{d+1}) for every constant d.

An (important) example for a promise problem is the CIRCUIT ACCEPTANCE PROBABILITY (CAP) promise problem. The CAP promise problem is the pair $(\text{CAP}_Y, \text{CAP}_N)$ where CAP_Y contains all circuits C such that $\text{Pr}_{x \in \{0,1\}^n}[C(x) = 1] > 2/3$ (where n is the number of inputs to the circuit C) and CAP_N contains all circuits C such that $\text{Pr}_{x \in \{0,1\}^n}[C(x) = 1] < 1/3$. Clearly the problem CAP is in **PromiseBPP**.

We say that a language L is *consistent* with a promise problem $\Pi = (\Pi_Y, \Pi_N)$ if for any $x \in \{0,1\}^*$ it holds that $x \in \Pi_Y \implies x \in L$ and $x \in \Pi_N \implies x \notin L$. We have the following Lemma:

Lemma 3.4. *Let L be a language consistent with the promise problem* CAP. *Then L is* **BPtime**-*hard (in the sense of Definition 3.1).*

Proof (Sketch). Every **BPtime**(t) language L' can be reduced to CAP (and therefore to L) in t^2 steps using the Cook-Levin reduction. □

We thus have the following corollary:

Corollary 3.5. *If there exists a language L such that*

1. *L is consistent with the promise problem* CAP.
2. *$L \in$ **BPP***

*Then there exists a **BPP**-complete language (in the sense of Definition 3.1).*

We remark that it is reasonable to believe that there exists a language L satisfying the conditions of Corollary 3.5. In particular, Impagliazzo and Wigderson [4] show that if there exists a language in $\mathbf{E} = \mathbf{Dtime}(2^{O(n)})$ that requires exponential-sized circuits, then the CAP promise problem can be solved in *deterministic* polynomial-time, and so there exists a language satisfying the conditions of Corollary 3.5 in $\mathbf{P} \subseteq \mathbf{BPP}$.

It is not known whether or not the reverse direction Corollary 3.5 also holds. That is, it is not known whether the existence of a **BPP**-complete language implies that there exists a language in **BPP** that is consistent with CAP.

3.3   The Hierarchy Theorem

We now turn to proving the hierarchy theorem of this section.

Theorem 1. *Suppose that* **BPP** *has a complete problem (by Definition 3.1). Then there exists a constant c such that for every time constructible $t : \mathbb{N} \to \mathbb{N}$ it holds that* **BPtime**$(t(n)) \subsetneq$ **BPtime**$((t(n))^c)$.

Note that by Corollary 2.5 this implies that under this assumption **BPtime**$(n^d) \subsetneq$ **BPtime**(n^{d+1}) for every constant $d \geq 1$.

Proof. Let L be the **BPP** complete problem and let M_L be a probabilistic machine that accepts L and runs in time n^a for some constant a. We know that there exists a constant b such that for every time-constructible function $t(\cdot)$, every language in **BPtime**(t) is reducible to L using a $t(n)^b$-time deterministic reduction. For a string $i \in \{0,1\}^*$, we denote by M_i the *deterministic* Turing machine whose description is i. We denote by M_i^t the machine M_i, restricted (using some timeout mechanism) to run for at most t steps.

We define the following language $Kx \in K$ iff $M_x^{t(|x|)^b}(x) \notin L$. We claim that

1. $K \in$ **BPtime**$(t(n)^{O(ab)})$
2. $K \notin$ **BPtime**$(t(n))$.

Once we prove both items we are done. Item 1 is true because deciding K can be done by simply returning $1 - M_L(M_x^{t(|x|)^b}(x))$. This value can be computed in $t(|x|)^{O(ab)}$ steps by using the universal Turing machine $\mathcal{U}$ to compute $M_x^{t(|x|)^b}(x)$ and then by running M_L on the result.

Suppose that Item 2 was false. That is, that $K \in$ **BPtime**$(t(n))$. By the completeness of L, there exists an i such that for every $x \in \{0,1\}^*$, $x \in K \iff M_i^{t(|x|)^b}(x) \in L$. In particular for $x = i$ it holds that $i \in K \iff M_i^{t(|i|)^b}(i) \in L$. Yet by the definition of K this happens if and only if $i \notin K$, and so we get a contradiction. $\square$

Remark 3.6. Suppose that we followed a more standard definition of **BPtime**-hardness by allowing a *different* polynomial running time of the reduction for every language $L \in$ **BPtime**. If we look at the proof of Theorem 3.3, we can see that we will still be able to derive a meaningful result from the existence of a **BPP**-complete language under this more relaxed notion. Indeed, the same proof will yield that under this assumption it holds that **BPtime**$(t(n)) \subsetneq$ **BPtime**$(f(t(n))$ for every time-constructible super-polynomial function $f:\mathbb{N} \to \mathbb{N}$ (Because the language K we defined will be in **BPtime**$(f(t(n))$ but not in **BPtime**$(t(n))$). In particular, under this assumption it holds that **BPtime**$(n) \subsetneq$ **BPtime**$(n^{\log n})$.

Remark 3.7. The proof of the Sipser-Lautemann Theorem [5,6], implies that the promise problem CAP is reducible to a language L in the polynomial hierarchy (actually in Σ_2). This implies that if **BPP** $\supseteq$ **NP** then **BPP** has a complete language (by Definition 3.1) since **BPP** $\supseteq$ **NP** implies that **BPP** $\supseteq$ **PH** and so that $L \in$ **BPP**. Combining this with Theorem 3.3, we get that **BPP** $\supseteq$ **NP** implies that **BPtime**$(n) \neq$ **BPP**. As a corollary we obtain that **BPtime**$(n) \neq$ **NP** unconditionally, since if **BPtime**$(n) =$ **NP** then by scaling up we get the equality **BPP** $=$ **NP** which would imply that both **BPP** has a complete language and **BPtime**$(n) =$ **BPP**, which is a contradiction. We note that this does not rule out the possibility that **NP** $\subsetneq$ **BPtime**(n).

4   A Hierarchy Theorem Using Small Advice

In this section we will prove an unconditional hierarchy theorem for "slightly non-uniform" probabilistic machines. Our results will have also a (rather weak) implication on standard (uniform) probabilistic machines (see Section 4.1). We start by defining what we mean by non-uniform probabilistic machines:

Definition 4.1. *Let $a:\mathbb{N} \to \mathbb{N}$ and $t:\mathbb{N} \to \mathbb{N}$ be two functions and let $L \subseteq \{0,1\}^*$ be a language. We say that $L \in \mathbf{BPtime}(t(n))_{/a(n)}$ if there exists a probabilistic Turing machine M and a sequence $\{s_n\}_{n\in\mathbb{N}}$ such that*

1. *For any $n \in \mathbb{N}$, $|s_n| \leq a(n)$.*
2. *For any $x \in \{0,1\}^*$, it holds that $\Pr[M^{t(|x|)}(x, s_{|x|}) = L(x)] > \frac{2}{3}$ (recall that M^t denotes the restriction of M to t steps).*

Note that the condition that for any $x \in \{0,1\}^*$ either $\Pr[M^{t(|x|)}(x, s) = 1] > \frac{2}{3}$ or $\Pr[M^{t(|x|)}(x, s) = 1] < \frac{1}{3}$ is guaranteed *only* if s is equal to the "good" advice string $s_{|x|}$. An almost equivalent formulation would be to say that $L \in \mathbf{BPtime}(t(n))_{/a(n)}$ if there exists a sequence $\{M_n\}_{n\in\mathbb{N}}$ of Turing machines, such that for every n it holds that $|M_n| \leq a(n)$ and for every $x \in \{0,1\}^n$, $\Pr[M_n^{t(n)}(x) = L(x)] > \frac{2}{3}$.

Our hierarchy theorem is:

Theorem 2. *For every constant $d \geq 1$, the class $\mathbf{BPP}_{/\log\log n}$ is not contained in the class $\mathbf{BPtime}(n^d)_{/\log n}$.*

This implies the following corollary:

Corollary 4.2. *Let $a : \mathbb{N} \to \mathbb{N}$ be a function that satisfies both $\omega(\log\log n) \leq a(n) \leq \log n$ and $a(n^{O(1)}) = \Theta(a(n))$. Then for every constant $d \geq 1$ there exists a constant γ such that*

$$\mathbf{BPtime}(n^d)_{/\gamma a(n)} \subsetneq \mathbf{BPtime}(n^{d+1})_{/\gamma a(n)}$$

Proof (Sketch). Let's look at the case $d = 1$. Suppose by contradiction that $\mathbf{BPtime}(n)_{/\gamma a(n)} = \mathbf{BPtime}(n^2)_{/\gamma a(n)}$ for all constants γ. By padding this implies that $\mathbf{BPtime}(t(n))_{/\gamma a(t(n))} = \mathbf{BPtime}(t(n)^2)_{/\gamma a(t(n))}$ for all time-constructible functions $t:\mathbb{N} \to \mathbb{N}$ and constants γ. In particular we get that

$$\mathbf{BPtime}(n)_{/a(n)} = \mathbf{BPtime}(n^2)_{/a(n)}$$
$$\supseteq \mathbf{BPtime}(n^2)_{/\gamma a(n^2)} = \mathbf{BPtime}(n^4)_{/\gamma a(n^2)}$$

For some constant $\gamma < 1$ that is chosen so that $a(n) \geq \gamma a(n^2)$. Continuing in the same way we obtain that for every constant c there exists a constant γ' such that

$$\mathbf{BPtime}(n)_{/a(n)} \supseteq \mathbf{BPtime}(n^c)_{/\gamma' a(n)}$$

But since $a(n) \leq \log n$ and $a(n) = \omega(\log\log n)$ we actually obtain that

$$\mathbf{BPtime}(n)_{/\log n} \supseteq \mathbf{BPtime}(n)_{/a(n)} \supseteq \mathbf{BPtime}(n^c)_{/\gamma' a(n)} \supseteq \mathbf{BPtime}(n^c)_{/\log\log n}$$

For every constant c, or in other words $\mathbf{BPtime}(n)_{/\log n} \supseteq \mathbf{BPP}_{/\log\log n}$, contradicting Theorem 4.   □

4.1   Proof Idea for Theorem 2

To illustrate the proof idea, we will show a flawed approach for proving the fully uniform version of Theorem 4. For simplicity, we'll restrict ourselves to the case $d = 1$. That is, we will try to prove that $\mathbf{BPP} \not\subseteq \mathbf{BPtime}(n)$. We will see that there is a point in which this proof direction fails, but this point can be overcome by introducing slight non-uniformity.

Suppose, towards a contradiction, that $\mathbf{BPtime}(n) = \mathbf{BPP}$. By Lemma 2.4, this means that $\mathbf{BPtime}(t(n)) = \mathbf{BPtime}(t(n)^c))$ for every time constructible function $t(\cdot)$ and every constant $c \geq 1$. This means that for every language L, if there exists a $\mathbf{BPtime}(T(n))$ algorithm for L (for some time-constructible function $T(\cdot)$), then there exists an algorithm that solves L in $\mathbf{BPtime}(T(n)^{1/c})$ for every constant $c \geq 1$ (assuming that $T(\cdot)$ is super-polynomial). This means that every language $L \notin \mathbf{BPP}$ has no "optimal algorithm", but rather that any algorithm for L can be improved by any polynomial factor.

We see that to get a contradiction it is sufficient to provide a language $L \notin \mathbf{BPP}$ that has an *optimal algorithm* in the following sense: there exists a $\mathbf{BPtime}(T(n))$ (for some super-polynomial function $T(\cdot)$) Turing machine A that solves L and a constant c such that every other probabilistic Turing machine A' that solves L takes *at least* $T(n)^{1/c}$ steps.

How can we do that? The idea is that our optimal algorithm A will have the following general form. Algorithm A will enumerate all Turing machines of size at most $f(n)$ (where $f(n)$ is some unbounded function, e.g. $f(n) = \log n$) and will try to use each machine to solve the language L. To prove that A is indeed an optimal algorithm we will use the fact that if there indeed exists a fast Turing machine A' that solves L, then for all but finitely many inputs, the machine A' will be one of the machines enumerated by A. Therefore, we will claim that A is able to use A' to solve L and so A will be at most polynomially slower than A' (we'll choose $f(\cdot)$ such that $2^{f(n)}$ is polynomial). Of course, one detail that is missing is how exactly will our algorithm A use the machines it enumerated to solve L. Although for sufficiently large x's, one of these machines will solve L, how can we know which one does so?

In general, this does seem like a hard problem. However, recall that we are free to choose L to be any language we want. In particular, we can choose L to be a language that has an *instance checker* [7]. Roughly speaking, an instance checker C for a language L is an efficient procedure that when given an input x, and an oracle $O(\cdot)$, outputs a value $v \in \{0, 1, \mathsf{fail}\}$, where we are guaranteed that if $O(\cdot)$ is an oracle for L then $v = L(x)$ and otherwise with high probability $v \in \{L(x), \mathsf{fail}\}$. Clearly, given an input x and m oracles $O_1, \ldots, O_m$ such that one of these oracles solves the language L, one can use the instance checker to compute $L(x)$. Therefore if we take L to be a language that has an instance checker, we can (almost) give an *optimal algorithm* A for L. Since all $\mathbf{EXP}$-complete languages have an instance checker [8], we can just pick L to be some $\mathbf{EXP}$-complete language. The resulting algorithm A will work as follows:

Algorithm A:

- Input: $x \in \{0,1\}^*$ (denote $n \stackrel{def}{=} |x|$).

1. For $m = 1, 2, \ldots$
2. For each probabilistic Turing machine M of size $\log n$ do:
   a) Let $a \leftarrow C^{M^m}(x)$ (where C is the instance checker for L and M^m denotes the machine M restricted to m steps).
   b) If $a \neq$ fail output a

Note that when A halts it will with high probability halt with the correct answer. It is not hard to prove the following claim:

Claim 4.3. *If there exists a* **BPtime**(t) *algorithm A' for L, then with high probability, on input x algorithm A stops after* $poly(|x|) \cdot t(l(|x|))^2$ *steps and outputs $L(x)$, where $l(n)$ is the length of the queries the instance checker C makes on inputs of length n.*

Define the function $T : \mathbb{N} \to \mathbb{N}$ in the following way: let $T(n)$ be the minimal number k such that for every $x \in \{0,1\}^n$, $A(x)$ halts with probability at least $\frac{2}{3}$ within k steps. We will assume that $T(\cdot)$ is super-polynomial (otherwise, we can show that **BPP** = **EXP** and then we can get a hierarchy theorem easily, e.g., by Theorem 3.3). It may seem like we're done since we have shown that L can be solved by a time $T(\cdot)$ probabilistic algorithm, but it is not in **BPtime**$(T(n)^{1/c})$ for some universal constant $c \geq 1$ (since if it was then A would have halted sooner).[5] However we still have two problems:

1. The definition of a **BPtime**$(T(n))$ Turing machine requires the machine to stop within $T(n)$ steps *with probability* 1 and does not allow a machine that only stops within $T(n)$ steps with high probability.
2. The function $T(\cdot)$ that we defined is not necessarily time-constructible. Therefore, even if we show that **BPtime**$(T(n)) \not\subseteq$ **BPtime**$(T(n)^{1/c})$ we will not be able to use Lemma 2.4 to show that this implies that **BPtime**$(n) \not\subseteq$ **BPP**.

It turns out that if we solve the second problem we will also solve the first. This is because if $T(\cdot)$ was time-constructible then we could implement a *time-out* mechanism and so ensure that algorithm A halts within $T(n)$ steps with probability 1. Note that it would have been enough if we found a time-constructible function $T'(\cdot)$ that approximates $T(\cdot)$ from above to a polynomial power (i.e., $T'(\cdot)$ should satisfy $T(n) \leq T'(n) \leq T(n)^d$ for some constant $d \geq 1$). The problem is that it may be the case that the function $T(\cdot)$ sits in a large "gap" between the time-constructible functions. That is, it may be that the smallest time-constructible function $T'(\cdot)$ that is larger than $T(\cdot)$ is actually *much* larger than $T(\cdot)$. That is, since the function $T(\cdot)$ is arbitrary and is not necessarily time-

---

[5] In this proof outline we make the simplifying assumption that $t(l(n)) = t(n)^{O(1)}$, which is the case if $l(n) = O(n)$ (e.g., if we use the instance checker of [9]) and $t(n)$ is a "nice" function (e.g., $t(n) = 2^{n^\epsilon}$). Note that the constant c is a universal constant that depends on the instance checker.

constructible it may be the case that it cannot even be roughly approximated by a time-constructible function.

This is where we need to resort to non-uniformity. It turns out that if we add even a very small amount of advice (i.e., $\log \log n$) then any function $T(\cdot)$ where $n \leq T(n) \leq 2^n$ can be approximated by a time-constructible function.[6] This is done by simply hardwiring the number $\lceil \log \log T(n) \rceil$. This number will take at most $\log \log n$ bits to specify, and thus we see that the function $T'(n) \overset{def}{=} 2^{2^{\lceil \log \log T(n) \rceil}}$, that approximates $T(n)$ from above (i.e., $T(n) \leq 2^{2^{\lceil \log \log T(n) \rceil}} \leq T(n)^2$), is time-constructible with $\log \log n$ bits of advice.

This shows that there exists a language L such that $L \in \mathbf{BPtime}(T(n))_{/\log \log n}$ but $L \notin \mathbf{BPtime}(T(n)^{1/c})$, which is not so impressive, because in fact even the stronger result $\mathbf{BPtime}(T(n))_{/1} \not\subseteq \mathbf{BPtime}(T(n))$ can be easily shown to hold due to non-uniformity. However, if we look again at our algorithm A we can see that in fact it works even against slightly *non-uniform* algorithms. Indeed, Algorithm A enumerated all machines with description size $\log n$. Therefore, even if there exists a $\mathbf{BPtime}(t(n))$ algorithm A' with $\log n$ bits of advice to solve L then A will solve L within $t(n)^c$ steps. It follows that

$$\mathbf{BPtime}(T(n))_{/\log \log n} \not\subseteq \mathbf{BPtime}(T(n)^{1/c})_{/\log n}$$

By a careful padding argument,[7] similar in spirit to Lemma 2.4, we can show that

$$\mathbf{BPP}_{/\log \log n} \not\subseteq \mathbf{BPtime}(n^d)_{\log n}$$

For every constant $d \geq 1$. We now proceed to the actual proof, that involves some issues and subtleties not mentioned in the above description.

4.2   The Actual Proof of Theorem 2

We define the class **i.o. BPP** to contain all languages that a probabilistic machines decides infinitely often. That is, $L \in \mathbf{i.o.\,BPP}$ if there exists a probabilistic polynomial-time machine such that for infinitely many n's it is the case that for all $x \in \{0,1\}^n$, $\Pr[M(x) = L(x)] > \frac{2}{3}$. We prove Theorem 4 by considering two cases according to whether or not $\mathbf{EXP} \subseteq \mathbf{i.o.\,BPP}$.

Lemma 4.4. *Suppose that* $\mathbf{EXP} \subseteq \mathbf{i.o.\,BPP}$. *Then for every* d, *and every* $a : \mathbb{N} \to \mathbb{N}$ *such that* $a(n) \leq n - \omega(1)$,

$$\mathbf{BPP}_{/1} \not\subseteq \mathbf{BPtime}(n^d)_{/a(n)}$$

Lemma 4.5. *Suppose that* $\mathbf{EXP} \not\subseteq \mathbf{i.o.\,BPP}$. *Then, for every* d,

$$\mathbf{BPP}_{/\log \log n} \not\subseteq \mathbf{BPtime}(n^d)_{/\log n}$$

Combining Lemmas 4.4 and 4.5 yields Theorem 4.

---

[6] We can assume that $T(n) \leq 2^n$ since we can choose the language L to be in $\mathbf{Dtime}(2^n)$.

[7] As we'll see in Section 4.2, to make things work we need to make a slight modification Algorithm A.

Proof of Lemma 4.4 Let $d \geq 1$ be some constant. We let L' be the following language: M is in L' (where M describes a probabilistic Turing machine) if $\Pr[M^{|M|^{d+1}}(M) = 1] < \frac{1}{2}$, where M^t denotes M restricted to running t steps. We claim that for every $\mathbf{BPtime}(n^d)$ Turing machine M and any large enough n and $s \in \{0,1\}^{a(n)}$ there exists a string $x \in \{0,1\}^n$ such that $\Pr[M^{n^d}(s,x) \neq L'(x)] > \frac{1}{3}$. Indeed, if we let x be the Turing machine that is M with the string s "hardwired" into it then we see that $L'(x) = 1 \iff \Pr[M^{n^d}(s,x) = 1] < \frac{1}{2}$.

Yet, since $L' \in \mathbf{EXP}$ and we assume that $\mathbf{EXP} \subseteq \mathbf{i.o.BPP}$, we *can* solve L' in $\mathbf{BPP}$ on an infinite set $I \subseteq \mathbb{N}$ of input sizes. We define L in the following way $L(x) = L'(x)$ if $|x| \in I$ and $L(x) = 0$ otherwise. We see that $L \in \mathbf{BPP}_{/1} \setminus \mathbf{BPtime}(n^d)_{/a(n)}$ (we use the advice bit to tell if $n \in I$). $\square$

Proof of Lemma 4.5 We will choose L to be a language that satisfies the following properties:

1. L is $\mathbf{EXP}$-complete.
2. L can be decided by a 2^n-time deterministic algorithm.
3. L has an *instance checker*. That is, there exists a probabilistic polynomial-time oracle machine C with output in $\{0, 1, \mathsf{fail}\}$ such that for every $x \in \{0,1\}^*$ and every oracle $O(\cdot)$, $\Pr[C^O(x) \notin \{L(x), \mathsf{fail}\}] < 2^{-\Omega(|x|)}$ and $\Pr[C^L(x) = L(x)] = 1$. Furthermore, we require that the instance checker will make only queries of fixed length that is linear in $|x|$. That is, all the queries C makes are of length $c|x|$ for some constant c.
4. For convenience we will assume that L allows padding. That is, given $x \in \{0,1\}^n$ and $m > n$ it is possible to (efficiently) extend x to a string $x' \in \{0,1\}^m$ such that $x' \in L$ iff $x \in L$.

The existence of such a language follows from Arora and Safra [9] (they strengthen the result of Babai, Fortnow and Lund [8] to obtain an instance checker with linear-sized queries, see [12]). We will also assume that L only contains strings whose length is a power of 2. (If L doesn't satisfy this assumption then it can be modified to do so by dropping all other strings; the padding property implies that the modified language will still satisfy the other properties.) We denote by $I \subseteq \mathbb{N}$ the set of "interesting" input lengths. That is, $I = \{2^k \mid k \in \mathbb{N}\}$. Note that the completeness of L implies that under the assumptions of the Lemma (that $\mathbf{EXP} \not\subseteq \mathbf{i.o.BPP}$) there is no probabilistic-polynomial-time algorithm that solves L on infinitely many $n \in I$. Note also that once we restrict L only to strings in I, we can assume that the constant c of the instance checker mentioned in Item 3 is a power of 2.

We define the following function $\hat{t}:\{0,1\}^* \to \mathbb{N}$. We let $\hat{t}(x)$ be the minimum number m such that there exists a probabilistic Turing machine M of size $\log m$ such that

$$\Pr[C^{M^m}(x) \neq \mathsf{fail}] > \tfrac{2}{3}$$

We define the function $t : I \to \mathbb{N}$ as follows: $t(n) = \frac{1}{n}\max_{x \in \{0,1\}^{n/c}} \hat{t}(x)$. Intuitively, we defined $t(n)$ in this way so that on one hand we will have that

$L \notin \mathbf{BPtime}(t(n))$ but on the other hand we will have that (a slightly modified version of) Algorithm A described in Section 4.1 will decide L in $poly(t(n))$ time.

We have the following two claims:

Claim 4.6. *There exists a constant e such that $L \in \mathbf{BPtime}(T'(n))$ for every time-constructible function $T' : \mathbb{N} \to \mathbb{N}$ such that $T'(n) \geq t(cn)^e$. (Recall that cn is the size of query the instance checker for L makes on input of size n).*

Proof (Sketch). If we take Algorithm A from Section 4.1 and change it so that in Step 2 it will go over all machines of size $\log m$ (instead of $\log n$) then on input $x \in \{0,1\}^n$ with high probability it will output $L(x)$ after at most $t(cn)^e$ steps for some constant e (e.g., $e = 7$). $\qquad\square$

As a corollary we get that under our assumption, it must be the case that $t(n)$ is super-polynomial (as otherwise we would have that $L \in \mathbf{i.o.\,BPP}$). That is, $t(n) = n^{\omega(1)}$.

Claim 4.7. $L \notin \mathbf{BPtime}(t(n))_{/\log t(n)}$. *Furthermore, this holds for almost all input lengths in I. That is, for every large enough $n \in I$ and every Turing machine M with $\log t(n)$ bits of advice, there exists a string $x \in \{0,1\}^n$ such that $\Pr[M(x) = L(x)] < 2/3$.*

Proof (Sketch). Suppose otherwise. Then it must hold that for some $n \in I$ there exists a $\log m + \log n$-sized[8] Turing machine M such that $\Pr[M^m(x) = L(x)] > 2/3$ for every $x \in \{0,1\}^n$, where $m \leq t(n) = \frac{1}{n}\max_{x \in \{0,1\}^{n/c}} \tilde{t}(x)$. Take $\tilde{M}$ to be the machine M with success probability amplified so that $\Pr[\tilde{M}^{\tilde{m}}(x) = L(x)] > 1 - 2^{-\Omega(n)}$, where $\tilde{m} = \frac{n}{2}m$ ($\tilde{M}$ can be described with less than $\log \tilde{m}$ bits). By the definition of the function $t(\cdot)$, there exists a string $x \in \{0,1\}^{n/c}$ such that $\tilde{t}(x) > \tilde{m}$ and so for every $\log \tilde{m}$ sized Turing machine, and in particular for the machine $\tilde{M}$, the probability that $C^{\tilde{M}^{\tilde{m}}}(x) \neq \mathsf{fail}$ is at most $\frac{2}{3}$. Yet this is a contradiction since $\Pr[C^L(x) \neq \mathsf{fail}] = 1$, and by the union bound, the probability that the checker will ask a query x' and get an answer from $\tilde{M}^{\tilde{m}}$ that is different from $L(x')$ is negligible (i.e., $2^{-\Omega(n)}$). $\qquad\square$

Will we now use the language L and a padding argument to construct for every constant d, a language L' such that L' will be in $\mathbf{BPP}_{\log\log n} \setminus \mathbf{BPtime}(n^d)_{/\log n}$. As a first observation note that since $t(n) \leq 2^n$, it is not hard to show that it must hold that for infinitely many $n \in I$ it holds that $t(cn) \leq t(n)^{2c}$. Therefore, $t(cn)^e \leq t(n)^{e'}$ for some constant e' for every $n \in J$ where J is an infinite subset of I. Let $d \geq 1$ be some constant. We consider the following language L':

$$L' = \{x\#1^l \ : \ x \in L \text{ and } |x| \in J \text{ and } l = t(|x|)^{1/d} - |x| - 1\}$$

(note that $t(|x|)^{1/d} > |x|$ since $t(\cdot)$ is super-polynomial). We claim that

1. $L' \in \mathbf{BPtime}(n^{de'})_{/\log\log n}$.
2. $L' \notin \mathbf{BPtime}(n^d)_{/\log n}$.

---

[8] We could have used any other unbounded function instead of $\log n$.

Once we prove both items we'll be done since we will have

$$\mathbf{BPP}_{/\log\log n} \not\subseteq \mathbf{BPtime}(n^d)_{/\log n}$$

Proof of Item 1. Suppose that we are given a string $y = x\#1^l$, where $n \overset{def}{=} |y| = |x| + l + 1$. We claim that we can check whether $|x| \in J$ and $n = t(|x|)^{1/d}$ using $\log\log n$ bits of advice. If we can do this then we're done since we can then run the modified Algorithm A on x for $n^{de'} = t(x)^{e'}$ steps to decide whether or not $x \in L$. Yet, we only need 1 bit of advice to check whether or not $n = t(k)^{1/d}$ for some $k \in J$. Then, we need to check if $|x| = k$.[9] Yet, this can be done by using $\lceil \log\log k \rceil$ bits of advice since k is a power of 2. Note that since $t(\cdot)$ is super-polynomial we know that $n = t(k)^{1/d} > k^2$ and so $\lceil \log\log k \rceil < \lfloor \log\log n \rfloor$.

Proof of Item 2. Suppose for contradiction that L' is in $\mathbf{BPtime}(n^d)_{/\log n}$. This means that for every $x \in \{0,1\}^*$ such that $|x| \in J$, we can decide whether or not x is a member of the language L using $(t(|x|)^{1/d})^d = t(|x|)$ steps and $\log(t(|x|)^{1/d}) = \frac{1}{d}\log t(|x|)$ bits of advice. Yet this means that the language L can be decided infinitely often in $\mathbf{BPtime}(t(n))_{/\log t(n)}$, contradicting Claim 4.2.
□

Acknowledgments. I thank Oded Goldreich and Avi Wigderson for helpful discussions. In particular, it was Avi who suggested to me the **BPP** hierarchy problem. I am also grateful to the anonymous RANDOM reviewers for their helpful comments.

References

1. Karpinski, M., Verbeek, R.: Randomness, provability, and the separation of monte carlo time and space. In E. Börger, ed: Computation Theory and Logic. Volume 270 of LNCS. Springer (1987) 189–207.
2. Allender, E., Beigel, R., Hertrampf, U., Homer, S.: Almost-everywhere complexity hierarchies for nondeterministic time. Theoretical Computer Science **115** (1993) 225–241.
3. Cai, J.-Y., Nerurkar, A., Sivakumar, D.: Hardness and hierarchy theorems for probabilistic quasi-polynomial time. In ACM, ed.: Proceedings of the thirty-first annual ACM Symposium on Theory of Computing: Atlanta, Georgia, May 1–4, 1999, New York, NY, USA, ACM Press (1999) 726–735
4. Impagliazzo, R., Wigderson, A.: $P = BPP$ if E requires exponential circuits: Derandomizing the XOR lemma. In: Proceedings of the Twenty-Ninth Annual ACM Symposium on Theory of Computing, El Paso, Texas (1997) 220–229
5. Sipser, M.: A complexity theoretic approach to randomness. In: Proceedings of the Fifteenth Annual ACM Symposium on Theory of Computing, Boston, Massachusetts (1983) 330–335

---

[9] We assume that there do not exist $k \neq k' \in J$ such that $t(k)^{1/d} = t(k')^{1/d}$. We can remove elements from J to ensure this condition while keeping it an infinite set.

6. Lautemann, C.: BPP and the polynomial hierarchy. Information Processing Letters **17** (1983) 215–217
7. Blum, M., Kannan, S.: Designing programs that check their work. Journal of the Association for Computing Machinery **42** (1995) 269–291
8. Babai, L., Fortnow, L., Lund, C.: Nondeterministic exponential time has two-prover interactive protocols. Computational Complexity **1** (1991) 3–40 (Preliminary version in Proc. 31st FOCS.).
9. Arora, S., Safra, S.: Probabilistic checking of proofs: A new characterization of NP. Journal of the ACM **45** (1998) 70–122 Preliminary version in FOCS' 92.
10. Levin, L.: Universal search problems (in russian). Problemy Peredachi Informatsii **9** (1973) 265–266 English translation in Trakhtenbrot, B. A.: A survey of Russian approaches to Perebor (brute-force search) algorithms. Annals of the History of Computing, 6 (1984), 384–400.
11. Schnorr, C. P.: Optimal algorithms for self-reducible problems. In Michaelson, S., Milner, R., eds.: Third International Colloquium on Automata, Languages and Programming, University of Edinburgh, Edinburgh University Press (1976) 322–337
12. Trevisan, L., Vadhan, S.: Pseudorandomness and average-case complexity via uniform reductions. In IEEE, ed.; Proceedings of 17th Conference on Computational Complexity, Montréal, Québec, May 21–24, IEEE (2002)

Derandomization That Is Rarely Wrong
from Short Advice That Is Typically Good

Oded Goldreich[1][*] and Avi Wigderson[2][**]

[1] Department of Computer Science, Weizmann Institute of Science (Rehovot, Israel).
oded@wisdom.weizmann.ac.il.
[2] Institute for Advanced Study (Princeton, NJ) and School of Computer Science of the Hebrew
University (Jerusalem, Israel). avi@ias.edu

Abstract. For every $\epsilon > 0$, we present a *deterministic* log-space algorithm that
correctly decides undirected graph connectivity on all but at most 2^{n^ϵ} of the n-
vertex graphs. The same holds for every problem in Symmetric Log-space (i.e.,
$\mathcal{SL}$).

Using a plausible complexity assumption (i.e., that $\mathcal{P}$ cannot be approximated by
$\mathrm{SIZE}(p)^{\mathrm{SAT}}$, for every polynomial p) we show that, for every $\epsilon > 0$, each problem
in $\mathcal{BPP}$ has a *deterministic* polynomial-time algorithm that errs on at most 2^{n^ϵ}
of the n-bit long inputs. (The complexity assumption that we use is not known to
imply $\mathcal{BPP} = \mathcal{P}$.)

All results are obtained as special cases of a general methodology that explores
which probabilistic algorithms can be derandomized by generating their coin
tosses *deterministically* from the input itself. We show that this is possible (for all
but extremely few inputs) for algorithms which take advice (in the usual Karp-
Lipton sense), in which the advice string is short, and most choices of the advice
string are good for the algorithm.

To get the applications above and others, we show that algorithms with short
and typically-good advice strings do exist, unconditionally for $\mathcal{SL}$, and under
reasonable assumptions for $\mathcal{BPP}$ and $\mathcal{AM}$.

1 Introduction

1.1 A Motivating Example

More than two decades ago, Aleliunas *et. al.* [3] presented a *randomized log-space
algorithm* for deciding undirected connectivity (**UCONN**). Their randomized algorithm
triggered (or maybe only draw attention to) the following open problem:

Can undirected connectivity be decided by a deterministic log-space algorithm?

Despite extensive study, the above question is still open. The lowest space-bound
currently known for deterministic algorithms (for undirected connectivity of n-vertex

---

[*] Supported by the MINERVA Foundation, Germany.
[**] Partially supported by NSF grants CCR-9987845 and CCR-9987077.

J.D.P. Rolim and S. Vadhan (Eds.): RANDOM 2002, LNCS 2483, pp. 209–223, 2002.
© Springer-Verlag Berlin Heidelberg 2002

graphs) is $(\log n)^{4/3}$ [4], which builds upon [19] (obtaining space $(\log n)^{3/2}$) and improves upon Savitch's [21] classical bound of $(\log n)^2$. We show that if a deterministic log-space algorithm for UCONN does not exist, then this is due to very few graphs. That is:

Theorem 1 *For every* $\epsilon > 0$, *there exists a deterministic log-space algorithm that correctly decides undirected connectivity on all but at most* 2^{n^ϵ} *of the* n-*vertex graphs. Furthermore, the algorithm never outputs a wrong answer; it either outputs a correct answer or a special ("don't know") symbol. Such algorithms exist for every problem in Symmetric Log-space* $(\mathcal{SL})$.[1]

Surprisingly enough, the proof of Theorem 1 is not difficult (see Sections 3 and 4). It is based on a new viewpoint of the high-level "derandomization" process of Nisan, Szemeredi and Wigderson [19]. Under a different setting of parameters, for every $\epsilon > 0$, this process may be viewed as a *deterministic log-space algorithm that takes advice* (for UCONN), denoted A_{NSW}, that satisfies the following two conditions:

1. The advice string is relatively short. Specifically, the length of the advice is $n^{\epsilon/2}$, where n denotes the number of vertices in the input graph.
2. Most choices of the advice string are "good" *for all* n-vertex graphs. More precisely, the algorithm works correctly (i.e., decides correctly whether the input n-vertex graph is connected) whenever the $n^{\epsilon/2}$-bit long advice string is a universal traversal sequence for $n^{\epsilon/10}$-vertex graphs. Moreover, by [3], most advice strings satisfy that property.

Note that we use the term "advice" in the standard sense of Karp and Lipton [16]; an advice string is good if it makes the algorithm using it give the correct answer on *all* inputs of the given length. The remarkable property of algorithm A_{NSW} is that it satisfies *both* conditions above: it has good advice strings that are *much shorter* than the input, and furthermore most strings of this length are good advice strings.

Remark 2 *Note that the Condition 2 (i.e., many good advice strings) is easy to obtain from every probabilistic algorithm. Indeed, Adleman's simulation [1] of probabilistic algorithms by nonuniform ones shows that any BPP-algorithm A can be modified into a deterministic A′ which takes advice, for which most advice strings are good. However, since this transformation requires amplification[2] in order to enable a union bound over all inputs of a given length, the advice is* necessarily longer *than the input length, which violates the Condition 1 (i.e., short advice string).*

---

[1] For a recent survey of this class, including a list of complete problems, see [2]

[2] Denoting by $A(x, r)$ the output of A on input x and coins r, we obtain A' by letting $A'(x, (r_1, ..., r_{O(|x|)}))$ output the most frequent value among $A(x, r_1), ..., A(x, r_{O(|x|)})$. Using sophisticated amplification methods, the sequence $(r_1, ..., r_{O(|x|)})$ can be encoded by a string of length $O(|x| + |r|)$, but we cannot hope for length shorter than $|x|$ (because we need to reduce the error to less than $2^{-|x|}$ in order to apply the union bound).

To demonstrate the impact of having *many good advice strings that are shorter than the input*, we sketch how to complete the proof of Theorem 1. The general claim is that an advice-taking algorithm for which at least $2/3$ of the possible advice strings are good can be transformed into a (standard) deterministic algorithm (of comparable complexity) that errs on a number of inputs that is roughly exponential in the length of the (original) advice. Thus, we obtain a meaningful result if and only if the length of the advice is significantly smaller than the length of the input.

The claimed transformation is presented in two stages. First, we derive a randomized (log-space) algorithm A' that uses a *logarithmic amount of randomness* (and errs on at most 2^{n^ϵ} inputs). Specifically, on input G, algorithm A' uses its randomness as a seed to an adequate extractor (cf. [26]), and extracts *out of its input* G an advice string, s, of length $n^{\epsilon/2}$. Then A' invokes A_{NSW} on input G and advice s, and outputs whatever the latter does. It is easy to show that there can be at most $2^{(n^{\epsilon/2})^2}$ inputs on which A' errs with probability greater than $1/2$. Applying a straightforward derandomization (and ruling by majority), we obtain a deterministic (log-space) algorithm A'' that decides correctly on all but at most of the 2^{n^ϵ} inputs.

1.2   The Underlying Principle

The above transformation can be applied to any advice-taking algorithm, and is meaningful if and only if most of the advice strings are good and the length of the advice is significantly smaller than the length of the input. That is:

Theorem 3 *Let A be an advice-taking polynomial-time (resp., log-space) algorithm for a problem Π, and let $\ell : \mathbb{N} \to \mathbb{N}$. Suppose that for every n it holds that at least a $2/3$ fraction of the $\ell(n)$-bit long strings are good advice stings for A; that is,*

$$\Pr_{r \in \{0,1\}^{\ell(n)}}[\forall x \in \{0,1\}^n \text{ it holds that } A(x,r) = \Pi(x)] \geq \frac{2}{3}$$

where $\Pi(x)$ denotes the correct answer for x. Then, for every $c > 1$, there exists a deterministic polynomial-time (resp., log-space) algorithm for Π that errs on at most $2^{\ell(n)^c}$ of the n-bit inputs. Furthermore, in case $\ell(n) = \Omega(n)$, the resulting algorithm errs on at most $2^{c \cdot \ell(n)}$ of the n-bit inputs.

The proof of Theorem 3 appears in Section 3. As hinted above, the proof proceeds by viewing the input itself as a source of randomness and extracting a random advice string via an adequate extractor (which uses logarithmically long seeds). The extracted advice string is sufficiently random (and thus is a good advice with probability greater than $1/2$) provided that the input comes from a source with min-entropy sufficiently larger than the length of the extracted advice. It follows that the number of inputs on which most extracted advice strings are not good can be bounded in terms of the min-entropy bound.

212   O. Goldreich and A. Wigderson

1.3   Other Applications

The question is how to obtain algorithms to which Theorem 3 can be (meaningfully) applied. We have seen already one such example (i.e., A_{NSW}), and in this subsection we will discuss some more.

Conditional derandomization of BPP and AM. As mentioned above, any randomized algorithm can be transformed into a (deterministic) advice-taking algorithm for which most possible advice strings are good. The problem is that the length of the advice will be longer than the length of the input. One natural idea is to use an adequate pseudorandom generator in order to shrink the length of the advice strings, while maintaining the fraction of good advice strings. We observe that the property of being a good advice string for a specific algorithm can be efficiently tested using a SAT-oracle. Thus, it suffices to have pseudorandom generators that withstand probabilistic polynomial-time distinguishers that use a SAT-oracle. (Actually, it suffices to have pseudorandom generators that withstand distinguishers that correspond to the class $\mathcal{MA}$.)

Recall that pseudorandom generators that withstand non-uniform polynomial-size distinguishers that use a SAT-oracle were constructed before (cf. [5,17]) under various non-uniform assumptions (which also refer to circuits with SAT-oracle gates). In particular, we will use the following result:

Theorem 4 (informal, implicit in [17,20]): *Suppose that for every polynomial p there exist a predicate in $\mathcal{P}$ that is hard to approximate[3] in $\mathrm{SIZE}(p)^{\mathrm{SAT}}$. Then, for every polynomial q, there exists a deterministic polynomial-time algorithm that expands k-bit long random seeds to $q(k)$-bit long sequences that are indistinguishable from random ones by any $q(k)^2$-size circuit that uses a SAT-oracle.*

Combining Theorem 4 with the observation that being a good advice string for a polynomial-time algorithm is testable in comparable time with the help of an SAT-oracle, we obtain (see proof in Section 5):

Theorem 5 (informal) *For every polynomial p and constant $\epsilon > 0$, under the assumption of Theorem 4, every probabilistic p-time algorithm can be converted into a functionally-equivalent polynomial-time advice-taking algorithm that uses advice strings of length n^ϵ, such that more than $2/3$ fraction of the possible advice strings are good.*

---

[3] Here, hard to approximate may mean that any machine in the class fails to guess the correct value of a random instance with success probability greater than $2/3$. Using Yao's XOR Lemma (cf. [27,8]), hardness to approximate can be amplified such that this class fails to guess the correct value of a random n-bit instance with probability greater than $(1/2) + (1/p(n))$. (Recall that such amplification requires only logarithmically many instances, and thus in our case it affects the resource bounds in a minor way.)

Combining Theorem 3 and 5, we obtain (see proof in Section 5):

Corollary 6 (informal) *Under the assumption of Theorem 4, for every $\epsilon > 0$, every language in $\mathcal{BPP}$ can be decided by a deterministic polynomial-time algorithm that errs on at most 2^{n^ϵ} of the n-bit long inputs.*

Interestingly, under the same assumption we can also derandomize $\mathcal{AM}$ (see proof in Section 5):

Theorem 7 (informal) *Under the assumption of Theorem 4, for every $\epsilon > 0$, every language in $\mathcal{AM}$ can be decided by a non-deterministic polynomial-time algorithm that errs on at most 2^{n^ϵ} of the n-bit long inputs.*

Comparison to previous results: When making the comparison, we refer to three issues:

1. The running-time of the derandomization,
2. For how many instances does the derandomization fail,
3. The intractability assumption used.

Our focus is on polynomial-time derandomization. Thus, Corollary 6 should be compared with Impagliazzo and Wigderson [13], who proved that $\mathcal{BPP} = \mathcal{P}$ under the assumption that *for some $c > 0$, the class $\mathcal{E}$ is not contained in* $\mathrm{SIZE}(2^{cn})$. Likewise, Theorem 7 should be compared with Kilvans and van Melkebeek [17], who proved that $\mathcal{AM} = \mathcal{NP}$ under the assumption *for some $c > 0$, the class $\mathcal{E}$ is not contained in* $\mathrm{SIZE}(2^{cn})^{\mathrm{SAT}}$. Both [13] and [17] present perfect derandomizations, whereas our derandomizations fail on very few inputs (which is of course weaker).

Comparing the assumptions is easier for the $\mathcal{AM}$ derandomization, as both Theorem 7 and [17] refer to lower bounds for circuits with SAT oracle gates. But as Theorem 7 refers to separation of classes with smaller resources, our assumption is weaker.[4] As for the $\mathcal{BPP}$ derandomization, it seems that the assumption made by [13] and Corollary 6 are incomparable. We still need the SAT oracles, but as above, our classes are lower.

There is another set of derandomization results, typically under uniform complexity assumptions (which are seemingly weaker than the analog non-uniform assumptions, cf. [14]). These results yield deterministic simulations that may make many errors; however, no efficient procedure can find inputs on which the simulation errs. This notion of imperfect simulation seems incomparable to ours; we seem to make much fewer errors, but the inputs for which errors occur may be easy to generate.

---

[4] For example, if $\mathcal{P}$ is contained in $\mathrm{SIZE}(n^2)$ then $\mathcal{E}$ is contained in $\mathrm{SIZE}(2^{cn})$ for every $c > 0$, but the converse is not known. Thus, "$\mathcal{E}$ is not contained in $\mathrm{SIZE}(2^{cn})$ for every $c > 0$" implies "$\mathcal{P}$ is not contained in $\mathrm{SIZE}(n^2)$" (but again the converse is not known). Also note that, in case of predicates in $\mathcal{E}$, hardness on the worst-case yields hardness to approximate.

About our assumption: To gain some intuition about the assumption used in Corollary 6, we consider a few plausible conjectures that imply it.

- *Given a Boolean circuit, determine whether it evaluates to 1 on more than half of its possible inputs.* Clearly, this problem can be decided in time that is exponential in the number of inputs to the circuit, but it seems hard to decide it in significantly less time. Specifically, suppose that the input circuit has size n and $\ell(n)$ input bits (e.g., $\ell(n) = O(\log n)$), then the problem is solvable in time $2^{\ell(n)} \cdot n$ but seems hard to decide in time $2^{\ell(n)/10}$. Furthermore, the problem seems hard even for $2^{\ell(n)/10}$-size circuit that use SAT-gates (i.e., an oracle to SAT). However, conjecturing that this problem is hard even to approximate (i.e., when given a random circuit as input) requires a suitable notion of the distribution of inputs (i.e., circuits). A conceivably good definition is that the input/circuit distribution is uniform over $\ell(n)$-variate polynomials over GF(2) having n monomials.
- For any polynomial q, we consider the following decision problem: *Given a description of a prime P and an natural number $x < P$, decide whether most of the natural numbers in the set $\{x + 1, ..., x + q(|P|)\}$ are quadratic residues modulo P.* Clearly, this problem can be decided in time $q(|P|) \cdot |P|^3$, but it seems hard to decide (or even approximate) it by (say) $q(|P|)^{1/3}$-size circuits even ones having SAT-gates.
- *Given a sequence of n-by-n matrices, $A_1, ..., A_\ell$, over a sufficiently large finite field, determine $\sum_{S \subseteq [\ell]} \det(\sum_{i \in S} A_i)$*, where $\det(M)$ denote the determinant of the matrix M. Again, this problem can be solved in time $2^\ell \cdot n^3$, but it seems hard to solve it by (say) $\min(2^{\ell/3}, \ell^{n/3})$-size circuits, even ones having SAT-gates.[5] (Again, we may use $\ell = O(\log n)$.)
  We comment that this problem is downward-self-reducible and random-self-reducible (see our Technical Report [10]). This is particularly interesting in light of the paper [14], which shows that for such functions *uniform, worst-case hardness* implies that they can be used in the NW-generator (obtaining derandomization on the average). More concretely, assume this function is not in BPTIME$(p)^{\text{SAT}}$, for any fixed polynomial p, for infinitely many input lengths. Then every language in BPP has a deterministic polynomial-time algorithm which, for infinitely many input lengths errs on at most 2^{n^ϵ} inputs, with $\epsilon > 0$ an arbitrarily small constant. To summarize, the special structure of this function, enables reducing our hardness assumption from non-uniform and average-case to uniform and worst-case. Thus, it will be interesting to try and substantiate (in direct or indirect ways), the conjecture that for this function is indeed difficult to compute in time $2^{\ell/O(1)}$.

---

[5] The $\ell^{n/3}$ term, which is meaningful only if $\ell = \Omega(n \log n)$, is introduced to account for a possible $(\ell^n \cdot n^3)$-time algorithm that computes the formal (degree n) polynomial $p(x_1, ..., x_\ell) \overset{\text{def}}{=} \det(\sum_{i=1}^\ell x_i A_i)$, and computes the sum $\sum_{x_1, ..., x_\ell \in \{0,1\}} p(x_1, ..., x_\ell)$ by computing separately the contribution of each of the $\binom{n+\ell}{n}$ terms of p.

Direct Product Problems. Direct product problems yield another interesting case where randomized algorithms can be transformed into ones having relatively short good advice strings. Specifically, any good advice string for the original problem constitutes a good advice string for the direct product problem. Applying Theorem 3, we obtain a deterministic algorithm (of related complexity) for the direct product problem that errs on a number of inputs that is independent of the arity of the (direct product) problem. For further details, see Section 6.

2 Preliminaries

In this section we recall some standard notions and notations. We will also recall some known results.

2.1 Randomness and Extractors

We consider random variables that are assigned binary values as strings. In particular, U_m will denote a random variable that is uniformly distributed over $\{0,1\}^m$. The min-entropy of a random variable X is the maximal (real number) k such that for every string x it holds that $\Pr[X = x] \leq 2^{-k}$.

The statistical difference between two random variables X and Y, denoted $\Delta(X,Y)$, is defined as $\frac{1}{2} \cdot \sum_z |\Pr[X = z] - \Pr[Y = z]|$. Clearly, $\Delta(X,Y) = \max_S \{\Pr[X \in S] - \Pr[Y \in S]\}$. We say that Y is ϵ-close to X if $\Delta(X,Y) \leq \epsilon$; otherwise, we say that Y is ϵ-far from X.

A function $E : \{0,1\}^n \times \{0,1\}^t \to \{0,1\}^\ell$ is called a (k,ϵ)-extractor if for every random variable X that has min-entropy at least k it holds that $E(X,U_t)$ is ϵ-close to U_ℓ. The following fact is well-known and easy to establish:

Fact 8 *Suppose that $E : \{0,1\}^n \times \{0,1\}^t \to \{0,1\}^\ell$ is a (k,ϵ)-extractor. Then, for every set $S \subseteq \{0,1\}^\ell$, all but at most 2^k of the x's in $\{0,1\}^n$ satisfy $\Pr[E(x,U_t) \in S] \geq (|S|/2^\ell) - \epsilon$.*

Proof: Let $B \subseteq \{0,1\}^n$ denote the set of x's that satisfy $\Pr[E(x,U_t) \in S] < (|S|/2^\ell) - \epsilon$. Consider a random variable X that is uniformly distributed on the set of the latter x's. Then,

$$\Delta(E(X,U_t), U_\ell) \geq \Pr[E(X,U_t) \in S] - \Pr[U_\ell \in S] > \epsilon$$

Thus, $\log_2 |B| < k$ must hold, and $|B| < 2^k$ follows. ∎

Uniform families of extractors: We actually consider families of (extractor) functions (of the above type). These families are parameterized by n, whereas t, ℓ, k and ϵ are all functions of n. Thus, when we say that $\{E_n : \{0,1\}^n \times \{0,1\}^{t(n)} \to \{0,1\}^{\ell(n)}\}_{n \in \mathbb{N}}$ is a (k, ϵ)-**extractor** we mean that for every n the function E_n is a $(k(n), \epsilon(n))$-extractor. We will use the following well-known result of Trevisan [26]:[6]

Theorem 9 *For every $c > 1$, $\epsilon > 0$ and every linear-space computable $\ell : \mathbb{N} \to \mathbb{N}$, there exists a polynomial-time computable family of functions $\{E_n : \{0,1\}^n \times \{0,1\}^{O(\log n)} \to \{0,1\}^{\ell(n)}\}_{n \in \mathbb{N}}$ that constitute a (k, ϵ)-extractor, where $k(n) \stackrel{\text{def}}{=} \ell(n)^c$. Furthermore, these functions are computable in log-space.*

2.2 Some Complexity Classes

We denote by $\text{SIZE}(p)$ the class of (non-uniform) families of p-size circuits, and by $\text{SIZE}(p)^{\text{SAT}}$ the class of such circuits augmented by SAT-gates (oracle gates to SAT).

   We refer to two restricted classes of interactive proof systems (cf. [11]), specifically $\mathcal{MA}$ and $\mathcal{AM}$. The class $\mathcal{MA}$ (resp., $\mathcal{AM}$) consists of all languages having a two-round public-coin interactive proof in which the prover (called Merlin) sends the first (resp., second) message (cf. [6]). In both cases, the verifier's (Arthur's) message consists of the entire contents of its random-tape (hence the term *public-coin*). By $\mathcal{MA}(p)$ (resp., $\mathcal{AM}(p)$) we denote a parameterized version of $\mathcal{MA}$ (resp., $\mathcal{AM}$) in which the verifier's complexity is bounded by p.

3   The Transformation: Proof of Theorem 3

We combine an algorithm A as in the hypothesis (of Theorem 3) with an adequate extractor to obtain the desired deterministic algorithm. Specifically, let use denote by S the set of good advice strings of algorithm A; that is, for every $r \in S$ and $x \in \{0,1\}^n$, it holds that $A(x, r) = \Pi(x)$. Then, by the theorem's hypotheses, $|S| \geq (2/3) \cdot 2^{\ell(n)}$. Let $E_n : \{0,1\}^n \times \{0,1\}^{t(n)} \to \{0,1\}^{\ell(n)}$ be a $(k(n), 0.1)$-extractor. Then, by Fact 8, for all but at most $2^{k(n)}$ of the n-bit long x's, the probability that $E_n(x, U_{t(n)}) \in S$ is at least $(2/3) - 0.1 > 1/2$. Thus, for all but at most $2^{k(n)}$ of the x's, for a strict majority of the (extractor seeds) $r' \in \{0,1\}^{t(n)}$ it is the case that $E_n(x, r') \in S$. Scanning all possible $r' \in \{0,1\}^{t(n)}$, an ruling by the majority of the $A(x, E_n(x, r'))$ values, we obtain the correct answer for all but at most $2^{k(n)}$ of the n-bit long inputs. (The resulting algorithm is depicted in Figure 1.)

---

[6] We mention that both the error-correcting code and the (weak) designs used by Trevisan's extractor can be constructed in log-space.

On input $x \in \{0,1\}^n$, scan all possible $r' \in \{0,1\}^{t(n)}$,
performing the following steps for each r':

   1. Obtain $r \leftarrow E_n(x, r')$.
   2. Invoke A on input x and advice r, and record the answer in $v(r')$.

Output the majority value in the sequence of $v(r')$'s.

Fig. 1. The resulting deterministic algorithm.

Clearly, the time complexity of the resulting algorithm is $2^{t(n)} \cdot (T_A(n) + T_E(n))$, where T_A (resp., T_E) denotes the time complexity of algorithm A (resp., of the extractor $E = \{E_n\}_{n \in \mathbb{N}}$). Similarly, the space complexity of the resulting algorithm is $t(n) + S_A(n) + S_E(n)$, where S_A (resp., S_E) denotes the space complexity of A (resp., of E).

Using Theorem 9, we may set $t(n) = O(\log n)$ and $k(n) = \ell(n)^c$, while using a log-space computable extractor. Thus, the main part of Theorem 3 follows. To establish the furthermore part of Theorem 3 (which refers to $\ell(n) = \Omega(n)$), we use Zuckerman's extractor [28] instead of Theorem 9.[7]

Remark 10 *If algorithm A never errs* (but may rather output a special "dont know" symbol on some (x, r) pairs) *then the same property is inherited by the resulting deterministic* (single-input) *algorithm. A similar statement holds for one-sided error of algorithms for decision problems.* (In both cases, we may actually use dispersers instead of extractors.)

4   Undirected Connectivity: Proof of Theorem 1

As explained in the introduction, Theorem 1 is proved by applying Theorem 3 to an advice-taking algorithm that is implicit in (or rather derived from) the work of Nisan, Szemeredi and Wigderson [19]. The latter algorithm refers to the notion of a universal traversal sequence for the set of all (3-regular) graphs of a certain size. Such sequences were defined by Cook (cf. [3]) and extensively studied since. There are many variants of this definition, and we choose one that is most convenient for our purpose.

Definition 11 (Universal traversal sequences for d-regular graphs):

---

[7] To handle log-space algorithms, we need a log-space computable extractor for this case (i.e., of $\ell(n) = \Omega(n)$ and $k(n) = c \cdot \ell(n)$). Such extractors do exist [22].

- *Let G be a d-regular graph, v be a vertex in G, and $\sigma = (\sigma_1, ..., \sigma_t)$ be a sequence over $[d] \stackrel{\text{def}}{=} \{1, ..., d\}$. The σ-directed G-walk starting at v is the vertex sequence $(v_0, v_1, ..., v_t)$, where $v_0 = v$ and v_{i+1} is the σ_i^{th} neighbor of v_i.*
- *We say that $\sigma \in [d]^*$ is (d, m)-universal if for every m-vertex d-regular graph G and every vertex v in G, the σ-directed G-walk starting at v visits all the vertices of the connected component of v in G.*
  We say that $\sigma \in \{0, 1\}^*$ is (d, m)-universal if when viewed as a sequence over $[d]$ it is (d, m)-universal.

The following result is implicit in the work of Nisan, Szemeredi and Wigderson [19].

Theorem 12 (following [19]): *Let $\ell, m : \mathsf{N} \to \mathsf{N}$ be space-constructible functions such that $m(n) \le \ell(n) \le n$ for all n's.[8] Then, there exists a deterministic $O((\log^2 n)/(\log m(n)))$-space advice-taking algorithm A_{NSW} for UCONN that on input an n-vertex graph uses an advice string of length $\ell(n)$, where the set of good advice strings contains every $(3, m(n))$-universal sequence.[9] Furthermore, algorithm A_{NSW} never errs, but may rather output a special ("dont know") symbol (in case the advice string is not good).*

Nisan *et. al.* [19] used the fact that (by [18]) $(3, m)$-universal sequences of length $\exp(\log^2 m)$ can be constructed in $O(\log^2 m)$-space. Setting $\ell(n) = n$ and $m(n) = \exp(\log^{1/2} m)$, their $O(\log^{3/2} m)$-space algorithm follows by combining Theorem 12 with the $O(\log n)$-space algorithm (of Nisan [18]) for constructing a $m(n)$-universal sequence of length n. Here, instead, we use the well-known fact that most sequences of length $\tilde{O}(m^3)$ are $(3, m)$-universal [3]. That is:

Theorem 13 (Aleliunas *et. al.* [3]): *More than a 2/3 fraction of the $O(m^3 \cdot \log m)$-bit long strings are $(3, m)$-universal sequences.*

Setting $\ell(n) = O(m(n)^3 \cdot \log m(n)) < m(n)^4$, it follows that a 2/3 fraction of the $\ell(n)$-bit long strings are $(3, \ell(n)^{1/4})$-universal, and thus are good advice strings for A_{NSW} (when applied to n-vertex graphs). Specifically, under this setting, A_{NSW} uses space $O((\log^2 n)/\log m(n)) = O((\log^2 n)/\log \ell(n))$ (and $\ell(n)$-bit long advice strings).

---

[8] The growth restriction on these functions is natural in the context of [19]. Of course $m(n) \le \ell(n)$ must hold, as otherwise the claim holds vacuously (because m-universal sequences must have length greater than m). For $\ell(n) > n$, algorithm A_{NSW} uses space $O((\log \ell(n))(\log n)/(\log m(n)))$ rather than $O((\log^2 n)/(\log m(n)))$.

[9] Nisan *et. al.* [19] work with an auxiliary 3-regular $O(n^2)$-vertex graph that is derived from the initial n-vertex graph in a straightforward manner. The algorithm works in iterations, where a good advice allows to shrink the size of the graph by a factor of $m(n)/4$ in each iterations. (In case the advice is not good, the algorithm may fail to shrink the graph, but will detect this failure, which justifies the furthermore clause.) Thus, there are $(\log n)/(\log(m(n)/4))$ iterations, and each iteration is implementable in $O(\log n)$ space.

For $\ell(n) = n^{1/O(1)}$, we obtain a deterministic log-space advice-taking algorithm for UCONN that uses $\ell(n)$-bit long advice strings such that at least a $2/3$ fraction of the possible advice strings are good.

We are now ready to invoke Theorem 3: given any desired constant $\epsilon > 0$, we set $\ell(n) = n^{\epsilon/c}$ (for any constant $c > 1$) and invoke Theorem 3. This yields a deterministic log-space algorithm for UCONN that errs on at most $2^{\ell(n)^c} = 2^{n^\epsilon}$ of the n-vertex graphs. The main part of Theorem 1 follows. By Remark 10, the fact that A_{NSW} never errs is inherited by the log-space algorithm that we derive, and thus the furthermore-part of Theorem 1 follows.

USTCONN and $\mathcal{SL}$. The arguments presented above apply also to USTCONN, where one is given an undirected graph G and a pair of vertices (u, v) and needs to determine whether or not these vertices are connected in G. Actually, the main algorithm presented in [19] is for that version, and so all the above holds. Since USTCONN is log-space complete for the class $\mathcal{SL}$ (symmetric log-space), it is easy to see that for every $\epsilon > 0$ every problem in $\mathcal{SL}$, has a deterministic log-space algorithm which is always correct, and answers "dont-know" on at most 2^{n^ϵ} inputs of length n. Indeed, one only has to observe that log-space reductions can only blow-up the input length polynomially, and since ϵ can be made arbitrarily small we get the same bound on the number of "dont-know"s. A compendium of interesting problems in $\mathcal{SL}$ can be found in [2].

5   Derandomizing BPP and AM

Comments on the proof of Theorem 4: With a minor modification (to be discussed), Theorem 3.2 in (the full version of) [17] yields Theorem 4. Theorem 3.2 in [17] assumes the existence of a predicate that is hard to approximate[10] by $\mathrm{SIZE}(p)^{\mathrm{SAT}}$, and conclude that a certain pseudorandom generator expanding k-bit strings to $q(k)$-bit strings exists, where $q(k) \leq p(\sqrt{k} \log q(k))^{1/2}$. The resulting generator withstand $q(k)$-sized circuits with SAT-gates, and operates in time related to the complexity of evaluating the predicate on $(\sqrt{k} \log q(k))$-bit long inputs and to $2^{O(k)}$. However, the latter additive term (of $2^{O(k)}$) is merely due to the fact that [17] use the brute-force design construction of [20] rather than their efficient (i.e., $\mathrm{poly}(k)$-time) construction, which can be used whenever p is a polynomial.[11] (Another minor detail is that we want to withstand $q(k)^2$-sized circuits rather than withstand $q(k)$-sized circuits.) Theorem 4 follows by setting, for any given polynomial q, the polynomial p such that $q(k)^2 < p(\sqrt{k} \log q(k))^{1/2}$ holds (e.g., $p(n) \stackrel{\mathrm{def}}{=} (q(n^2))^4$). ■

---

[10] See Footnote 3.

[11] Indeed, Theorem 3.2 in [17] is stated for arbitrary p's, and the focus is actually on super-polynomial p's.

Proof of Theorem 5: For each set in $\mathcal{BPP}$, by using straightforward amplification, we obtain a randomized polynomial-time algorithm that errs with probability at most $2^{-(n+2)}$ (on each n-bit input). This algorithm yields an advice-taking algorithm for which at least $3/4$ of the possible advice strings are good. We call such an algorithm **canonical**.

For any polynomial q and $\epsilon > 0$, we construct a generator $G_{\epsilon,q}$ as in Theorem 4 such that n^{ϵ}-bit long strings are stretched into sequences of length $q(n)$ that pass all $q(n)^2$-size distinguishers (i.e., circuits with SAT-gates).

We claim that for every q-time canonical algorithm, A, at least a $2/3$ fraction of the sequences generated by $G_{\epsilon,q}$ are good advice strings. Otherwise, we consider an $\mathcal{MA}$-proof system for **bad** (i.e., non-good) advice strings. On input a string $r \in \{0,1\}^m$, where $m = q(n)$, the prover (i.e., Merlin) sends $x \in \{0,1\}^n$, and the verifier accepts if and only if $A(x,r)$ differs from the majority vote of $A(x,s)$ taken over a sample (of say 100) uniformly selected $s \in \{0,1\}^m$. Note that if r is a bad advice string then the verifier accepts with probability at at least 0.99 (provided Merlin acts optimally), whereas if r is a good (i.e., not bad) advice string then the verifier accepts with probability at most 0.01 (no matter what Merlin does). Thus, the probability that the verifier accepts a string produced by $G_{\epsilon,q}$ is at least $(1/3) \cdot 0.99 = 0.33$, whereas the probability that the verifier accepts a uniformly distributed $q(n)$-bit long string is at most $(1/4) \cdot 1 + (3/4) \cdot 0.01 < 0.3$. Observe that the verifier's running-time is bounded by $O(m) = O(q(n))$. Applying known transformations from $\mathcal{MA}$ to $\mathcal{AM}$, and from the latter to $\text{BPTIME}^{\text{SAT}}$, yields a distinguisher in $\text{BPTIME}(m^2)^{\text{SAT}}$ for m-bit inputs (provided that $q(n) > n\log n$).[12] We derive a contradiction to the security of $G_{\epsilon,q}$, and the theorem follows. ∎

Proof of Corollary 6: By applying Theorem 5 (with ϵ replaced by $\epsilon/2$), we can derive for any set in $\mathcal{BPP}$ a polynomial-time advice-taking algorithm with advice strings of length $n^{\epsilon/2}$ such that at least $2/3$ of the advice strings are good. Then applying Theorem 3 (with $c = 2$), the current claim follows. ∎

Proof of Theorem 7: We just follow the proof of Corollary 6, adapting the notion of a good advice to the AM setting, and observing that the proof of Theorem 5 still applies. Specifically, a **good advice for an AM-game** is a verifier message that is good for all inputs of a certain length (i.e., for inputs in the language there exist an acceptable prover response, whereas no such response exists in case the input is not in the language). The $\mathcal{MA}$-proof system described in the proof of Theorem 5 extends in the straightforward

---

[12] The transformation of $\mathcal{MA}$ to $\mathcal{AM}$ increases the running-time by a factor related to the length of the original Merlin's message, which equals n in our case. In the second transformation (i.e., from $\mathcal{AM}$ to $\text{BPTIME}^{\text{SAT}}$), we need to make a SAT-query that is answered with an optimal Merlin message. This amounts to encoding the accepting predicate of Arthur as a SAT-instance, where the length of that instance is almost linear in the verifier's running-time (on a multi-tape Turing machine). Thus, $\mathcal{MA}(m) \subseteq \mathcal{AM}(mn) \subseteq \text{BPTIME}(mn\log(nm))^{\text{SAT}}$.

manner (i.e., Merlin now sends an adequate x along with an adequate prover message for the AM-game). Indeed, moving to $\mathcal{AM}$ will blow-up the complexity by a factor related to the prover message, and so we should start (w.l.o.g.) with a AM-game in which the verifier's messages are longer than the prover messages.[13] But otherwise, the argument remains intact. ∎

6 Direct Product Problems

The following observation is due to Noam Nisan: A natural domain where one may obtain advice-taking algorithms with good advice strings that are much shorter than the input is the domain of direct product problems. That is, suppose that Π is a problem having a (polynomial-time) randomized algorithm. As we have commented in the introduction, by straightforward amplification [1] we may obtain an advice-taking algorithm for Π such that for some polynomial ℓ, the algorithm uses $\ell(n)$-bit long advice for n-bit long inputs such that at least a $2/3$ fraction of all possible advice are good. The key point is that such an advice-taking algorithm for Π yields an algorithm with similar performance for the direct product of Π. That is, for any n and t, given input $(x_1, ..., x_t) \in \{0,1\}^{t \cdot n}$ and an $\ell(n)$-bit long advice, the latter algorithm invokes the single-instance algorithm on each x_i using the same advice in all invocations. Clearly, if the advice is good for the single-instance algorithm then it is also good for the multiple-instance algorithm. Applying Theorem 3, we obtain

Theorem 14 *For every problem Π in $\mathcal{BPP}$, there exist a polynomial p and a deterministic polynomial-time algorithm A such that for every n and t, for all but at most $2^{p(n)}$ of the sequences $\overline{x} = (x_1, ..., x_t) \in \{0,1\}^{t \cdot n}$ it holds that $A(\overline{x}) = (\Pi(x_1), ..., \Pi(x_t))$.*

We stress that the number of inputs on which A may err depends only on the length of the individual Π-instances (i.e., n), and not on their number (i.e., t). We comment that this is superior to what could be obtained by straightforward considerations.[14]

---

[13] Specifically, assuming that the prover's messages in the AM-game are shorter than m, we get an $\mathcal{MA}(m)$ proof system for bad advice strings, which is transformed to an $\mathcal{AM}(m^2)$ proof system, which in turn resides in $\mathrm{BPTIME}(m^2 \log(m^2))^{\mathrm{SAT}}$. So we should use a generator as in Theorem 4 such that n^ε-bit long strings are stretched into sequences of length $q(n)$ that pass all $q(n)^3$-size distinguishers (rather than $q(n)^2$-size ones).

[14] For example, suppose that Π has a BPP-algorithm of randomness complexity ρ (such that $\rho(n) \geq n$ or else we can derandomize Π itself). Then, by straightforward amplification, we obtain a 2/3-majority of good advice strings for advice length $\ell(n) = O(\rho(n) \cdot n)$ (or even $\ell(n) = O(\rho(n) + n)$ by using expander-based amplification). Thus, in Theorem 14, we obtain $p(n) = \ell(n)^c$, for any desired $c > 1$. In contrast, if we use $x_2, ..., x_t$ to generate $m \approx t/\rho(n)$ disjoint random pads for invocations of the BPP-algorithm on input x_1 then we may err on $\exp(-m) \cdot 2^{tn} = 2^{(1-o(1)) \cdot tn}$ of the sequences. Using $x_2, ..., x_t$ to generate $tn/(\log d)$ related pads (by using a d-regular optimal expander) may yield error on $(1/\sqrt{d})^{tn/(\log d)} \cdot 2^{tn} = 2^{tn/2}$ sequences. For large t, this is inferior to the upper bound of $2^{O(\rho(n) \cdot n)^{3/2}} \leq 2^{\rho(n)^3}$ sequences (not to mention $2^{\tilde{O}(\rho(n)+n)^c} = 2^{\rho(n)^c}$, for any $c > 1$) obtained by using Theorem 14.

Acknowledgments. We are grateful to Noam Nisan for suggesting the application to direct product presented in Section 6. We also thank Eric Allender, Shien Jin Ong, Salil Vadhan and the anonymous reviewers for their helpful comments.

References

1. L. Adleman. Two theorems on random polynomial time. In *10th FOCS*, pages 75–83, 1978.
2. C. Alvarez and R. Greenlaw. A compendium of problems complete for symmetric logarithmic space. ECCC report TR96-039, 1996.
3. R. Aleliunas, R.M. Karp, R.J. Lipton, L. Lovász and C. Rackoff. Random walks, universal traversal sequences, and the complexity of maze problems. In *20th FOCS*, pages 218–223, 1979.
4. R. Armoni, M. Saks, A. Wigderson and S. Zhou. Discrepancy sets and pseudorandom generators for combinatorial rectangles. In *37th FOCS*, pages 412-421, 1996.
5. V. Arvind and J. Köbler. On pseudorandomness and resource-bounded measure. In *17th FSTTCS*, Springer-Verlag, LNCS 1346, pages 235–249, 1997.
6. L. Babai. Trading Group Theory for Randomness. In *17th STOC*, pages 421–429, 1985.
7. M. Blum and S. Micali. How to Generate Cryptographically Strong Sequences of Pseudo-Random Bits. *SICOMP*, Vol. 13, pages 850–864, 1984. Preliminary version in *23rd FOCS*, 1982.
8. O. Goldreich, N. Nisan and A. Wigderson. On Yao's XOR-Lemma. *ECCC*, TR95-050, 1995.
9. O. Goldreich, D. Ron and M. Sudan. Chinese Remaindering with Errors. TR98-062, available from *ECCC*, at http://www.eccc.uni-trier.de/eccc/, 1998.
10. O. Goldreich and A. Wigderson. Derandomization that is rarely wrong from short advice that is typically good. TR02-039, available from *ECCC*, 2002.
11. S. Goldwasser, S. Micali and C. Rackoff. The Knowledge Complexity of Interactive Proof Systems. *SICOMP*, Vol. 18, pages 186–208, 1989. Preliminary version in *17th STOC*, 1985.
12. R. Impagliazzo, V. Kabanets and A. Wigderson. In search of an easy witness: Exponential versus probabilistic time. In proceedings of *16th CCC*, pages 2–12, 2001.
13. R. Impagliazzo and A. Wigderson. P=BPP if E requires exponential circuits: Derandomizing the XOR Lemma. In *29th STOC*, pages 220–229, 1997.
14. R. Impagliazzo and A. Wigderson. Randomness vs. Time: De-randomization under a uniform assumption. In *39th FOCS*, pages 734–743, 1998.
15. V. Kabanets. Easiness assumptions and hardness tests: Trading time for zero error. *Journal of Computer and System Sciences*, 63(2):236–252, 2001.
16. R.M. Karp and R.J. Lipton. Some connections between nonuniform and uniform complexity classes. In *12th STOC*, pages 302-309, 1980.
17. A. Klivans and D. van Melkebeek. Graph Nonisomorphism has Subexponential Size Proofs Unless the Polynomial-Time Hierarchy Collapses. In *31st STOC*, pages 659–667, 1998. To appear in *SICOMP*.
18. N. Nisan. Pseudorandom Generators for Space Bounded Computation. *Combinatorica*, Vol. 12 (4), pages 449–461, 1992.
19. N. Nisan, E. Szemeredi, and A. Wigderson. Undirected connectivity in $O(log^{1.5} n)$ space. In *33rd FOCS*, pages 24-29, 1992.
20. N. Nisan and A. Wigderson. Hardness vs Randomness. *JCSS*, Vol. 49, No. 2, pages 149–167, 1994.
21. W.J. Savitch. Relationships between nondeterministic and deterministic tape complexities. *JCSS*, Vol. 4 (2), pages 177-192, 1970.

22. R. Shaltiel. A log-space extractor for high values of min-entropy. Personal communication, June 2002.
23. M. Sudan. Decoding of Reed-Solomon codes beyond the error-correction bound. *Journal of Complexity*, Vol. 13 (1), pages 180–193, 1997.
24. M. Sudan, L. Trevisan and S. Vadhan. Pseudorandom Generators without the XOR Lemma. *JCSS*, Vol. 62, No. 2, pages 236–266, 2001.
25. A. Ta-Shma. Almost Optimal Dispersers. In *30th STOC*, pages 196–202, 1998.
26. L. Trevisan. Constructions of Near-Optimal Extractors Using Pseudo-Random Generators. In *31st STOC*, pages 141–148, 1998.
27. A.C. Yao. Theory and Application of Trapdoor Functions. In *23rd FOCS*, pages 80–91, 1982.
28. D. Zuckerman. Randomness-Optimal Sampling, Extractors, and Constructive Leader Election. In *28th STOC*, pages 286–295, 1996.

Is Constraint Satisfaction Over Two Variables Always Easy?

Lars Engebretsen[1,*] and Venkatesan Guruswami[2,**]

[1] Department of Numerical Analysis and Computer Science
Royal Institute of Technology
SE-100 44 Stockholm
SWEDEN
[2] University of California at Berkeley
Miller Institute for Basic Research in Science
Berkeley, CA 94720
USA

Abstract. By the breakthrough work of Håstad, several constraint satisfaction problems are now known to have the following *approximation resistance* property: satisfying more clauses than what picking a random assignment would achieve is **NP**-hard. This is the case for example for Max E3-Sat, Max E3-Lin and Max E4-Set Splitting. A notable exception to this extreme hardness is constraint satisfaction over two variables (2-CSP); as a corollary of the celebrated Goemans-Williamson algorithm, we know that every Boolean 2-CSP has a non-trivial approximation algorithm whose performance ratio is better than that obtained by picking a random assignment to the variables. An intriguing question then is whether this is also the case for 2-CSPs over larger, non-Boolean domains. This question is still open, and is equivalent to whether the generalization of Max 2-SAT to domains of size d, can be approximated to a factor better than $(1 - 1/d^2)$.

In an attempt to make progress towards this question, in this paper we prove, firstly, that a slight restriction of this problem, namely a generalization of linear inequations with two variables per constraint, *is not* approximation resistant, and, secondly, that the Not-All-Equal Sat problem over domain size d with three variables per constraint, *is* approximation resistant, for every $d \geq 3$. In the Boolean case, Not-All-Equal Sat with three variables per constraint is equivalent to Max 2-SAT and thus has a non-trivial approximation algorithm; for larger domain sizes, Max 2-SAT can be reduced to Not-All-Equal Sat with three variables per constraint. Our approximation algorithm implies that a wide class of 2-CSPs called *regular 2-CSPs* can all be approximated beyond their random assignment threshold.

* Research partly performed while the author was visiting MIT with support from the Marcus Wallenberg Foundation and the Royal Swedish Academy of Sciences.
** Supported by a Miller Research Fellowship.

J.D.P. Rolim and S. Vadhan (Eds.): RANDOM 2002, LNCS 2483, pp. 224–238, 2002.

1   Introduction

In a breakthrough paper, Håstad [6] studied the problem of giving approximate solutions to maximization versions of several constraint satisfaction problems. An instance of a such a problem is given as a set of variables and a collection of constraints, i.e., functions from some domain to $\{0,1\}$, on certain subsets of variables, and the objective is to find an assignment to the variables that satisfies as many constraints as possible. An approximate solution of a constraint satisfaction program is simply an assignment that satisfies roughly as many constraints as possible. In this setting we are interested in proving either that there exists a polynomial time algorithm producing approximate solutions, i.e., solutions that are at most some constant factor worse compared to the optimum, or that no such algorithms exist.

Typically, each individual constraint depends on a fixed number k of the variables—this case is usually called the Max k-CSP problem. The complexity of the constraint satisfaction problem (CSP) is determined by the precise set of constraints that may be posed on subsets of k variables, and accordingly we get various families of Max k-CSP problems. For each such CSP, there exists a very naive algorithm that approximates the optimum within a constant factor: The algorithm that just guesses a solution at random. In his paper, Håstad [6] proved the very surprising fact that this algorithm is essentially the best possible efficient algorithm for several constraint satisfaction problems, unless $\mathbf{P} = \mathbf{NP}$. Håstad [6] suggests that predicates with the property that the naive randomized algorithm is the best possible polynomial time approximation algorithm should be called *non-approximable beyond the random assignment threshold*; we also use the phrase *approximation resistant* to refer to the same phenomenon.

Definition 1. *A solution to a maximization problem is α-approximate if it is feasible and has weight at least α times the optimum. An approximation algorithm has* performance ratio α *if it delivers α-approximate solutions in polynomial time.*

Definition 2. *A CSP is said to be* approximation resistant *or* non-approximable beyond the random assignment threshold *if, for any constant $\varepsilon > 0$, it is $\mathbf{NP}$-hard to compute a $(\rho + \varepsilon)$-approximate solution, where ρ is the expected fraction of constraints satisfied by a solution guessed uniformly at random.*

Clearly, understanding which predicates are approximation resistant is an important pursuit. The current knowledge is that for Boolean CSPs, which understandably have received the most attention so far, there is a precise understanding of which CSPs on *exactly three* variables are approximation resistant: All predicates that are implied by parity have this property [6,11]. It is known that *no* Boolean CSP over two variables is approximation resistant; this is a corollary of the breakthrough Goemans-Williamson algorithm [4]. For the case of four or more variables, very little is known; therefore it seems to be a good approach to first understand the situation for two and three variables.

Accordingly, we are interested in the situation for CSPs with two and three variables over larger, non-Boolean, domains. In particular, it is a really intriguing question whether every CSP over two variables can be approximated better than random, no matter what the domain size is. The central aim of this paper is to study this question. We are not able to resolve it completely, but we conjecture that the answer to the question is yes.

1.1   Formal Definitions of Some CSPs

Before discussing our results, we will need to define some of the CSPs that we will be concerned with in this paper. A specific k-CSP problem is defined by the family of constraints that may be imposed on subsets of k variables. Allowing arbitrary constraints gives the most general problem, which we call Max Ek-CSP(d). In this paper, d refers to the domain size from which the variables may take values, with $d = 2$ corresponding to the Boolean case. Over domain size d, a constraint is simply a function $f: [d]^k \rightarrow \{0,1\}$, where $[d] = \{0,1,\ldots,d-1\}$. Equivalently, a constraint f can be viewed as a subset of $[d]^k$ consisting of all inputs which it maps to 1.

The Max Ek-Sat(d) problem is defined by the constraint family $\{f \subseteq [d]^k : |f| = d^k - 1\}$, i.e., the family of all constraints having just one non-satisfying assignment. Max Ek-NAE-Sat(d) is the problem where the constraints assert that the specific variables are not all equal, except that we also allow translates of variables, e.g., for the two variable case, a constraint can assert $x_1 + 1 \neq x_2 + 3$ (the addition being done modulo d); this is the analog of complementation of Boolean variables. In the Max Ek-Lin(d) problem, the constraint family is given by all linear constraints: $\{\text{Lin}(\alpha_1,\ldots,\alpha_k,c) : \alpha_i, c \in [d]\}$ where $\text{Lin}(\alpha_1,\ldots,\alpha_k,c) = \{(x_1,\ldots,x_k) : \sum_i \alpha_i x_i = c \bmod d\}$. The Max E$k$-LinInEq($d$) problem is defined by the family of all linear inequations: $\{f \subseteq [d]^k : f^c \text{ is a linear constraint}\}$—here $f^c = [d]^k \setminus f$ denotes the complement of the constraint f.

For the two variable case, we define the constraint satisfaction problems Max BIJ(d) and Max Co-BIJ(d) which are generalizations of Max E2-Lin(d) and Max E2-LinInEq(d) respectively. Let S_d be the set of all bijections from $[d]$ to $[d]$. For each $\pi \in S_d$, define the 2-ary constraint $f_{\pi,d} = \{(a,b) \in [d]^2 : b = \pi(a)\}$. Now define the family BIJ(d) $= \{f_{\pi,d} : \pi \in S_d\}$; we call the CSP associated with this family Max BIJ(d). The problem Max Co-BIJ(d) is obtained by constraints which are complements of those in BIJ(d), i.e., a constraint is of the form $\pi(x_1) \neq x_2$ for some bijection π defined over $[d]$. It is clear that these problems generalize Max E2-Lin(d) and Max E2-LinInEq(d) respectively.

For the three variable case, we define the problem Max E3-NAE-Sat(G) for finite Abelian groups G. For each triple $(g_1,g_2,g_3) \in G^3$ define the constraint $N_{g_1,g_2,g_3} = \{(x_1,x_2,x_3) \in G^3 : \neg(g_1 x_1 = g_2 x_2 = g_3 x_3)\}$. Now define the family $\text{NAE}(G) = \{N_{g_1,g_2,g_3} : (g_1,g_2,g_3) \in G^3\}$; we denote by Max E3-NAE-Sat(G) the CSP associated with this family of constraints. Note that the group structure is indeed present in the problem since the constraints involve multiplication by elements from G. In fact, we are able to prove in this paper that Max E3-

NAE-Sat($\mathbf{Z}_4$) is approximation resistant while we are unable to determine the approximability of Max E3-NAE-Sat($\mathbf{Z}_2 \times \mathbf{Z}_2$).

It is an interesting open question to determine what kind of hardness holds for the restricted version of Max E3-NAE-Sat(G) where group multipliers are not allowed; for this problem the group structure is, of course, not present at all. Recently, Khot [8] has shown that Max E3-NAE-Sat($\mathbf{Z}_3$) is approximation resistant even without group multipliers.

1.2   Our Results

Preliminaries: First, we make explicit the easily seen result that an approximation algorithm for Max E2-Sat(d) with performance ratio better than $1 - 1/d^2$, i.e., better than the random assignment threshold, implies that any CSP over 2 variables can be approximated to within better than *its respective* random assignment threshold. In other words, Max E2-Sat(d) is the hardest problem in this class, and if there is some Max E2-CSP(d) which is approximation resistant, then Max E2-Sat(d) has to be approximation resistant.

Consequently, our interest is in the approximability of Max E2-Sat(d), specifically to either find a polynomial time approximation algorithm with performance ratio greater than $1 - 1/d^2$ or to prove a tight hardness result that the trivial $1 - 1/d^2$ is the best one can hope for. While we are unable to resolve this question, we consider and prove results for two predicates whose difficulty sandwiches that of solving Max E2-Sat(d): namely Max Co-BIJ(d) and Max E3-NAE-Sat($\mathbf{Z}_d$). The former problem is (in a loose sense) the natural 2-CSP which is next in "easiness" after Max E2-Sat(d) as far as approximating better than the random assignment threshold is concerned. There is an approximation preserving reduction from Max E2-Sat(d) to Max E3-NAE-Sat($\mathbf{Z}_d$), implying that Max E3-NAE-Sat($\mathbf{Z}_d$) is a harder problem than Max E2-Sat(d).

Algorithms: For the Max Co-BIJ(d) problem, we prove that it is *not* approximation resistant by presenting a polynomial time approximation algorithm with performance ratio $1 - d^{-1} + 0.07d^{-4}$. This result implies that a large class of 2-CSPs, called *regular 2-CSPs* (defined below), are *not* approximation resistant. Viewing a 2-ary constraint C over domain size d as a subset of $[d] \times [d]$, the constraint is said to be *r-regular* if for each $a \in [d]$, $|\{x : (x, a) \in C\}| = |\{y : (a, y) \in C\}| = r$ (the term regular comes from the fact that the bipartite graph defined by C is regular). The constraint is *regular* if it is *r*-regular for some $1 \le r < d$. A 2-CSP is regular if all the constraints in the CSP are regular and it is *r*-regular if all the constraints are *r*-regular.

Our result for regular 2-CSPs includes as a special case the result of Frieze and Jerrum [3] that Max d-Cut can be approximated to better than its random threshold. Our performance ratio is weaker, but our analysis is simpler and gives a more general result. Another special case is the result for Max E2-Lin(d) where our result actually improves the approximation ratio of Andersson *et al* [2]. Recently, Khot [9] gave a simpler algorithm that beats the random assignment threshold for Max E2-Lin(d) as well as the more general Max BIJ(d) problems—his result is actually more general and can find a near-satisfying assignment given a near-satisfiable instance, i.e., an instance where the optimum solution satisfies a fraction $1 - \varepsilon$ of constraints. Our approximation algorithm for Max Co-BIJ(d) is based on a semidefinite programming relaxation, similar to that used for Max BIJ(d) by Khot [9], combined with a rounding scheme used by Andersson [1] to construct an approximation algorithm for Max d-Section, the generalization of Max Bisection to domains of size d. Technically, we view this algorithmic result as the main contribution of this paper.

Inapproximability results: For the Boolean case, $d = 2$, it is known that Max E3-NAE-Sat can be approximated to better than random. The GW-algorithm [4] for Max E2-Sat essentially gives such an algorithm, and the performance ratio was later improved by Zwick [12]. We prove that for larger domains, the problem becomes approximation resistant; in other words, it is **NP**-hard to approximate Max E3-NAE-Sat($\boldsymbol{Z}_d$) to better than $(1 - 1/d^2 + \varepsilon)$ for any $d \geq 3$ and any $\varepsilon > 0$. This result rules out the possibility of a non-trivial algorithm for Max E2-Sat(d) that works by reducing it to Max E3-NAE-Sat($\boldsymbol{Z}_d$). In fact, we prove that for any finite group G which is not of the form $\boldsymbol{Z}_2 \times \boldsymbol{Z}_2 \times \cdots \times \boldsymbol{Z}_2$, Max E3-NAE-Sat($G$) is hard to approximate within a factor $(1 - 1/|G|^2 + \varepsilon)$.

We remark that the above hardness results hold with *perfect completeness*; in other words, the stated approximation factors are hard to obtain even on satisfiable instances of the concerned constraint satisfaction problems.

Conclusions: We are not able to completely resolve the status of the 2-CSP problem over larger domains. Using reductions, we prove a hardness result of $1 - \Omega(1/d^2)$ for Max E2-Sat(d), which compares reasonably well with the $(1 - 1/d^2)$ random assignment threshold. For satisfiable instances of Max E2-Sat(d), we prove a hardness result of $1 - \Omega(1/d^3)$. Nevertheless, we conjecture that there is an approximation algorithm beating the random assignment threshold for Max E2-Sat(d), and hence for all instances of 2-CSP.

Organization: We begin with a brief Section 2 highlighting why Max E2-Sat(d) is the hardest Max E2-CSP(d) problem in terms of beating the random assignment threshold. Next, in Section 3 we prove that every Max Co-BIJ(d) problem admits an algorithm that beats the random assignment threshold, and record some of its consequences. In Section 4, we prove that Max E3-NAE-Sat(G) is approximation resistant for most groups, including $G = \boldsymbol{Z}_d$ (the case of most interest in the context of Max E2-Sat(d)). Finally, we record results that directly apply to Max E2-Sat(d) in Section 5.

2 The "Universality" of Max E2-Sat(d)

We note that the existence of an approximation algorithm that beats the random threshold for every 2-CSP is equivalent to the existence of such an algorithm for

Max E2-Sat(d). Thus, an algorithm for Max E2-Sat(d) with performance ratio better than $(1 - 1/d^2)$ will imply that no 2-CSP is approximation resistant, thus resolving our conjecture that every 2-CSP is "easy".

This claim is seen by a "gadget" reducing an arbitrary CSP(d) to Max E2-Sat(d). Given an instance of any 2-CSP, construct an instance of Max E2-Sat(d) by repeating the following for every constraint C in the original 2-CSP: For every non-satisfying assignment to C, add one 2SAT(d) constraint which has precisely this non-satisfying assignment. If an assignment satisfies C then it also satisfies all the 2SAT(d) constraints in the gadget, and otherwise it satisfies precisely all but one of the 2SAT(d) constraints. Using this fact, it is straightforward to show that if Max E2-Sat(d) can be approximated beyond the random threshold, the above procedure gives an approximation algorithm that approximates an arbitrary 2-CSP beyond the random threshold. Conversely, if any 2-CSP at all is approximation resistant, then Max E2-Sat(d) must be approximation resistant.

3   Approximation Algorithm for Max Co-BIJ(d)

To construct an approximation algorithm for Max Co-BIJ(d) we combine a modification of the semidefinite relaxation used by Khot [9] for the Max BIJ(d) problem with a modification of the randomized rounding used by Andersson [1] for the Max d-Section problem. Recall that a specific clause in the Max Co-BIJ(d) problem is of the form (x, x', π), where x and x' are variables in the Max Co-BIJ(d) instance and π is a permutation, and that the clause is satisfied unless $x = j$ and $x' = \pi(j)$ for some j. In our semidefinite relaxation of Max Co-BIJ(d) there are d vectors $\{u_0^x, \dots, u_{d-1}^x\}$ for every variable x in the Max Co-BIJ(d) instance. Intuitively, the vector u_j^x sets the value of the variable x to j. It is straightforward to see that the semidefinite program in Fig. 1 and that the *barycenter* $b = \frac{1}{d} \sum_{j=0}^{d-1} u_j^x$ is independent of x for any feasible solution to the program. To establish a bound on the performance ratio of the algorithm in Fig. 2, we use local analysis:

Lemma 1. *For any clause (x, x', π) in the Max Co-BIJ(d) instance, the algorithm in Fig. 2 satisfies (x, x', π) with probability at least*

$$\left(1 - \sum_{j=0}^{d-1} \langle u_j^x, u_{\pi(j)}^{x'} \rangle\right) \int_B \left(1 - \frac{1}{d} + \frac{K^2 r_1^2}{d}\right) dP(r)$$

where K is any positive constant, the vectors u_j^x and $u_{j'}^{x'}$ are as described in the algorithm, $B = \{r \in \mathbf{R}^{2d} : |r| \leq 1/Kd\}$, and P is the probability distribution of a $2d$-dimensional Gaussian with mean zero and identity covariance matrix.

Proof. Consider an arbitrary clause (i, i', π) and the corresponding values q_{xj} computed by the algorithm. Let $B = \{r \in \mathbf{R}^{2d} : |r| \leq 1/Kd\}$. When $r \in B$,

$$
\begin{aligned}
\text{maximize} \quad & \sum_{x,x',\pi} w_{x,x',\pi}\left(1 - \sum_{j=0}^{d-1}\langle u_j^x, u_{\pi(j)}^{x'}\rangle\right) \\
\end{aligned}
$$

$$
\begin{aligned}
\text{such that} \quad & \langle u_j^x, u_{j'}^{x'}\rangle \geq 0 && \forall x \in X, x' \in X, j \in \mathbf{Z}_d, j' \in \mathbf{Z}_d \\
& \langle u_j^x, u_{j'}^x\rangle = 0 && \forall x \in X, (j,j') \in \mathbf{Z}_d^2 : j \neq j' \\
& \sum_{j=0}^{d-1}\langle u_j^x, u_j^x\rangle = 1 && \forall x \in X \\
& \sum_{j=0}^{d-1}\sum_{j'=0}^{d-1}\langle u_j^x, u_{j'}^{x'}\rangle = 1 && \forall (x,x') \in X^2 : x \neq x'
\end{aligned}
$$

Fig. 1. Semidefinite relaxation of Max Co-BIJ(d) with variable set X. A clause in the Max Co-BIJ(d) instance is denoted by (x, x', π) where $x \in X$ and $x' \in X$ are variables and $\pi\colon \mathbf{Z}_d \to \mathbf{Z}_d$ is a permutation. The clause is satisfied unless $x = j$ and $x' = \pi(j)$ for some $j \in \mathbf{Z}_d$. Each clause (x, x', π) has a non-negative weight $w_{x,x',\pi}$ associated with it.

1. Solve the semidefinite program in Fig. 1.
2. Denote by u_j^x the vectors obtained from the solution.
3. For every $(x,j) \in X \times \mathbf{Z}_d$, let $v_j^x = u_j^x - \frac{1}{d}\sum_{j=0}^{d-1} u_j^x$.
4. Select r from a dn-dimensional Gaussian distribution.
5. Set $q_{xj} = \frac{1}{d} + K\langle r, v_j^x\rangle$ for all $(x,j) \in X \times \mathbf{Z}_d$.
6. For each $x \in X$,
   – set $p_{xj} = q_{xj}$ if $q_{xj} \in [0, 2/d]$ for all $j \in \mathbf{Z}_d$;
   – set $p_{xj} = 1/d$ for all $j \in \mathbf{Z}_d$ otherwise.
7. For each $x \in X$, let $x = j$ with probability p_{xj}.

Fig. 2. Approximation algorithm for Max Co-BIJ(d) with variable set X. The algorithm is parameterized by the positive constant K.

both q_{xj} and $q_{x'j}$ are in the interval $[0, 2/d]$; hence the clause (i, i', π) is satisfied with probability

$$
1 - \sum_{j=0}^{d-1} p_{xj} p_{x',\pi(j)} = 1 - \sum_{j=0}^{d-1}\left(\frac{1}{d} + K\langle r, v_j^x\rangle\right)\left(\frac{1}{d} + K\langle r, v_{\pi(j)}^{x'}\rangle\right)
$$

$$
= 1 - \frac{1}{d} - K^2 \sum_{j=0}^{d-1}\langle r, v_j^x\rangle\langle r, v_{\pi(j)}^{x'}\rangle
$$

given r in this case. Using the definition of v_j^x from the algorithm, and integrating over B, we can lower bound the probability that the clause (i, i', π) is accepted by

$$
\int_B \left(1 - \frac{1}{d} + K^2\left(d\langle r, b\rangle\langle r, b\rangle - \sum_{j=0}^{d-1}\langle r, u_j^x\rangle\langle r, u_{\pi(j)}^{x'}\rangle\right)\right) dP(r).
$$

where b is the barycenter of the vectors $\{u_j^x : j \in Z_d\}$, which is indepen-dent of x. To compute the integral of $\langle r, b \rangle \langle r, b \rangle$, introduce an orthonormal basis $\{e_k\}$ such that $b = e_1/d$ and write $r = \sum_k r_k e_k$ in this basis. Then $\int_B \langle r, b \rangle \langle r, b \rangle \, dP(r) = \frac{1}{d^2} \int_B r_1^2 \, dP(r)$, where the last integral is actually inde-pendent of the basis since both P and B are spherically symmetric. To compute the integral of $\langle r, u_j^x \rangle \langle r, u_{\pi(j)}^{x'} \rangle$, we proceed similarly: Introduce an orthonormal basis $\{e_k\}$ such that $u_j^x = x_1 e_1$ and $u_{\pi(j)}^{x'} = y_1 e_1 + y_2 e_2$. Then

$$\int_B \langle r, u_j^x \rangle \langle r, u_{\pi(j)}^{x'} \rangle \, dP(r) = x_1 y_1 \int_B r_1^2 \, dP(r) = \langle u_j^x, u_{\pi(j)}^{x'} \rangle \int_B r_1^2 \, dP(r).$$

To conclude, we can write the probability that the clause is satisfied as $a - cx$ where

$$a = \int_B \left(1 - \frac{1}{d} + \frac{K^2 r_1^2}{d} \right) dP(r),$$

$$c = K^2 \int_B r_1^2 \, dP(r),$$

$$x = \sum_{j=0}^{d-1} \langle u_j^x, u_{\pi(j)}^{x'} \rangle.$$

Since $a \geq c > 0$, $a - cx > a(1-x)$; therefore the clause is satisfied with probability at least $a(1-x)$, which equals

$$\left(1 - \sum_{j=0}^{d-1} \langle u_j^x, u_{\pi(j)}^{x'} \rangle \right) \int_B \left(1 - \frac{1}{d} + \frac{K^2 r_1^2}{d} \right) dP(r).$$

Using standard calculus, it can be shown that

$$\int_B \left(1 - \frac{1}{d} + \frac{K^2 r_1^2}{d} \right) dP(r) \geq 1 - \frac{1}{d} + \frac{0.07}{d^4}$$

with the parameter choice $K = 1/\sqrt{13d^3}$. Together with the above lemma, this proves our main theorem:

Theorem 1. *The algorithm in Fig. 2 with $K = 1/\sqrt{13d^3}$ is a randomized poly-nomial time approximation algorithm for Max Co-BIJ(d) with expected perfor-mance ratio $1 - d^{-1} + 0.07d^{-4}$ and thus better than the random assignment threshold.*

3.1   An Approximation Algorithm for Max E2-Lin(d)

We can use the above algorithm for Max Co-BIJ(d) to construct an algorithm also for Max E2-Lin(d): Simply replace an equation $ax + by = c$ with the $d - 1$ inequations $ax + by \neq c_i$ for all $c_i \neq c$. Then an assignment that satisfies a linear equation satisfies all of the corresponding linear inequations and an assignment that does not satisfy a linear equation satisfies $d - 2$ of the $d - 1$ corresponding linear equations.

Run the following two algorithms and take the assignment producing the largest weight as the result: The first algorithms selects a random assignment to the variables; the second algorithm runs the above algorithm for Max E2-Lin(d). This algorithm gives a performance ratio of $1/d + \Omega(1/d^4)$ which improves significantly on the previously best known ratio of $1/d + \Omega(1/d^{14})$ [2]. The formal proof is omitted from this extended abstract.

Theorem 2. *For all $d \geq 4$, the above algorithm is a randomized polynomial time approximation algorithm for Max E2-Lin(d) with expected performance ratio $d^{-1} + 0.05d^{-4}$.*

Corollary 1. *For all $d \geq 2$, there is a polynomial time approximation algorithm for Max E2-Lin(d) with expected performance ratio $d^{-1} + 0.05d^{-4}$ and thus better than the random assignment threshold.*

Proof. Algorithms for $d = 2$ and $d = 3$ have been provided by Goemans and Williamson [4,5], for $d \geq 4$ the result follows by Theorem 2

3.2   An Approximation Algorithm for Regular 2-CSPs

We can obtain an approximation algorithm for all regular CSPs by a straightforward generalization of the ideas from the previous section. Given an r-regular 2-CSP, we proceed as follows for every relation R defining the CSP: Decompose R^c, the "bipartite complement" of the graph defined by R, into $(d - r)$ perfect matchings $\pi_R^1, \pi_R^2, \ldots, \pi_R^{d-r}$. Then let these matchings define the Max Co-BIJ(d) instance. An assignment that satisfies R satisfies all of the $d - r$ matchings while an assignment that does not satisfy R satisfies $d - r - 1$ of them. Run the following two algorithms and take the assignment producing the largest weight as the result: The first algorithms selects a random assignment to the variables; the second algorithm runs the above algorithm for Max Co-BIJ(d). The formal proof is, again, omitted.

Theorem 3. *For all $d \geq 2$ and all $1 \leq r \leq d - 1$, the above algorithm is a randomized polynomial time approximation algorithm for r-regular CSPs with expected performance ratio $r/d + \Omega(d^{-4})$.*

It is not necessary that the various constraints be r-regular for the same r, and a similar argument also shows that every regular 2-CSP can be approximated in polynomial time beyond its random assignment threshold.

4   Hardness Results for Max E3-NAE-Sat(Z_d)

In this section, our aim is to prove that unlike the Boolean case, for every $d \geq 3$, Max E3-NAE-Sat(Z_d) is approximation resistant. We will actually prove that Max E3-NAE-Sat(G) is approximation resistant for pretty much every finite Abelian group. Specifically, we will prove:

Theorem 4. *For every constant $\varepsilon > 0$ and every finite Abelian group G that is not isomorphic to $\mathbf{Z}_2^m$ for any positive integer m, it is **NP**-hard to distinguish instances of Max E3-NAE-Sat(G) that are satisfiable from instances where at most a fraction $(1 - |G|^{-2} + \varepsilon)$ of the constraints are simultaneously satisfiable.*

Inevitably, the proof of the above theorem will involve the machinery of probabilistically checkable proofs (PCP). We first give a rapid introduction to the necessary background, and then indicate the kind of PCP verifiers that yield Theorem 4.

4.1   Background on PCP Constructions

Our hardness of approximation results are proved by constructions of suitable probabilistically checkable proofs (PCPs) for **NP**. Our PCP constructions follow the by now standard paradigm used by Håstad; we refer the reader to Håstad's paper [6] for the complete details. The basic construction methodology is to start with an instance of 3SAT, called μ-gap E3-Sat(5), where there is a gap in the optimum (either the formula is satisfiable or at most a fraction μ of clauses are satisfied by any assignment) and each variable occurs in exactly five clauses. One then uses Raz's parallel repetition theorem to obtain a two-prover one-round (2P1R) system with completeness 1 and arbitrarily low soundness. The verifier in the 2P1R system picks a set W of u clauses of the 3SAT instance at random and also picks a set U of u variables, one in each of the picked clauses, at random. The first prover P_1 is expected to give an assignment to the variables in U and the second prover P_2 is expected to give an assignment that satisfies all the clauses in W. The verifier accepts if the answer of the P_2 is consistent with the answer of P_1. This protocol has soundness at most c^u for some absolute constant c.

In order to prove that several constraint satisfaction programs are non-approximable beyond the random assignment threshold, we use a verifier whose acceptance predicate is closely related to the particular CSP we want to analyze. The final verifier expects as proof encodings of the answers of P_1 and P_2 in the Raz 2P1R, and then checks very efficiently, by making very few queries, that the proof is close to valid encodings of answers that would have made the 2P1R verifier accept with good probability. To get hardness results for CSPs over domain size d, the specific encoding used is the *Long G-Code* where G is an Abelian group of order d.

Definition 3. *Let U be a set of variables and denote by $\{-1,1\}^U$ the set of assignments to the variables in U. The* long G-code *of some assignment x to the variables in U is a function $A_{U,x} : \{-1,1\}^U \to G$ defined by $A_{U,x}(f) = f(x)$.*

Definition 4. *Let W be a set of clauses and denote by SAT^W the set of satisfying assignments to the clauses in W. The* long G-code *of some satisfying assignment y to the clauses in W is a function $A_{W,y} : \mathrm{SAT}^W \to G$ defined by $A_{W,y}(h) = h(y)$.*

The proof is a standard written G-proof with parameter u:

The verifier acts as follows:

1. Select a sequence W of clauses, each clause uniformly and independently at random from Φ.
2. Select a sequence U of variables by selecting one variable from each clause in W, uniformly and independently.
3. Let $\pi\colon \mathrm{SAT}^W \to \{-1,1\}^U$ be the function that creates an assignment in $\{-1,1\}^U$ from an assignment in SAT^W.
4. Let $F = G^{\{-1,1\}^U}$ and $H = G^{\mathrm{SAT}^W}$.
5. Select $f \in F$ and $h \in H$ uniformly at random.
6. Select $e \in H$ such that independently for every $y \in \mathrm{SAT}^W$, $e(y)$ is uniformly distributed in $G \setminus \{\mathbf{1}\}$.
7. Accept if $A_U(f)$, $A_W(h)$ and $A_W((f \circ \pi)^{-1} h^2 e)$ are not all equal; Reject otherwise.

Fig. 3. The PCP used to prove optimal approximation hardness for Max E3-NAE-Sat($\mathbf{Z}_d$) for odd d. The PCP is parameterized by the constant u and tests if a μ-gap E3-Sat(5) formula Φ is satisfiable.

The proof expected by the PCP verifier now consists of purported Long G-Codes of the assignments to the u variables in U and the u clauses in W for each possible choice U, W of the 2P1R verifier. We call such a proof in such a format the *Standard Written G-proof with parameter u*.

The PCP design task now reduces to designing an "inner" verifier to check if two purported Long G-Codes encode assignments which are consistent answers for P_1 and P_2 in the 2P1R. One designs such a verifier with an acceptance predicate closely tied to the problem at hand, and its performance is analyzed using Fourier analysis. The basic strategy here is to show how proofs that make the "inner" verifier accept with high probability can be used to extract good strategies for P_1 and P_2 in the 2P1R protocol.

4.2   Intuition Behind Our PCP Constructions

For lack of space, we defer the analysis of our PCPs to the full version of our paper and only give the intuition behind the constructions here. The first construction turns out to work for all Abelian groups of odd order. A verifier in a PCP typically first selects sets U and W uniformly at random and then checks a small number of positions in tables corresponding to U and W. Specifically, the standard way to get a PCP with three queries is to query one position in a table corresponding to U and two positions in a table corresponding to W. The three values obtained are then tested to see if they satisfy some given constraint—such a construction gives a hardness result for the CSP corresponding to the type of constraint checked. To get a hardness result for Max E3-NAE-Sat(G) the constraint checked by the verifier therefore has to be a not-all-equal constraint. Moreover, we want the verifier to have perfect completeness, i.e., to always verify

> The proof is a standard written $\mathbf{Z}_d$-proof with parameter u:
>
> The verifier acts as follows:
>
> Steps 1–5 are as in Fig. 3 applied to $G = \mathbf{Z}_d$.
>
> 6. Select $e_1 \in H$ such that independently for every $y \in \mathrm{SAT}^W$,
> $e_1(y)$ is uniformly distributed in $\{\omega^{4i}, \omega^{4i+1}\}_{i=0}^{d/4-1}$.
>   Select $e_2 \in H$ such that independently for every $y \in \mathrm{SAT}^W$,
> $e_2(y)$ is uniformly distributed in $\{\omega^{4i+1}, \omega^{4i+2}\}_{i=0}^{d/4-1}$.
>   Let $e = e_1 e_2$.
> 7. Accept if $A_U(f)$, $A_W(h)$ and $A_W((f \circ \pi)^{-1} h^2 e)$ are not all equal;
>   Reject otherwise.

Fig. 4. The PCP used to prove optimal approximation hardness for Max E3-NAE-Sat($\mathbf{Z}_d$) where $d = 2^m$ for integers $m \geq 2$. The group $\mathbf{Z}_d$ is represented by $\{\omega^i\}_{i=0}^{d-1}$ where $\omega = e^{2\pi i/d}$ and multiplication is the group operator. The PCP is parameterized by the constant u and tests if a μ-gap E3-Sat(5) instance Φ is satisfiable.

> The proof is a standard written G-proof with parameter u:
>
> The verifier acts as follows:
>
> Steps 1–5 are as in Fig. 3.
>
> 6. Select e by selecting independently the components of $e(y)$
>   according to Step 6 in Figs. 3 and 4, respectively.
> 7. Accept if $A_U(f)$, $A_W(h)$ and $A_W((f \circ \pi)^{-1} h^2 e)$ are not all equal;
>   Reject otherwise.

Fig. 5. The PCP used to prove optimal approximation hardness for Max E3-NAE-Sat(G) for any finite Abelian group $G \cong G_o \times \mathbf{Z}_{2^{\alpha_1}} \times \mathbf{Z}_{2^{\alpha_2}} \times \cdots \times \mathbf{Z}_{2^{\alpha_s}}$ where G_o is a finite Abelian group of odd order and $\alpha_i > 1$ for all i. The PCP is parameterized by the constant u and tests if a μ-gap E3-Sat(5) formula Φ is satisfiable.

a correct proof. We accomplish this by querying the positions $A_U(f)$, $A_W(h)$ and $A_W(f^{-1} h^2 e)$ where f and h are selected uniformly at random and e is selected such that $e(y)$ is selected independently and uniformly at random from $G \setminus \{\mathbf{1}\}$ (see Figure 3). Here, the function $f^{-1} h^2 e$ is the map $y \mapsto (f(y|_U))^{-1}(h(y))^2 e(y)$. For a correct proof of a satisfying assignment, the answers to these queries will be $f(y|_U)$, $h(y)$ and $(f(y|_U))^{-1}(h(y))^2 e(y)$ where y is a satisfying assignment to the clauses in W. These three values can never be all equal, since $f(y|_U) = h(y)$ implies that $(f(y|_U))^{-1}(h(y))^2 e(y) = h(y)e(y) \neq h(y)$. Therefore the verifier always accepts a correct proof and we prove in the full version of our paper that the verifier accepts a proof corresponding to an unsatisfying assignment with probability at most $1 - |G|^{-2} + \varepsilon$, where $\varepsilon > 0$ is an arbitrary constant.

To obtain results for the Abelian groups $\mathbf{Z}_d$, where $d = 2^m$ for some $m \geq 2$, we need to change the verifier slightly. The reason that $|G|$ being odd was very useful in the above mentioned analysis is the following: For a random function h with range in G, h^2 is also a random function when $|G|$ is odd. This is no longer the case when $|G|$ is even. We get around this obstacle by a different, clever choice of the error function e. The PCP verifier is described in Figure 4. This verifier also has perfect completeness and we prove in the full version of our paper that it has soundness $1 - |G|^{-2} + \varepsilon$, where $\varepsilon > 0$ is an arbitrary constant.

Finally, in order to obtain a hardness result for every Abelian group except powers of Z_2, we construct a verifier that essentially runs several of the verifiers described previously in parallel. This verifier is described in Fig. 5. We prove in the full version of our paper that the analysis of the previous two types of verifiers can be combined into an analysis that works in this case, which establishes the proof of Theorem 4.

5   The Status of Max E2-Sat(d)

For the Max E2-Sat(d) problem itself, we present some hardness results below. We first prove a result for domain size 3 and then prove a result for general domains.

Lemma 2. *For every constant $\varepsilon > 0$, the predicate $(x \neq a) \vee (y = b)$ over domains of size 3 is hard to approximate within $(23/24 + \varepsilon)$ with perfect completeness.*

Proof. Consider the following 2P1R interactive proof system for 3SAT: The first prover is given a variable and returns an assignment to that variable, the second prover is given a clause and returns an index of a literal that makes the clause satisfied. The verifier selects a clause at random, then a variable in the clause at random, sends the variable to P_1, the clause to P_2 and accepts unless P_2 returns the index of the variable sent to P_1 and P_1 returned an assignment that does not satisfy the literal. It is known that there are satisfiable instances of 3SAT such that it is **NP**-hard to satisfy more than $7/8 + \varepsilon$ of the clauses, for any constant $\varepsilon > 0$ [6]. When the above protocol is applied to such an instance of 3SAT, the test has perfect completeness and soundness $(1 - 1/24 + \varepsilon)$. To obtain the hardness for the claimed constraint satisfaction problem, we just use the following reduction: x specifies the name of a clause, y specifies a variable in this clause, a specifies the location of y in x (encoded as 0,1 or 2), and b specifies a Boolean assignment to y (encoded over 0,1,2, where 2 is meaningless).

Theorem 5. *For every constant $\varepsilon > 0$, it is **NP**-hard to approximate Max E2-Sat(3) within $47/48 + \varepsilon$ with perfect completeness.*

Proof. Follows from Lemma 2 since $(x \neq a) \vee (y = b)$ can be written as a two E2-Sat(3) clauses.

Theorem 6. *For every $d \geq 3$ and every constant $\varepsilon > 0$, Max E2-Sat(d) is hard to approximate within a factor of $(1 - d^{-4} + \varepsilon)$ with perfect completeness.*

Proof. We reduce Max E3-Sat(d), which is known to be hard to approximate within $(1 - d^{-3} + d\varepsilon)$ with perfect completeness to Max E2-Sat(d). A constraint SAT(x, y, z), which requires that at least one of x, y, z does not equal 0, is replaced with the constraints SAT(x, t), SAT($x, t + 1$), SAT($x, t + 2$), $\ldots$, SAT($x, t + d - 3$), SAT($y, t + d - 2$), SAT($z, t + d - 1$), where t is an auxiliary variable specific to this constraint and the additions are done modulo d. If all d 2SAT clauses are satisfied, the 3SAT clause has to be satisfied; if the 3SAT clause is not satisfied we can satisfy $d - 1$ of the 2SAT clauses. Therefore it is hard to distinguish the case when all the constraints are satisfied from the case when a fraction $\frac{1}{d}(d(1 - d^{-3} + d\varepsilon) + (d-1)(d^{-3} - d\varepsilon)) = (1 - d^{-4} + \varepsilon)$ of the constraints are satisfied.

Theorem 7. *Max E2-Sat(d) is hard to approximate within a factor $1 - \Omega(d^{-2})$ with non-perfect completeness. It is also hard to approximate within a factor $1 - \Omega(d^{-3})$ with perfect completeness for all $d \geq 3$.*

Proof. We reduce d-CUT, which is hard to approximate within $1 - 1/34d + \varepsilon$ with non-perfect completeness [7], to Max E2-Sat(d). A clause CUT(x, y) is replaced with the clauses SAT($x + i, y + i$) for all i from 0 to $d - 1$. A d-CUT instance with n constraints corresponds to a 2SAT(d) instance with dn constraints and an assignment satisfying all but k 2SAT(d) constraints satisfies all but k d-CUT constraints. The hardness result with perfect completeness follows by a reduction from Max 3-CUT on 3-colorable graphs (i.e., Max 3-CUT with perfect completeness), which is known to hard to approximate within a factor γ for some absolute constant $\gamma < 1$ [10], to Max d-CUT for $d \geq 3$. Such a reduction which preserves perfect completeness and shows the hardness of approximating Max d-CUT within $1 - \Omega(1/d^2)$ exists. Combining with the above gadget that reduces d-CUT to 2SAT(d), we get the claimed $1 - \Omega(1/d^3)$ hardness for satisfiable instances of Max E2-Sat(d) for all $d \geq 3$. We omit the details.

The result of Theorem 6 is not entirely subsumed by the result of Theorem 7 for satisfiable instances, since the constant in front of $1/d^3$ implies that the result of Theorem 6 will actually be stronger for small values of d. An interesting question is whether a factor $(1 - \Omega(d^{-2}))$ hardness can be shown for satisfiable instances of Max E2-Sat(d).

Conjecture. Although we did not completely resolve the status of Max E2-Sat(d), we conjecture at this point that the problem is *not* approximation resistant.

Acknowledgments. We would like to thank Subhash Khot for providing us with a copy of [9]; his algorithm for Max BIJ(d) directly inspired our algorithm for Max Co-BIJ(d).

References

1. Gunnar Andersson. *Some New Randomized Approximation Algorithms.* Doctoral dissertation, Department of Numerical Analysis and Computer Science, Royal Institute of Technology, May 2000.
2. Gunnar Andersson, Lars Engebretsen, and Johan Håstad. A new way of using semidefinite programming with applications to linear equations mod p. *Journal of Algorithms*, 39(2):162–204, May 2001.
3. Alan Frieze and Mark Jerrum. Improved approximation algorithms for MAX k-CUT and MAX BISECTION. *Algorithmica*, 18:67–81, 1997.
4. Michel X. Goemans and David P. Williamson. Improved approximation algorithms for maximum cut and satisfiability problems using semidefinite programming. *Journal of the ACM*, 42(6):1115–1145, November 1995.
5. Michel X. Goemans and David P. Williamson. Approximation algorithms for Max-3-Cut and other problems via complex semidefinite programming. In *Proceedings of the 33rd Annual ACM Symposium on Theory of Computing*, pages 443–452. Hersonissos, Crete, Grece, 6–8 July 2001.
6. Johan Håstad. Some optimal inapproximability results. *Journal of the ACM*, 48(4):798–859, July 2001.
7. Viggo Kann, Sanjeev Khanna, Jens Lagergren, and Alessandro Panconesi. On the hardness of approximating Max k-Cut and its dual. *Chicago Journal of Theoretical Computer Science*, 1997(2), June 1997.
8. Subhash Khot. Hardness results for coloring 3-colorable 3-uniform hypergraphs. To appear in *Proceedings of the 43rd IEEE Symposium on Foundations of Computer Science*. Vancouver, Canada, 16–19 November 2002.
9. Subhash Khot. On the power of unique 2-prover 1-round games. In *Proceedings of the 34th Annual ACM Symposium on Theory of Computing*, pages 767–775. Montréal, Québec, Canada, 19–21 May 2002.
10. Erez Petrank. The hardness of approximation: Gap location. *Computational Complexity*, 4(2):133–157, 1994.
11. Uri Zwick. Approximation algorithms for constraint satisfaction programs involving at most three variables per constraint. In *Proceedings of the Ninth Annual ACM-SIAM Symposium on Discrete Algorithms*, pages 201–210. San Francisco, California, 25–27 January 1998.
12. Uri Zwick. Outward rotations: a tool for rounding solutions of semidefinite programming relaxations, with applications to MAX CUT and other problems. In *Proceedings of the Thirty-First Annual ACM Symposium on Theory of Computing*, pages 679–687. Atlanta, Georgia, 1–4 May 1999.

Dimensionality Reductions That Preserve Volumes and Distance to Affine Spaces, and Their Algorithmic Applications

Avner Magen

NEC Research Institute, Princeton, NJ.
`avner@research.nj.nec.com`

Abstract. Let X be a subset of n points of the Euclidean space, and let $0 < \varepsilon < 1$. A classical result of Johnson and Lindenstrauss [JL84] states that there is a projection of X onto a subspace of dimension $O(\varepsilon^{-2} \log n)$, with distortion $\leq 1 + \varepsilon$. Here we show a natural extension of the above result, to a stronger preservation of the geometry of finite spaces. By a k-fold increase of the number of dimensions used compared to [JL84], a good preservation of volumes and of distances between points and affine spaces is achieved. Specifically, we show it is possible to embed a subset of size n of the Euclidean space into a $O(\varepsilon^{-2} k \log n)$- dimensional Euclidean space, so that no set of size $s \leq k$ changes its volume by more than $(1+\varepsilon)^{s-1}$. Moreover, distances of points from affine hulls of sets of at most $k - 1$ points in the space do not change by more than a factor of $1 + \varepsilon$. A consequence of the above with $k = 3$ is that angles can be preserved using asymptotically the same number of dimensions as the one used in [JL84]. Our method can be applied to many problems with high-dimensional nature such as *Projective Clustering* and *Approximated Nearest Affine Neighbor Search*. In particular, it shows a first poly-logarithmic query time approximation algorithm to the latter. We also show a structural application that for volume respecting embedding in the sense introduced by Feige [Fei00], the host space need not generally be of dimensionality greater than polylogarithmic in the size of the graph.

1 Introduction

The dimension of a normed space that accommodates a finite set of points plays a critical role in the way this set is analyzed. The running time of most geometric algorithms is at least linear in the dimensionality of the space, and in many cases exponential in it. To represent the metric of n points in the Euclidean space, one clearly needs no more than $n - 1$ dimensions. By relaxing the notion of isometry to near-isometry, the underlying structure of these points is well represented in a space with a much smaller number of dimensions; In their seminal paper [JL84], Johnson and Lindenstrauss show that far more efficient representations capture almost precisely the metric nature of such sets. They present a simple and elegant principle that allows one to embed such an n-point set into a t-dimensional Euclidean space, with t merely $O(\varepsilon^{-2} \log n)$, while preserving the

J.D.P. Rolim and S. Vadhan (Eds.): RANDOM 2002, LNCS 2483, pp. 239–253, 2002.

pairwise distances to within a relative error of ε. One simply needs to project the original space onto a random t-dimensional subspace (and scale accordingly) to obtain a low-distortion embedding with high probability. Their argument therefore supplies a probabilistic algorithm for producing such an embedding.

Modifications that relate to the exact way the randomization is applied, and further improvements in the parameters were later proposed. For example it was shown that by using a projection onto t independent random unit vectors, a similar result can be achieved. In [Ach01], Achlioptas shows an even simpler probability space for the desired embedding, by projecting the space onto random vectors in $\{-1, 1\}^N$. It is interesting to note that these simplifications do not require higher dimensionality to satisfy the same quality of embeddings. In [EIO02], a derandomization of the probabilistic projection is given, leading to an efficient deterministic algorithm for finding such low-dimensional embeddings. Among the other simpler proofs to the result of Johnson and Lindenstrauss (which we sometimes call *JL-lemma*) are [FM88,IM98,LLR95,DG99,AV99].

For metric spaces (X, d_X) and (Y, d_Y), and an embedding $f : X \to Y$, we define the distortion of f by $\sup_{x,y \in X} \frac{d_Y(f(x), f(y))}{d_X(x,y)} \cdot \sup_{x,y \in X} \frac{d_X(x,y)}{d_Y(f(x), f(y))}$. By this definition a good embedding is one that preserves the pairwise distances. However, a set X of points in the Euclidean space has many more characteristics in addition to the metric they represent, such as the center of gravity of a set of points and its average distance to them, the angles defined by triplets of points, volumes of sets, and distances between points to lines, planes and higher dimensional affine spaces that are spanned by subsets of X. Regarding (some of) these characteristics as part of the structure of X, we redefine the quality of an embedding to be one that preserves both the volumes of (certain) subsets of X and the distances of points from affine hulls of subsets of X.

We define the volume of a set of k points in the Euclidean space as the $(k-1)$-dimensional volume (Lebesgue measure) of its convex-hull. For $k = 2$ this is just the distance between the points. For $k = 3$, this is the area of the triangle with vertices that are the three points of the set, etc. Throughout this paper we denote the volume of a set S in the Euclidean space by $\text{Vol}(S)$. When considering a general metric space (not necessarily Euclidean), it is not a-priori clear whether there is a reasonable way to define a volume. In [Fei00] Feige defined a notion of volumes for general metric spaces, and measured the quality of an embedding from general metric spaces into Euclidean spaces (he calls such embeddings *volume respecting embeddings*. The volume preservation there applied to two different definition of volumes, the one in general metric spaces, and the one in Euclidean space. This line of work led to important algorithmic applications, most notably a polylogarithmic approximation algorithm for the bandwidth problem [Fei00], and an approximation algorithm to a certain VLSI layout problem [Vem98]. Our attention focuses on the case where both the original and the image space are Euclidean, and consequently the volume preservation notion is a well defined one and need not use the more involved definition of [Fei00]. Accordingly, our result should not be confused with results in the aforementioned framework, most notably Rao's result [Rao99], that deals with Euclidean metric, but with respect to its metric structure alone.

Consider an embedding $f : \mathbb{R}^N \to \mathbb{R}^t$ that does not expand distances in $X \subset \mathbb{R}^N$. We say f distorts the volume of a set $S \subseteq X$ of size k by $\left(\frac{\text{Vol}(S)}{\text{Vol}(f(S))}\right)^{\frac{1}{k-1}}$. The exponent in this expression should be thought of as a natural normalization measure (and was introduced in [Fei00]).

Our Result: Let $\varepsilon \leq \frac{1}{4}$, X be an n-point subset of $\mathbb{R}^N$, and let $t = O(\varepsilon^{-2}k \log n)$. We show that there is a mapping of $\mathbb{R}^N$ into $\mathbb{R}^t$ that (i) does not distort the volume of subsets of X of size at most k to by more than a factor of $1 + \varepsilon$. (ii) preserves the distance of points from affine hulls of subsets of X of size at most $k - 1$ to within a relative error of ε.

To see how our result is achieved, we take a closer look into JL-lemma. The JL-lemma is based on the following lemma that can be found (in slightly different formulations) in [Ach01,DG99,IM98].

Lemma 1. *Let* $\varepsilon \leq \frac{1}{3}$, $v \in \mathbb{R}^N$, *and let* f *be a random projection onto t dimensions, multiplied by* $\sqrt{\frac{N}{t(1+\varepsilon)}}$. *Then*

$$\Pr\left[\frac{\|v\|}{1+\varepsilon} \leq \|f(v)\| \leq \|v\|\right] \geq 1 - \exp(\frac{2}{15}t\varepsilon^2)$$

For a low distortion embedding, $\binom{n}{2}$ vectors (the unsigned pairwise differences between the points) should maintain their norms approximately, and so t is chosen so that the above probability is smaller than $1/\binom{n}{2}$ resulting in $O(\varepsilon^{-2} \log n)$ needed dimensions.

In this paper we show that in order for a linear embedding to preserve volumes and affine-distances of sets of size at most k to within a relative error of ε, it is sufficient to preserve the norms of a certain set of $\exp(O(k \log n))$ vectors, to within a relative error of $\varepsilon/3$. The result is then achieved by taking t to be $O(\varepsilon^{-2}k \log n)$ which guarantees a positive probability for the preservation of the norms of that many vectors.

Applications. In [IM98], Indyk and Motwani describe the way projections can be used for designing efficient algorithms for the *Approximate Nearest Neighbor* problem. The generalization of this problem from a set of points to a set of k-dimensional affine spaces is the *Approximate Nearest affine neighbor*. For $k = 1$, i.e., this is the problem of finding the closest line to a query point. This is a natural proximity problem, that has appeared few times in the literature. Using Meiser's result for point location in arrangements of hyperplanes together with our result yields a randomized poly-logarithmic query-time approximation to the *Approximate Nearest affine neighbor* problem, which is to the best of our knowledge the first. In Section 5.1 we show how exactly this can be achieved. Our result can be applied to another classical problem in computational geometry that stems from data mining. Consider a data set in $\mathbb{R}^N$, where the "true dimensionality" is anticipated to be much smaller than N. In other words, it is assumed to be possible to cover (up to proximity) the set by a small number of k-dimensional affine spaces. The problem is known to be NP-hard even for dimension 2. Currently, there are approximation algorithms for the cases of two and three dimensions. For the higher dimensions, our method provides a way to reduce the problem to $\varepsilon^{-2}k\log n$ dimensions.

There are also straight forward applications of our result to other problems which extend metrical questions to ones that consider volumes. In particular, the problem of finding the diameter of a set of n points, can be extended to finding the biggest volume subset of size k. It is immediate by our result that this can be approximated when dimensionality is reduced to $O(k\log n)$.

Can Feige's volume-respecting embeddings benefit from our result? We briefly describe the general framework in those embeddings. Consider an embedding of a graph on n vertices into the Euclidean space that does not increase distances. Such an embedding is good if the (Euclidean) volumes of sets of size at most k are big. Typically the value of k is $O(\log n)$. Our result shows that by combining a volume-respecting embeddings with a random projection onto a low dimension, the distortion is asymptotically the same. Specifically, for $k = O(\log n)$ as often is the case, we get that the restriction to $O(\log^2 n)$-dimensional embeddings entails only an extra constant factor to volume-distortion of the embedding.

2   Preliminaries and Notation

The norm $\|\cdot\|$ always stands for the Euclidean norm. We say that an embedding $\phi : X \to \mathbb{R}^N$ is a *contraction* if for every $x, y \in X$, $\|\phi(x) - \phi(y)\| \leq \|x - y\|$. Whenever the dimensionality of an Euclidean space is immaterial, we call it $\mathbb{R}^N$ without explicitly defining N. We will sometime refer to an affine space as a *flat*. An affine subspace of $\mathbb{R}^N$ that is spanned by points of $X \subseteq \mathbb{R}^N$ is called X-flat (analogously, we define X-lines, X-planes, X-k-dimensional flats, etc.).

For a set $S \subseteq \mathbb{R}^N$ of size ν, we denote by $\mathcal{L}(S)$ the affine-hull of S, that is $\mathcal{L}(S) = \{\sum_{i=1}^{\nu} \lambda_i a_i | \sum_i \lambda_i = 1\}$. For a set $S \subseteq \mathbb{R}^N$ and $x \in \mathbb{R}^N$ we define $P(x, S)$ to be the projection of x onto $\mathcal{L}(S)$. The *affine distance* of x to S, $\mathrm{ad}(x, S)$, is defined to be the distance of x to the affine-hull of S, or equivalently $\|x - P(x, S)\|$. The affine distance will occasionally referred to as *height*. Let $r_1, r_2 \ldots, r_{\nu-1}$ be an arbitrary set of orthonormal vectors in $\mathcal{L}(S)^\perp$. We now define the *corner* points of the pair (x, S). The i-th corner point $c_i(x, S)$ is defined as $P(x, S) + \mathrm{ad}(x, S) \cdot r_i$, for $1 \leq i \leq \nu - 1$. When $s = 2$ we let $c(x, S)$ denote $c_1(x, S)$. See figure 3.

---

[1] We exclude the case where S is affinely dependent: Since we consider only linear mappings, the image of affinely dependent set will also be affinely dependent, and so degenerated case will remain degenerated in the image. Also notice that $\mathrm{ad}(x, S) = \mathrm{ad}(x, S')$ where $S' \subseteq S$ is an affinely independent set for which $\mathcal{L}(S') = \mathcal{L}(S)$, and that if $\mathrm{ad}(x, S) = 0$ then $\mathrm{ad}(f(x), f(S)) = 0$.

3   Preserving Distances to Lines and Preserving Areas of Triangles

Consider the problem of finding a low-dimensional Euclidean embedding of a finite subset X of the Euclidean space, such that pairwise distances, areas of triangles, and distance of points from X-lines do not change by much. Note that this is a special case of the the general problem we consider (here $k = 3$), as volumes of triplets of points are simply areas. This case is easier to analyze and moreover, the analysis of the general case uses some of the structure of the two dimensional case. The current case also gives a preservation result for angles as we later note.

The natural thing to try in order to reduce dimensionality and preserve geometrical features is simply to apply JL-lemma: after all, in an isometry not merely distances are preserved, but also volumes, affine distances and angles. One might expect that when f is nearly an isometry (i.e. f has small distortion) it will follow that volumes and affine distances are being quite reasonably preserved. We show that in general, this is very far from the truth: Consider a triangle with a very small angle. A low-distortion embedding can be applied to it, so that it is changed to a triangle with dramatically different angles, area and heights (see Figure 1).

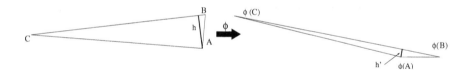

Fig. 1. Areas and affine distances can dramatically change and practically vanish, even under a low-distortion embedding. Here, the distortion of Φ is very small, but $h' \ll h$

We next show a certain class of triangles for which small distance-distortion *does* imply small heights-distortion.

Lemma 2. *Let A, B, C be the vertices of a right angle isosceles triangle, where the right angle is at A, and let Φ be a contracting embedding of its vertices to a Euclidean space, such that the edges do not contract by more than $1 + \varepsilon$, where $\varepsilon \le \frac{1}{6}$. Let h be the length of $[AC]$ ($h = \mathrm{ad}(C, \{A, B\})$), $\boldsymbol{b}$ be the vector $\Phi(B) - \Phi(A)$, and $\boldsymbol{c}$ be the vector $\Phi(C) - \Phi(A)$. Then*

1. $|\langle \boldsymbol{b}, \boldsymbol{c} \rangle| \le 2\varepsilon \cdot h^2$
2. $h/(1 + 2\varepsilon) \le \mathrm{ad}(\Phi(C), \{\Phi(A), \Phi(B)\}) \le h$.

Remark 1. The first assertion of the lemma says that the images of the perpendicular edges of the triangle are almost orthogonal too. This fact is used later in the analysis of the general case.

Proof. Let $a = \|\boldsymbol{b} - \boldsymbol{c}\|$, and let θ be the angle between $\boldsymbol{b}$ and $\boldsymbol{c}$. Now, $|\langle \boldsymbol{b}, \boldsymbol{c} \rangle| = |\|\boldsymbol{b}\|^2 + \|\boldsymbol{c}\|^2 - a^2|/2$. A simple analysis shows that this quantity is maximized (while satisfying the conditions on Φ) when $\|\boldsymbol{b}\| = \|\boldsymbol{c}\| = h/(1+\varepsilon)$ and $a = \sqrt{2}h$. Hence $|\langle \boldsymbol{b}, \boldsymbol{c} \rangle| \le \frac{1}{2}h^2 \cdot (2 - 2/(1+\varepsilon)^2) \le 2\varepsilon \cdot h^2$. As can be easily verified, these values of $a, \|\boldsymbol{b}\|, \|\boldsymbol{c}\|$ also maximize $|\cos\theta| = |\|\boldsymbol{b}\|^2 + c^2 - a^2|/(2\|\boldsymbol{b}\|\|\boldsymbol{c}\|)$. Hence $|\cos\theta| \le |(2 - 2(1+\varepsilon)^2)/2| = 2\varepsilon + \varepsilon^2$, and accordingly $\sin\theta = \sqrt{1 - \cos^2\theta} \ge \sqrt{1 - 4\varepsilon^2 - 4\varepsilon^3 - \varepsilon^4} > \frac{1+\varepsilon}{1+2\varepsilon}$ for $\varepsilon \le \frac{1}{6}$. Finally,

$$h/(1+2\varepsilon) = \frac{h}{1+\varepsilon} \cdot \frac{1+\varepsilon}{1+2\varepsilon} \le \mathrm{ad}(\Phi(C), \{\Phi(A), \Phi(B)\}) = c\sin\theta \le h.$$

In order to conclude that under a low-distortion embedding of the set X, areas and affine distances do not change by much, one would like to eliminate bad cases such as those in Figure 1. Think of the following physical model: Edges are rubber rods that can slightly contract due to a shock. Figure 1 demonstrates that this shock may very well change the areas and heights of triangles significantly. The remedy we propose is to supplement this rubber triangle with some additional rods to keep it stable. If the contraction of these rods is also limited, then placing them in appropriate locations, will eliminate cases such as the mapping Φ in Figure 1. This mental experience translates to an additional set of vectors whose norms must be approximately preserved.

Our strategy is therefore choosing "rods" such that nice triangles as in Lemma 2 emerge. This, together with the fact the embeddings we consider are linear, enables us to bound the changes of the heights. Area preservation then immediately follows.

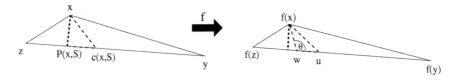

Fig. 2. The triangle $f(x), w, u$ is the f-image of a right angle isosceles triangle. Height-contraction is estimated via these triangle (dashed lines)

Theorem 1. *Let $\varepsilon \le \frac{1}{3}$ and let n, t be integers for which $t \ge 60\varepsilon^{-2}\log n$. Then for any n-point subset X of the Euclidean space $\mathbb{R}^N$, there is a linear contracting embedding $f : X \to \mathbb{R}^t$, under which the areas of triangles in X are preserved to within a factor of $(1+\varepsilon)^2$, the distances of points from X-lines are preserved to within a factor of $1 + \varepsilon$, and angles (of triplets of points from X) are preserved to within a (double-sided) factor of $1 + \frac{8}{\pi} \cdot \sqrt{\varepsilon}$.*

Proof. For every pair $S = \{y, z\}$ of elements of X, and every element $x \in X - S$, we consider the right angle isosceles triangle $\{x, P(x, S), c(x, S)\}$. Let V

be the collection of the unsigned vectors corresponding to these triangles $(x - P(x, S), c(x, S) - P(x, S)$ and $x - c(x, S))$ over all choices of S and x, together with all pairwise differences between the points of X. Let f be a random projection onto t dimensions, multiplied by $\sqrt{\frac{N}{t(1+\varepsilon/2)}}$. By Lemma 1, the probability that f does not expand norms of vectors in V, and that it does not contract them by more than $1 + \varepsilon/2$ is at least $1 - |V| \exp(-\frac{2}{15} t(\frac{\varepsilon}{2})^2)$. A little closer look shows that V contains merely $3\binom{n}{2}$ different directions : the directions of X-lines, and their rotations by $\pi/4$ and by $\pi/2$. Since f is linear, vectors in the same direction are preserved simultaneously, and therefore to establish the existence of f as above we need

$$1 - 3\binom{n}{2} \exp\left(-\frac{2}{15} t(\varepsilon/2)^2\right) > 0$$

which is satisfied when $t \geq 60\varepsilon^{-2} \log n$. Next we show that f satisfies the preservation statements of the theorem. Consider three different points $x, y, z \in X$. Let $w = f(P(x, \{y, z\})$ and $u = f(c(x, \{y, z\}))$. Now let θ be the angle between $f(x) - w$ and $f(y) - w$ (see Figure 2). Since $z, P(x, \{y, z\})$ are collinear and f is linear, $f(z), w, u, f(y)$ are also collinear. We now apply the second assertion of Lemma 2 with $P(x, \{y, z\}), c(x, \{y, z\})$ and x for A, B and C and $\varepsilon/2$ the error parameter. It follows that

$$\mathrm{ad}(x, \{y, z\})/(1 + \varepsilon) \leq \mathrm{ad}(f(x), \{f(z), f(y)\}) \leq \mathrm{ad}(x, \{y, z\}).$$

For the area estimates, we get

$$1/(1 + \varepsilon)^2 \leq \frac{\mathrm{Vol}(f(S))}{\mathrm{Vol}(S)} = \frac{\|f(y) - f(z)\|}{\|y - z\|} \cdot \frac{\mathrm{ad}(f(x), \{f(y), f(z)\})}{\mathrm{ad}(x, \{y, z\})} \leq 1.$$

Finally, let $\alpha = \angle(xyz)$ and $\alpha' = \angle(f(x)f(y)f(z))$. We have that

$$\frac{1}{1 + \varepsilon} \leq \frac{\sin \alpha'}{\sin \alpha} = \frac{\|y - x\|}{\|f(y) - f(x)\|} \cdot \frac{\mathrm{ad}(f(x), \{f(y), f(z)\})}{\mathrm{ad}(x, \{y, z\})} \leq 1 + \varepsilon. \qquad (1)$$

We now turn to analyze the relative change of the angles themselves. For the case $\alpha, \alpha' \leq \pi/2$, we use the following fact that holds for any $0 < \beta, \gamma \leq \pi/2$: if $\sin \beta / \sin \gamma \leq 1 + \varepsilon$ then $\beta/\gamma \leq 1 + \sqrt{\varepsilon}$. This fact together with inequality 1 implies that $1/(1 + \sqrt{\varepsilon}) \leq \alpha'/\alpha \leq 1 + \sqrt{\varepsilon}$. The case where $\alpha, \alpha' \geq \pi/2$ can be easily reduced to the previous case, by replacing α, α' by $\pi - \alpha, \pi - \alpha'$. Last, assume that $\alpha \geq \pi/2$ and $\alpha' \leq \pi/2$. Using the fact that a linear embedding is an isometry times a scalar when restricted to a line, and in particular that the order of points along a line does not change under a linear embedding, we get $\|f(x) - w\| \geq \|f(x) - f(y)\|$. Now,

$$\|f(x) - w\| \leq \|x - P(x, \{y, z\})\| = \|x - y\| \cdot \sin \alpha \leq \frac{\|f(x) - f(y)\|}{1 + \varepsilon/2} \cdot \sin \alpha,$$

and therefore $\sin \alpha \geq 1/(1 + \frac{\varepsilon}{2})$ which means $\alpha \leq \pi/2 + \sqrt{\varepsilon}$. Analogously, we can show that $\alpha' \geq \pi/2 - \sqrt{\varepsilon}$. Therefore $\alpha - \alpha' \leq 2\sqrt{\varepsilon}$. When the angle changes from acute to obtuse we similarly obtain that $\alpha' - \alpha \leq 2\sqrt{\varepsilon}$. Summing it all up, we get

$$1/(1 + \varepsilon') \leq \alpha'/\alpha \leq 1 + \varepsilon',$$

where $\varepsilon' = \frac{8}{\pi} \cdot \sqrt{\varepsilon}$ always holds.

4   The General Case

Analogously to the $k = 3$ case, we look for a set of vectors with the following properties: (i) it is not too big and (ii) if a linear embedding does not change the norms of the vectors by much, then it also does not change affine distances and volumes by much. We first restrict our attention to one affine distance, $\mathrm{ad}(x, S)$ with $|S| = s < k$. We construct a set of vectors in a way which is determined by the relation between ε and s. When s is small with respect to $1/\varepsilon$ we extend the previous construction (for the case $k = 3$) in the natural way. The right angle isosceles triangle is substituted by a simplex spanned by $[x, P(x, S)]$ and by orthogonal vectors in $\mathcal{L}(S)$ of size $\mathrm{ad}(x, S)$ (call this a *nice simplex*). When s is bigger than $1/\varepsilon$ we use, in addition to this set of vertices, a dense enough set of points. The analysis is mostly algebraic one. It relates to the geometry of interest in the following simple way: Let V_{s-1} be a simplex on s points, V_{s-2} be the facet of V_{s-1} opposite one of the points, and H be the distance from that point to the facet, then $\mathrm{Vol}(V_{s-1}) = H \cdot \mathrm{Vol}(V_{s-2})/(s-1)$. The simplices we take are the images of the nice simplices under the linear low distortion embedding. The actual estimate $\mathrm{Vol}(V_{s-1})/\mathrm{Vol}(V_{s-2})$ is then achieved by means of analysis of determinants of matrices with constraints resulting from the low distortion embedding on the set of auxiliary points we added.

Similarly to the case $k = 3$, we make use of the linearity of our embeddings to claim that it is enough to bound the contraction of heights in nice simplices to get the actual bound for all the required affine distances X (refer again to Figure 3 for the exposition). Eventually, the guarantee for the volume preservation is achieved by an iterative use of the relation $\mathrm{Vol}(V_{s-1}) = H \cdot \mathrm{Vol}(V_{s-2})/(s-1)$.

Proposition 1. *Let $\varepsilon \leq \frac{1}{12}$, let $S \subset \mathbb{R}^n$ be a set of size $s < k$ points, and let $x \in \mathbb{R}^n - S$. Then there is a subset $W = W_{x,S,\varepsilon}$ of $\mathbb{R}^n$ of size $\leq (5s)^{\frac{3}{2}s}$, such that if $f : \mathbb{R}^n \to \mathbb{R}^t$ is a linear embedding that does not expand distances in W and does not contract them by more than $1 + \varepsilon$, then $\mathrm{ad}(x, S)/(1 + 3\varepsilon) \leq \mathrm{ad}(f(x), f(S)) \leq \mathrm{ad}(x, S)$.*

Proof. Let $W_0 = \{x, P(x, S), c_1(x, S), \ldots, c_{s-1}(x, S)\}$. Recall by definition that

- $\forall i, c_i(x, S) \in \mathcal{L}(S)$.
- The vectors $\{x - P(x, S), c_1(x, S) - P(x, S), \ldots, c_{s-1}(x, S) - P(x, S)\}$ are orthogonal and are of the same length, namely $\mathrm{ad}(x, S)$.

Clearly, $\mathrm{ad}(x, S) = \mathrm{ad}(x, \{P(x, S), c_1(x, S), \ldots, c_{s-1}(x, S)\})$. Now, by linearity of f, $\mathrm{ad}(f(x), f(S)) = \mathrm{ad}(f(x), \{f(P(x, S)), f(c_1(x, S)), \ldots, f(c_{s-1}(x, S))\})$, and so it is enough to prove that

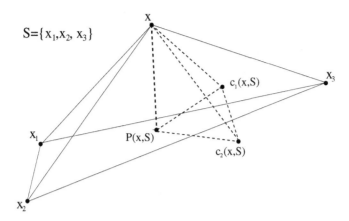

S={x_1, x_2, x_3}

Fig. 3. The set W_0 consists of the vertices of the 'nice' simplex (dashed lines). Notice that the height $[x, P(x, S)]$ is common to the original simplex (solid lines) and to the nice simplex, and that the linear images of these two simplices also share the same height

$$\frac{1}{1+3\varepsilon} \leq \frac{\text{ad}(f(x), f(\{P(x,S), c_1(x,S), \ldots, c_{s-1}(x,S)\})}{\text{ad}(x, \{P(x,S), c_1(x,S), \ldots, c_{s-1}(x,S)\})} \leq 1.$$

Since f is linear, we may simplify and assume that $P(x, S) = 0$ (the zero vector), $\text{ad}(x, S) = 1$ and that $\mathcal{L}(S)$ is spanned by the first $s - 1$ standard vectors (so $c_i(x, S) = e_i$) and also that $x = e_s$. We now need to show that

$$1/(1 + 3\varepsilon) \leq H \leq 1. \tag{2}$$

where H is the affine distance $\text{ad}(f(e_s), \{0, f(e_1), \ldots, f(e_{s-1})\})$.

The right inequality is immediate, since $H \leq \|f(e_s)\| \leq 1$. We proceed to the more interesting left side of inequality 2. The operation of f on the first s coordinates can be described as a $t \times s$ matrix U. Now, the volume of the the the simplex P which is the convex hull of $0, f(e_1), \ldots, f(e_s)$ is $\sqrt{\det(U^t U)}/(s-1)!$. Let $\hat{P}$ be the facet of P opposite $f(e_s)$. Similarly, $\hat{P}$ has volume $\sqrt{\det(\hat{U}^t \hat{U})}/(s-2)!$, where $\hat{U}$ is obtained be removing the last column of U. Consequently

$$H = (s-1) \cdot \text{Vol}(P)/\text{Vol}(\hat{P}) = \sqrt{\det(U^t U)/\det(\hat{U}^t \hat{U})}.$$

We let $A = U_s^t U_s$ and $B = \hat{U}^t \hat{U}$, and note that $A_{ij} = \langle f(e_i), f(e_j) \rangle$, and that B is the principal minor of A that is obtained by removing its last row and column. Our aim is therefore to bound $\det A / \det B$ from below. At this point we divide our analysis depending on the relation between $1/\varepsilon$ and k. The definition of W is dependent on this relation too.

Case 1: $1/\varepsilon \geq 4s$. We start with the following algebraic lemma.

248   A. Magen

Lemma 3. *Let* $\mu \leq \frac{1}{2(s-1)}$, *and let* A *be a real* $s \times s$ *matrix, such that* [2] $\|A - I\|_\infty \leq \mu$. *Denote by* B *the principal minor as described above; then* $\det(A)/\det(B) \geq 1 - 2\mu$.

The proof of Lemma 3 will be given shortly. We now show that for the case $1/\varepsilon >$ $4s$ it implies proposition 1. Indeed, we take $W = W_0$. Since f is a contraction with distortion $\leq 1+\varepsilon$ on W, for all i we have that $\sqrt{1 - 2\varepsilon} \leq 1/(1+\varepsilon) \leq \|f(e_i)\| \leq 1$. Now for $i \neq j$ consider the triangle $e_i, 0, e_j$. The first statement of Lemma 2 says that $|\langle f(e_i), f(e_j)\rangle| \leq 2\varepsilon$. We take A and B to be the matrices described above, and we take $\mu = 2\varepsilon$. We have just established that $\|A - I\|_\infty \leq \mu$, and since $1/\varepsilon \geq 4s$ it follows that $\mu \leq \frac{1}{2(s-1)}$. By the proceeding analysis, $H = \sqrt{\det A/\det B} \geq \sqrt{1 - 2\mu} = \sqrt{1 - 4\varepsilon} \geq \frac{1}{1+3\varepsilon}$ for $\varepsilon \leq \frac{1}{12}$. To conclude, notice that $|W| = s + 1 \leq (5s)^{\frac{3}{2}s}$. We now prove Lemma 3.

Proof. (Lemma 3) We first observe that if $y \in \mathbb{R}^s$ and $v = By$, then $\|v\|_\infty \geq \frac{1}{2}\|y\|_\infty$: Assume without loss of generality that $|y_1| = \|y\|_\infty$. Now,

$$|v\|_\infty \geq |v_1| = |\sum_{i<s} y_i a_{1,i}| = |y_1 + \sum_{i<s} y_i(a_{1,i} - \delta_{1,i})|$$

$$\geq |y_1| - \mu \sum_{i<s} |y_i| \geq |y_1| - (s-1)\mu|y_1| \geq \frac{1}{2}|y_1| = \frac{1}{2}\|y\|_\infty$$

($\delta_{i,j}$ denotes the Kronecker Delta, which is 1 if $i = j$ and 0 otherwise). One immediate consequence of the above is that B is nonsingular, since it implies that if v is the zero vector then so is y.

Now, let v be the vector $(a_{1,s}, a_{2,s}, \ldots, a_{s-1,s})$. For every $j < s$, let $B^{(j)}$ be the matrix B with the j-th column replaced by v. We next argue that $|\frac{\det(B^{(j)})}{\det(B)}| \leq$ 2μ for all $1 \leq j < s$. Let $y_j = \frac{\det(B^{(j)})}{\det(B)}$. Recall that by Cramer's rule, $y =$ $(y_1, y_2, \ldots, y_{s-1})$ is the (unique) solution to $By = v$. We now get $|\frac{\det(B^{(j)})}{\det(B)}| =$ $|y_j| \leq \|y\|_\infty \leq 2\|v\|_\infty = 2\max_{i<s} a_{i,s} \leq 2\mu$.

Now,

$$\det(A) = a_{s,s}\det(B) + \sum_{i<s}(-1)^{i+s}a_{s,i}\det(B^{(i)}).$$

Therefore

$$\frac{\det(A)}{\det(B)} = a_{s,s} + \sum_{i<s}(-1)^{i+s}a_{s,i}\frac{\det(B^{(i)})}{\det(B)} \geq 1 - \mu - (s-1)\mu \cdot 2\mu \geq 1 - 2\mu.$$

Case 2: $1/\varepsilon < 4s$. In this case W_0 alone is too sparse to provide us with the needed bound on the contraction of H. The approach we take here follows an argument used by Feige in [Fei00][3]: Instead of taking just W_0 we add to it a

---
[2] $\|M\|_\infty$ denote the maximal absolute value of the elements of the matrix M.
[3] although considerably modified for the present use

much denser set, namely an $O(\eta)$-net in the unit ball, where $\eta = \varepsilon/\sqrt{s}$. Such a net in the present case where $1/\varepsilon < 4s$ is not too big as a function of s, but still good enough for the bound on the contraction of H.

We turn to the details of this construction. Call a set η-*separated* if the distance between any two distinct points in the set is greater than η. We take W to be an inclusion-maximal η-separated subset of the unit ball. By standard arguments, for any point v in the unit ball there is a point $v' \in W$ such that $\|v - v'\| \le \eta$ (otherwise, $W \bigcup \{v\}$ would also be a η-separated subset of the ball). Without loss of generality we can assume that $W \supset W_0$. We now assume that f does not contract distances in W by more than $1 + \varepsilon$. We first bound the norm of U (as a linear operator). Since f is a contraction on W_0, the columns of U are of (Euclidean) norm at most 1. This means that $\|U\|_2 = \max\{\|Uv\| : \|v\| = 1\} \le \sqrt{s}$. We next show that if v is a unit vector then $\|Uv\| \ge 1 - 3\varepsilon$. Indeed, let v' be a vector in W such that $\|v - v'\| \le \eta$. Now

$$\|Uv\| = \|Uv' + U(v - v')\| \ge \|Uv'\| - \|U(v - v')\| \ge \frac{1}{1+\varepsilon} - \sqrt{s}\|v - v'\| \ge$$

$$\frac{1}{1+\varepsilon} - \eta\sqrt{s} = \frac{1}{1+\varepsilon} - \sqrt{s} \cdot \frac{\varepsilon}{\sqrt{s}} \ge 1 - 2\varepsilon.$$

Now, let $0 \le \lambda_1 \le \lambda_2 \le \ldots \le \lambda_s$ be the eigenvalues of A, and let $\sigma_1 \le \sigma_2 \le \ldots \le \sigma_{s-1}$ be the eigenvalues of B. It is known that $\lambda_i \le \sigma_i \le \lambda_{i+1}$, and so

$$\frac{\det A}{\det B} = \frac{\prod_{i=1}^{s} \lambda_i}{\prod_{i=1}^{s-1} \sigma_i} \ge \lambda_1.$$

We now use the standard fact that $\lambda_1 = \min\{\|Uv\|^2 : \|v\| = 1\}$. It follows that

$$H = \sqrt{\det A / \det B} \ge \sqrt{\lambda_1} = \min\{\|Uv\| : \|v\| = 1\} \ge 1 - 2\varepsilon \ge 1/(1 + 3\varepsilon)$$

for $\varepsilon < 1/12$. It remains to show that W is not too big. Since W is η-separated, by a volume argument $|W| \le (\frac{2+\eta}{\eta})^s$. Therefore $|W| \le (\frac{2+\eta}{\eta})^s = (1 + \frac{2\sqrt{s}}{\varepsilon})^s \le (5s)^{\frac{3}{2}s}$.

Corollary 1. *Let* $\varepsilon \le \frac{1}{4}$, *let* $S \subset \mathbb{R}^n$ *be a set of size k points, and let* $x \in \mathbb{R}^n - S$. *Let f be a random projection onto t dimension multiplied by* $\sqrt{\frac{N}{t(1+\varepsilon)}}$. *Then*

$$\Pr\left[\mathrm{ad}(x,S)/(1+\varepsilon) \le \mathrm{ad}(f(x), f(S)) \le \mathrm{ad}(x,S)\right] \ge 1 - \exp(3k(2 + \log k) - \frac{2}{135}t\varepsilon^2)$$

Proof. Let $W = W_{x,S,\varepsilon/3}$. By Lemma 1 we get that the probability that f is a contraction with distortion $\le 1 + \varepsilon/3$ on W is at least $1 - \binom{|W|}{2}\exp(-\frac{2}{15}t(\frac{\varepsilon}{3})^2)$. By proposition 1 such an embedding satisfies $\mathrm{ad}(x,S)/(1+\varepsilon) \le \mathrm{ad}(f(x), f(S)) \le \mathrm{ad}(x,S)$. Now $|W| \le (5k)^{\frac{3}{2}k}$ and so $\binom{|W|}{2} \le |W|^2 \le (5k)^{3k} \le \exp(3k(2 + \log k))$, and the bound in the lemma follows.

Remark 2. The different approaches we use for the two cases in Proposition 1 seem to be necessary, in the sense that no one of them can be applied to the other case: suppose we apply the approach of case 1 when $\varepsilon = \frac{1}{2s}$, then the linear embedding $f(e_j) = e_j - \frac{1}{s}\sum_i f(e_i)$ has distortion $\leq 1 + \varepsilon$, but in the same time makes H zero!. If, on the other hand, we use a dense net when $1/\varepsilon \gg n$ it means that $|W| \gg n^s$, which in turn leads to a dimensionality $O(\varepsilon^{-2} k \log(1/\varepsilon))$ rather than $O(\varepsilon^{-2} k \log n)$.

4.1   The Main Theorem

Theorem 2. *Let* $\varepsilon \leq \frac{1}{4}$ *and let* k, n, t *be integers greater than 1, for which* $t \geq 70\varepsilon^{-2}(k \log n + 3k(2 + \log k))$. *Then for any* n-*point subset* X *of the Euclidean space* $\mathbb{R}^n$, *there is a linear mapping* $f : X \rightarrow \mathbb{R}^t$, *such that for all subsets* S *of* X, $1 < |S| < k$,

$$\mathrm{Vol}(S)/(1 + \varepsilon) \leq \mathrm{Vol}(f(S)) \leq \mathrm{Vol}(S),$$

and for $x \in X - S$,

$$\mathrm{ad}(x, S)/(1 + \varepsilon) \leq \mathrm{ad}(f(x), f(S)) \leq \mathrm{ad}(x, S).$$

Proof. We apply Corollary 1 to all choices of S and x as above, and then use union bound. We get that as long as

$$\sum_{s=2}^{k} s \binom{n}{s} \exp\left(3k(2 + \log k) - \frac{2}{135} t\varepsilon^2\right) < 1,$$

the probability that a random projection onto t dimensions is a contraction on X and all relevant affine distances are preserved to within $1 + \varepsilon$, is positive. Now

$$\sum_{s=2}^{k} s \binom{n}{s} \exp\left(3k(2 + \log k) - \frac{2}{135} t\varepsilon^2\right) \leq n^k \exp\left(3k(2 + \log k) - \frac{2}{135} t\varepsilon^2\right) =$$

$$\exp\left(k \log n + 3k(2 + \log k) - \frac{2}{135} t\varepsilon^2\right).$$

Setting $t = 70\varepsilon^{-2}(k \log n + 3k(2 + \log k)) = O(\varepsilon^{-2} k \log n)$ then, guarantees that with positive probability f preserves all affine distance of sets of size at most k to within relative error of ε.

We turn to the other part of the theorem, namely volume preservation. Let $S_r = x_1, x_2, \ldots, x_r$. It is known that $\mathrm{Vol}(S_r) = \frac{1}{(r-1)!} \cdot \prod_{i=1}^{r-1} \mathrm{ad}(x_{i+1}, S_i)$, and so the volume distortion of S, $(\mathrm{Vol}(S)/\mathrm{Vol}(f(S)))^{\frac{1}{r-1}}$ is simply the geometric mean of $\{\mathrm{ad}(x_{i+1}, S_i)/\mathrm{ad}(f(x_{i+1}), f(S_i))\}_{i=1}^{r-1}$. We now conclude that $1 \leq (\mathrm{Vol}(S)/\mathrm{Vol}(f(S)))^{\frac{1}{r-1}} \leq 1 + \varepsilon$.

5   Applications

5.1   The Approximated Nearest Neighbor to Affine Spaces Problem

Let $\mathcal{F}_1, \mathcal{F}_2, \ldots, \mathcal{F}_n$ be k-dimensional flats in $\mathbb{R}^d$, and let $x \in \mathbb{R}^d$ be a query point. To answer a Nearest Neighbor to Affine-Spaces is to find the flat closest (in the Euclidean sense) to x. In [Mei93] Meiser presents a solution to the point location problem in arrangements of n hyperplanes in $\mathbb{R}^d$ with running time $O(d^5 \log n)$ and space $O(n^{d+1})$. That paper was a breakthrough in that it was the first time (and to the best of our knowledge also the last) where an algorithm to the problem was presented, that is not exponential neither in d nor in $\log n$. We show that by combining this result with ours, a considerable improvement to the problem of approximating the Nearest Affine Neighbor can be achieved. Let $\rho_i : \mathbb{R}^d \to \mathbb{R}$ be the functions that are the squares of distances from the flats, i.e. $\rho_i(x) = dist(x, \mathcal{F}_i)$. Note that ρ_i^2 are polynomials of degree 2 in $x_1, \ldots, x_d$. We now use the following standard linearization of such polynomials. We use the transformation $\xi : \mathbb{R}^d \to \mathbb{R}^{d'}$ which is the assignment transformation to all the monomials of degree at most two in $x_1, \ldots, x_d$. For example, if $d = 2$, $\xi(x) = \xi((x_1, x_2)) = (1, x_1, x_2, x_1 x_2)$. Clearly, $d' = \binom{d}{2} + 2d + 1$. Next, we introduce the functions ρ_i' defined by $\rho_i'(\xi(x)) = \rho_i^2(x)$. It is easy to see now that the ρ_i' are linear functionals [4]. We can now reduce the *Nearest Affine Neighbor* problem with query point x to the Vertical Ray Shooting problem from x: 'Shoot' a ray from $(\xi(x), -\infty)$ 'upwards', in other words to the positive direction of the last coordinate. Let $\mathcal{S}_i$ be (one of) the first surface this ray hits. Then $\mathcal{F}_i$ is the closest of $\mathcal{F}_1, \mathcal{F}_2, \ldots, \mathcal{F}_s$ to x. This vertical ray-shooting query in linear arrangements can be easily answered using the data-structure of Meiser.

The Preprocessing time, as well as space needed for the above are $O(n^{d'+1}) = O(n^{d^2})$, and the query time is $O(d'^5 \log n) = O(d^{10} \log n)$. We now apply our result in the following way.

For each flat $\mathcal{F}_i$, take an affinely-independent subset of $k + 1$ points. We can embed these $O(kn)$ points to $O(\varepsilon^{-2} k \log(kn)) = O(\varepsilon^{-2} k \log n)$ dimensions, and for k constant this is just $O(\varepsilon^{-2} \log n)$. Concatenation of the above yields a query time of $O(\varepsilon^{-20} \log^{11} n)$, and a preprocessing time (and space) of $n^{O(\varepsilon^{-4} \log^2 n)}$.

The standard problem in this approach remains: we need to answer queries on any point of the space, and not on a predetermined set. Looking into the proof of Theorem 2 we easily notice that by taking $t = \varepsilon^{-2} Q k \log n$ we get that the probability that a fixed projection is good in the sense that it does not distort affine distances from a random point x on the sphere to sets of size at most $k - 1$ in X is $n^{-\Omega(Q)}$. We can therefore approximate the answer to all but a small fraction of the query points, with only a constant factor sacrifice in the number of dimensions needed.

---

[4] The ρ_i' are no longer defined on the whole space, but this should not be of any concern to us.

5.2   Projective Clustering

Projective Clustering is a well known problem which has important applications in data-mining. This special variant of clustering, relates very closely to the geometric structures that are discussed in this paper. Here is the problem: Given an n-point set X in $\mathbb{R}^N$ and an integer $s > 0$, find s k-dimensional flats (k-flats), $h_1, \ldots, h_k$, so that the greatest distance of any point in X from its closest flat is minimized. In other words, this is a s-clustering problem, where a cluster is defined by a k-flat, and the quality of the clustering, its *width*, is the maximal distance of any point from its corresponding flat [5]. The "regular" clustering is in fact a special case, where $k = 0$. There are different ways to define the solution for the *projective clustering* problem: the optimal width, the clustering, and the k-flats themselves. In [PC00], Agarwal and Procopiuc give an efficient approximation algorithm for the planar case where the flats are lines, i.e. $N = 2$ and $k = 1$. In [HV02], Har-Peled and Varadarajan give a $dn^{O(\frac{kg}{\varepsilon^5} \log(1/\varepsilon))}$ algorithm that approximates the solution to within $1 + \varepsilon$. Here we show that under certain restrictions one can reduce the problem to one in which the space is of dimension that depends logarithmically in n, such that only a small inaccuracy incurs.

In [PC00], the original problem is reduced to the variant where the candidate flats are only the X-k-flats. Namely only affine-subspaces that are spanned by $k + 1$ points from X are considered. We call this variant the *Median Projective Clustering*. Agarwal et al. show that by this reduction an additional factor of at most 2 is added to the approximation. We claim

Theorem 3. *An instance of the* Median Projective Clustering *with* $X \subseteq \mathbb{R}^N$ *and* k *the dimensionality of the flats, can be reduced to a t-dimensional space instance with a $1 + \varepsilon$ approximation to the optimal width, where* $t = O(\varepsilon^{-2} k \log n)$,.

Proof. Using Theorem 2, we can map X to $\mathbb{R}^t$, such that the distance of points in X to k-dimensional X-flats do not expand, and do not contract by more than a factor of $1 + \varepsilon$ Therefore any solution w^* in the reduced problem corresponds to a solution w in the original problem, with $w^* \leq w \leq 2w^*$, and we immediately get that $\tau^* \leq \tau \leq 2\tau^*$, where τ, τ^* are the optimal solution to the original problem, and the optimal solution to the reduced problem respectively. Note that the clustering for that width can also be given from the reduced problem. It is not clear, however, how to reconstruct the k flats from the ones in the reduced problem, as the k-flats in the smaller space do not map back to the original space (a projection is not injective). □

Acknowledgments. I thank Robert Krauthgamer and Nati Linial for discussions that led to the question of this paper, to Uri Feige for his suggestion to use a dense net in case 2 of Proposition 1, to Piotr Indyk and Sariel Har-Peled for the Computational-Geometry relevant Background and references, and many suggestions for Applications. In particular to Sariel, who pointed out the way to use Meiser's result in Section 5.1.

---

[5] Any other variant, such as average distance, sum and sum-of-squares is probably as applicable here.

References

[Ach01]   Dimitris Achlioptas. Database-friendly random projections. In *roceedings of PODS 01*, pages 274–281, 2001.

[AV99]   R.I. Arriaga and S. Vempala. An algorithmic theory of learning: Robust concepts and random projections. In *Proceeding of the 31th Symposium on the Theory of Computing*, New York, 1999.

[DG99]   S. Dasgupta and A. Gupta. An elementary proof of the johnson-lindenstrauss lemma. In *Technical Report TR-99-06*, Computer Scinece Institute, Berkeley, CA, 1999.

[EIO02]   L. Engebretsen, P. Indyk, and R. O'Donnell. Derandomized dimensionality reduction with applications. In *Proceedings of the 13th Symposium on Discrete Algorithms*. IEEE, 2002.

[Fei00]   U. Feige. Approximating the bandwidth via volume respecting embeddings. *J. Comput. System Sci.*, 60(3):510–539, 2000.

[FM88]   P. Frankl and H. Mahera. The johnson lindenstrauss lemma and the sphericity of some graphs. *J. Combin. Theory Ser. B*, 44(2):355–362, 1988.

[HV02]   S. Har-Peled and K. R. Varadarajan. Projective clustering in high dimensions using core-sets. In *Proc. 18th Annu. ACM Sympos. Comput. Geom.*, pages 312–318, 2002.

[IM98]   P. Indyk and R. Motwani. Approximate nearest neighbor: towards removing the curse of dimensionality. In *Proceedings of the Thirty Second Symposium on Theory of Computing*, pages 604–613. ACM, 1998.

[JL84]   W. B. Johnson and J. Lindenstrauss. Extensions of Lipschitz mappings into a Hilbert space. In *Conference in modern analysis and probability (New Haven, Conn., 1982)*, pages 189–206. Amer. Math. Soc., Providence, RI, 1984.

[LLR95]   N. Linial, E. London, and Yu. Rabinovich. The geometry of graphs and some of its algorithmic applications. *Combinatorica*, 15(2):215–245, 1995.

[Mei93]   S. Meiser. Point location in arrangements of hyperplanes. *Information and Computation*, 106:286–303, 1993.

[PC00]   Agarwal P. and Procopiuc C. Approximation algortihms for strip cover in the plane. In *Proceeding of the 11th ACM-SIAM Symposium on Discrete Algorithms*, pages 373–382, 2000.

[Rao99]   S. Rao. Small distortion and volume preserving embeddings for planar and Euclidean metrics. In *Proceedings of the Fifteenth Annual Symposium on Computational Geometry (Miami Beach, FL, 1999)*, pages 300–306 (electronic), New York, 1999. ACM.

[Vem98]   S. Vempala. Approximating vlsi layout problems. In *Proceedings of the thirty ninth Symposium on Foundations of Computer Science*. IEEE, 1998.

On the Eigenvalue Power Law

Milena Mihail[1] and Christos Papadimitriou[2]

[1] College of Computing,
Georgia Institute of Technology
Atlanta, GA
{mihail}@cc.gatech.edu
[2] Computer Science Department,
Berkeley, CA
{christos}@cs.berkeley.edu

Abstract. We show that the largest eigenvalues of graphs whose highest degrees are Zipf-like distributed with slope α are distributed according to a power law with slope $\alpha/2$. This follows as a direct and almost certain corollary of the degree power law. Our result has implications for the singular value decomposition method in information retrieval.

1 Introduction

There has been a recent surge of interest in graphs whose degrees have very skewed distributions, with the ith largest degree of the graph about $ci^{-\alpha}$, for some positive constants c and α. Such distributions are called *Zipf-like distributions* (the Zipf distribution being the one with $\alpha = 1$) or *power laws*. In contrast, the degrees of random graphs in the traditional $G_{n,p}$ model [12] are, by the law of large numbers, exponentially distributed around the mean. It had been observed for some time that the graph of documents and hyperlinks in the worldwide web follow such degree distributions; in fact, there are several papers proposing plausible models (based on "preferential attachment" [5,6,2,3,9], or "copying"[21]) for explaining, with varying degrees of persuasiveness and rigor, this phenomenon.

More recently, in [14] it was pointed out that the Internet graph (both the graph of the routers and that of the autonomous systems) also has degrees that are power law distributed, with an exponent α between .85 and .93. This created much interest in the Internet research community, because the graph generators used by researchers had theretofore lacked this property; generators that are realistic in this sense have since appeared [18,23,24,26]. In [13] a theoretical explanation of this phenomenon was proposed, in terms of a model of network growth driven by the trade-off of two optimization criteria (connection costs and communication delays), predicting a power law degree distribution.

Another very interesting, intriguing, and as of yet unexplained observation in [14] is that the (twenty or so) largest eigenvalues of the Internet graph (that is, the largest eigenvalues of its adjacency matrix) *are also power law distributed,* with α between .45 and .5. This is in line with similar observations in physics with $a = .5$ (see [15,16] where a heuristic explanation is described). In fact,

J.D.P. Rolim and S. Vadhan (Eds.): RANDOM 2002, LNCS 2483, pp. 254–262, 2002.

all graph generators aiming to accurately simulate Internet topologies use the eigenvalue power law as a performance measure [18,23,24,26].

The distribution of the largest eigenvalues of Internet-related graphs is of additional special interest for the following reason: Spectral techniques [19,27, 4,1] based on the analysis of the largest eigenvalues and eigenvectors of the web graph have proven algorithmically successful in detecting "hidden patterns" such as semantics and clusters in the worldwide web. Is the Internet graph also amenable to such analysis?

In this note we provide a very simple, intuitive, and rigorous explanation of the eigenvalue power law phenomenon: We point out that *it is a rather direct and almost certain corollary of the degree power law*. In particular, we consider a random graph model whose degrees are, in expectation, $d_1, \ldots, d_n$ and show that, if these degrees are power-law distributed, then, with high probability, the few largest eigenvalues of the graph are close to $\sqrt{d_1}, \sqrt{d_2}, \ldots$ —and therefore follow a power law with exponent half of that of the degrees. This is in good agreement with the findings of [14], where the eigenvalue exponent is a little larger than half that of the degree exponent.

There is a negative implication of our result: By being essentially determined by the largest degrees (a very "local" aspect of a graph), the largest eigenvalues are unlikely to be helpful in analyzing and understanding the structure of the internet topology (the corresponding eigenvectors are highly concentrated on the largest degrees). In [25] we show experimental evidence that spectral analysis of the Internet topology becomes useful only after the high degrees have been suitably normalized. A similar problem in the use of spectral methods in "term-document" contexts is known as the "term norm distribution problem" [17]; it is considered the main bottleneck in the use of spectral filtering for information retrieval. Extending our study from the context of undirected graphs (symmetric matrices) to the context of terms and documents (general matrices) is an interesting technical problem with direct practical significance.

The rest of the paper is organized as follows: In Section 2 we review some basics from algebraic graph theory and matrix perturbation. We state a first theorem that indicates the effect of high degrees on the spectrum. In Section 3 we show that for a rich class of random graphs whose high degrees are Zipf distributed follow a power law on their highest eigenvalues, almost surely. In Section 4 we discuss implications of our results for the singular value decomposition method.

2   Eigenvalues and Degrees

We begin by recalling certain basic facts from algebraic graph theory and matrix perturbation. For a symmetric graph G with n nodes, we denote by $\lambda_1(G) \geq \lambda_2(G) \geq \cdots \geq \lambda_n(G)$ the eigenvalues of its adjacency matrix in non increasing order. E denotes the set of edges, and d_1 the highest degree of the graph.

Fact 1. (See Lovász [22], pages 70-73.)

1. For any graph G, $|\lambda_i(G)| \leq \min\{d_1, \sqrt{|E|}\}$.

2. *If G is a star with $n-1$ leaves, then $\lambda_1(G) = \sqrt{n-1}, \lambda_n(G) = -\sqrt{n-1}$ and $\lambda_i(G) = 0, i = 2, \ldots, n-1$.*
3. *The multiset of the eigenvalues of a graph G is the union of the eigenvalue multisets of its connected components.*

Now let A and B be symmetric $n \times n$ matrices and let $\lambda_1(A) \geq \lambda_2(A) \geq \cdots \geq \lambda_n(A)$ and $\lambda_1(B) \geq \lambda_2(B) \geq \cdots \geq \lambda_n(B)$ be their eigenvalues in non increasing order.

Fact 2. (See Wilkison [30], page 101.)
$$\lambda_i(A) + \lambda_n(B) \leq \lambda_i(A + B) \leq \lambda_i(A) + \lambda_1(B).$$

Now the following theorem is immediate:

Theorem 1. *Suppose that an undirected graph G can be decomposed into*

- *Vertex disjoint stars S_i with degrees d_i, $i = 1, \ldots, k$.*
- *Vertex disjoint components G_j with corresponding maximum degrees $d(G_j)$ and number of edges $e(G_j)$, such that $\min\{d(G_j), \sqrt{e(G_j)}\} = o(d_k)$, $j = 1, \ldots, m$. In addition all the components G_j are disjoint from all the stars S_i.*
- *A graph H with maximum degrees d and E edges such that $\min\{d, \sqrt{E}\} = o(d_k)$, where H can have arbitrary intersections with the S_i's and the G_j's.*

Then the largest eigenvalues of G are $\sqrt{d_i}(1 - o(1)) \leq \lambda_i \ \sqrt{d_i}(1 + o(1))$, $i = 1, \ldots, k$.

Remark 1: It is clear that the spectrum of G is dominated by the spectrum of the highest degree stars. It is worth noticing how much information this dominance can hide. In particular, H could be *any* sparse graph: connected, disconnected, with or without clusters, a tree, an expander. However, we would not be able to retrieve the structure of H from the spectrum of G. If G was the topology of the Internet, H could be the network backbone, and yet, all information about this structure would be lost. Indeed, in experiment, we have been able to decompose the Internet topology analyzed in [14] precisely along the lines of Theorem 1. The mere numbers are striking: For highest degree vertices in November 2000 d(UUNET)=2034, d(Sprint)=1079, d(C&WUSA)=793, d(AT&T)=742, d(BBN)=529, d(QWest)=483, d(AboveNet)=405, d(Verio)=363, d(BusInter)=347, d(GlobCros)=311 and d(Level3)=274, the highest eigenvalues squared were $\lambda_1^2 = 3113$, $\lambda_2^2 = 1135$, $\lambda_3^2 = 787$, $\lambda_4^2 = 676$, $\lambda_5^2 = 590$, $\lambda_6^2 = 515$, $\lambda_7^2 = 424$, $\lambda_8^2 = 395$, $\lambda_9^2 = 289$, $\lambda_{10}^2 = 277$, $\lambda_{11}^2 = 268$.

Remark 2: A technical statement analogous to Theorem 1 can be made about the stability of eigenvectors that correspond to largest eigenvalues (along the lines of Stewart [29]). As expected, the statement is that these eigenvectors of G are very "close" to the eigenvectors of the stars, and are hence highly concentrated on the vertices with the highest degrees.

3 Random Graphs

We next consider a distribution of graphs with prescribed degrees. Let $\mathbf{d} = (d_1, d_2, \ldots, d_n)$ be a vector of integers between 0 and n, in decreasing order. Denote $\sum_{l=1}^{i} d_l$ by D_i, and assume that $d_1^2 \leq D_n$. Define now $G_n(\mathbf{d})$ to be the distribution of graphs with n nodes generated by the following experiment: Edges are added by independent draws, where, for $i, j = 1, \ldots, n$ the probability that the edge $[i, j]$ is added is $\frac{d_i \cdot d_j}{D_n}$. Notice that we allow self-loops, an analytical convenience that does not affect the highest degrees much. Notice also that, by definition, node i has degree d_i, in expectation. This random graph model has been also considered in [8] and is known to have robust connectivity properties. Our main result is the following:

Theorem 2. *For any constant γ, with $0 < \gamma < 1$, for any constant α, with $\frac{1}{2} < \alpha < 1$, for any constant β, with $0 < \beta < \frac{1-\gamma}{2\alpha}$, and for any positive integer c, if $\mathbf{d} = (d_1, d_2, \ldots, d_n)$, with*

$$d_1^2 = D_n = \Theta(n^{1+\gamma}) \tag{1}$$

and

$$d_i = \frac{d_1}{i^\alpha}, \quad \text{for } i = 1, \ldots, k = \Theta(n^\beta), \tag{2}$$

then, for any constant β', with $0 < \beta' < \beta$, the eigenvalues of $G_n(\mathbf{d})$ satisfy

$$\sqrt{d_i}(1 - o(1)) \leq \lambda_i \leq \sqrt{d_i}(1 + o(1)), \quad \text{for } i = 1, \ldots, k' = \Theta(n^{\beta'}), \tag{3}$$

with probability at least $1 - O(n^{-c})$, for large enough n $(n \geq n_0(\alpha, \gamma, c))$.

Proof. We decompose G into the following graphs:

- G_1 is a union of vertex disjoint stars $S_1, \ldots, S_k$ where, for $i = 1, \ldots, k$, S_i has node i as its center and leaves those nodes from among $k+1, \ldots, n$ which are adjacent to i and not adjacent to any node in $\{1, \ldots, i-1\}$.
- G_1' contains all edges of G with one endpoint in $\{1, \ldots, k\}$ and the other in $\{k+1, \ldots, n\}$, except those in G_1.
- G_2 is the subgraph of G induced by $\{1, \ldots, k\}$.
- G_3 is the subgraph of G induced by $\{k, \ldots, n\}$.

We will show that the spectrum of G_1 dominates and that each star S_i has degree very close to its expectation d_i. Let s_i be the expected degree of S_i in G_1. To get a lowerbound, define F_i as the subset of vertices $\{k+1, \ldots, n\}$ not adjacent to $\{1, \ldots, i-1\}$ and notice:

$$\begin{aligned} s_i &= \sum_{l=k+1}^{n} \frac{d_i d_l}{D_n} - \sum_{l \in F_i} \frac{d_i d_l}{D_n} \\ &= d_i \sum_{l=k+1}^{n} \frac{d_l}{D_n} - \frac{d_i}{D_n} \sum_{l \in F_i} d_l \\ &\geq d_i \sum_{l=k+1}^{n} \frac{d_l}{D_n} - \frac{d_i d_k E[|F_i|]}{D_n} \end{aligned} \tag{4}$$

In the above expression we need to argue about the quantities $\sum_{l=k+1}^{n} \frac{d_l}{D_n}$ and $E[|F_i|]$. For the first sum first notice:

$$
\begin{aligned}
\sum_{j=1}^{k} d_j &= d_1 \sum_{j=1}^{k} j^{-\alpha} \\
&\simeq d_1 \frac{k^{1-\alpha}}{1-\alpha} \qquad \text{, by approx with an integral} \\
&= \frac{n^{\frac{1+\gamma}{2}} n^{\beta(1-\alpha)}}{1-\alpha} \qquad \text{, by equations (1) and (2)} \\
&= n^{1+\gamma} \frac{n^{-\frac{1+\gamma}{2}} n^{\beta(1-\alpha)}}{(1-\alpha)} \qquad \text{, which for } \beta < \frac{1-\gamma}{2\alpha} \text{ becomes} \\
&< n^{1+\gamma} \frac{1}{(1-\alpha)n^{1-\frac{1-\gamma}{2\alpha}}} \\
&= \Theta(D_n) \frac{1}{(1-\alpha)n^{1-\frac{1-\gamma}{2\alpha}}} \qquad \text{, with } \alpha > \frac{1}{2}.
\end{aligned}
\tag{5}
$$

Now (1) and (5) imply

$$
\sum_{l=k+1}^{n} d_l \simeq D_n .
\tag{6}
$$

It can be seen that equation (6) above can be satisfied provided the average degree of nodes $k+1$ through n is $\Omega(D_n/n)$. But the maximum expected degree of these nodes is d_k, which implies that $nd_k = \Omega(D_n)$. From equations (1) and (2) this is equivalent to $n \cdot n^{\frac{1+\gamma}{2}} \cdot n^{-\beta\alpha} = \Omega(n^{1+\gamma})$, which is indeed satisfied for β as in the statement of Theorem 2.

For $E[|F_i|]$ we have:

$$
\begin{aligned}
E[|F_i|] &= \sum_{j=1}^{i-1} \sum_{l=k+1}^{n} \frac{d_j d_l}{D_n} \\
&= \sum_{j=1}^{i-1} d_1 j^{-\alpha} \sum_{l=k+1}^{n} \frac{d_l}{D_n} \\
&\simeq \frac{d_1 i^{1-\alpha}}{1-\alpha} \qquad \text{, by approx with an integral and equation (6).} \\
&\leq \frac{d_1 k^{1-\alpha}}{1-\alpha} .
\end{aligned}
\tag{7}
$$

Now combining (4), (6) and (7) we get:

$$
\begin{aligned}
s_i &\geq d_i \left(1 - \frac{d_k d_1 k^{1-\alpha}}{D_n(1-\alpha)}\right) \\
&= d_i \left(1 - \frac{d_1^2 n^{-\beta\alpha} n^{\beta(1-\alpha)}}{D_n(1-\alpha)}\right) \quad \text{,by substitution} \\
&= d_i \left(1 - n^{\beta(1-2\alpha)}\right) \qquad \text{,with } \alpha > \frac{1}{2}.
\end{aligned}
\tag{8}
$$

Combining 8 with the obvious upper bound we get

$$
d_i(1 - n^{\beta(1-2\alpha)}) \leq s_i \leq d_i \quad \text{,with } \alpha > \frac{1}{2}.
\tag{9}
$$

To argue about sharp concentration of the degrees of the S_i's around their means we will use the standard Chernoff bounds for small probabilities of success [[28], Lecture 4]: For independent random variables $X_1, \ldots, X_N$, such that such that $\Pr[X_i=1]=p_i$ and $\Pr[X_i=0]=1-p_i$, and where $p=(\sum_{i=1}^{N} p_i)/N$:

$$
\Pr[|\sum_{i=1}^{N} X_i - pN| > s] < e^{-\frac{s^2}{2pN} + \frac{s^3}{2(pN)^3} + 1}
\tag{10}
$$

It can be readily checked from (9) and (10) that for some constant c', the actual degrees $\tilde{s}_i$ are concentrated as follows:

$$d_i - \sqrt{c'd_i \log n} \le \tilde{s}_i \le d_i + \sqrt{c'd_i \log n} \tag{11}$$

and the probability that (11) fails even for one $i = 1, \ldots, k$ is at most $n^{-c}/4$. Now Fact 1 implies that the largest eigenvalues of G_1 are

$$\sqrt{d_i}(1 - o(1)) \le \lambda_i(G_1) \le \sqrt{d_i}(1 + o(1)), \quad i = 1, \ldots, k \tag{12}$$

and the probability that (12) fails even for one $i = 1, \ldots, k$ is at most $n^{-c}/4$.

Let m_i be the expected degree of vertex i in the graph G_1', $i = 1, \ldots, k = n^\beta$. Then using the calculations of (7) and straightforward substitutions we get:

$$\begin{aligned}
m_i &= \sum_{l \in F_i} \frac{d_i d_l}{D_n} \\
&= \frac{d_i}{D_n} \sum_{l \in F_i} d_l \\
&\le d_i \cdot \frac{d_k E[|F_i|]}{D_n} \\
&\simeq d_i \cdot \frac{n^{-\alpha\beta} i^{1-\alpha}}{1-\alpha} \\
&\le d_i \cdot \frac{n^{\beta(1-2\alpha)}}{1-\alpha} \\
&= d_i \cdot \frac{n^{\beta(2\alpha-1)}}{1-\alpha} \quad , \text{ with } a > \tfrac{1}{2}.
\end{aligned} \tag{13}$$

For the expected degree of vertex i in the graph G_1' when $i = k+1, \ldots, n$ we have the obvious bound $d_i \le d_k$. This together with (13) and the Chernoff bound (10) suggest that, for some constant c'' all actual degrees $\tilde{t}_i$ of G_1' satisfy:

$$\tilde{t}_i \le d_k + \sqrt{c''d_k \log n} \tag{14}$$

and the probability that (14) fails even for one $i = 1, \ldots, n$ is at most $n^{-c}/4$.

The total number of edges for the graph G_2 is:

$$\begin{aligned}
\sum_{i=1}^{k} \sum_{j=1}^{k} \frac{d_i d_j}{D_n} &= \frac{d_1^2}{D_n} \sum_{i=1}^{k} i^{-\alpha} \sum_{j=1}^{k} j^{-\alpha} \\
&\le \frac{d_1^2}{D_n} k^{2(1-\alpha)} \\
&= n^{2\beta(1-\alpha)}
\end{aligned} \tag{15}$$

The above together with (10) suggest that, for some constant c''' the actual total number of edges $e(G_2)$ satisfy

$$\Pr[e(G_2) > n^{2\beta(1-\alpha)} + \sqrt{c'''n^{2\beta(1-\alpha)} \log n}] < n^{-c}/4. \tag{16}$$

For the graph G_3 we have the obvious bound that all its degrees are in expectations bounded by d_k, and hence are at most $d_k + \sqrt{c''d_k \log n}$ with probability as in (14). Combining this with (14), (16) and Fact 1 we get that the largest eigenvalues of each one of the graphs G_1', G_2 and G_3 are

$$\lambda_i(G_1'), \lambda_i(G_2), \lambda_i(G_3) \le \sqrt{d_k}(1 + o(d_k)), \quad i = 1, \ldots, k \tag{17}$$

and the probability that (17) fails even for one $i = 1, \ldots, k$ is at most $3n^{-c}/4$. We may now combine (12) and (17) and see that, for any $\beta' < \beta$, we have

$$d_k = o(d_i), \quad i = 1, \ldots, k' = n^{\beta'}$$

hence the statement of the Theorem follows.

Remark: We stated our result for the case in which the highest d_i's follow an exact power law; obviously, essentially the same conclusion holds if the degrees follow a less precise law (e.g., if the degrees are within constant multiples of the bounds). Finally, the $d_1^2 \leq D_n$ assumption is useful for keeping the $G_n(\mathbf{d})$ model simple; unfortunately, it does not hold for the Internet topology. However, the Internet, as measured in [14], does satisfy the assumption, if its few (5 or 6) highest-degree nodes are removed. These high-degree nodes do not affect the other degrees much, and do not harm our argument, no matter how adversely they may be connected.

4   Implication on SVD Method for Information Retrieval

Spectral filtering and, in particular the singular value decomposition (SVD) method is repeatedly invoked in information retrieval and datamining. It has also been ameanable to theoretical analysis and has yielded a remarkable set of elegant algorithmic tools [19,27,4,1]. However, in practice, SVD is weakened by the so-called "term norm distribution problem": this arises when terms are used in frequencies disproportionately higher than their relative significance, and several heuristics (so-called "inverse frequency normalizations") are known, however, none of them is known to perform adequately in theory or in practice (see [17] for a nice exposition). The term norm distribution problem appears very similar to the problem of high degrees that we treated here. It would be interesting to study the term norm distribution problem in a theoretical framework and quantify the proposed heuristics to overcome it. For the Internet topology of [14], in practice we solved the problem of high degrees by pruning small ISP's (leaves and a few more nodes) [25]. However, we do not have a formal framework for this method, and we do not know how it would extend in the case of term-documents or directed graphs.

The effectiveness of several of the SVD-based algorithms [4,1] requires that the underlying space has "low rank", that is, a relatively small number of significant eigenvalues. Power laws on the statistics of these spaces, including eigenvalue power laws, have been observed [7,20,10] and are quoted as evidence that the involved spaces are indeed low rank and hence spectral methods should be efficient. In view of the fact that the corresponding eigenvalue power law on the Internet topology was essentially a restatement of the high degrees and thus revealing no "hidden" semantics (see Remark 2 in Section 2), it is intriguing to understand what kind of information the corresponding power laws on the spectra of term-document spaces convey (or hide...)

Acknowledgements. The first author was supported by NSF under grant CCR-9732746 and by a Georgia Tech Edenfield Faculty Fellowship. The second author was supported by NSF under grant CCR-0121555.

References

1. Achlioptas, D., Fiat, A., Karlin, A. and McSherry, F., "Web Search via Hub Synthesis", *Proceedings of the 42nd Annual IEEE Symposium on Foundations of Computer Science*, (FOCS 2001), pp 500–509.
2. Aiello, W., Chung, F.R.K. and Lu, L., "A random graph model for power law graphs", *Proceedings of the Thirtysecond Annual ACM Symposium on Theory of Computing* (STOC 2000), pp 171–180.
3. Aiello, W., Chung, F.R.K. and Lu, L., "Random Evolution in Massive Graphs", Proceedings of the Fourty-Second Annual IEEE *Symposium on Foundations of Computer Science*, (FOCS 2001), pp. 510–519.
4. Azar, Y., Fiat, A., Karlin, A., McSherry, F. and J. Saia, "Spectral Analysis for Data Mining", Proceedings of the Thirty-Third Annual ACM Symposium on Theory of Computing, (STOC 2001), pp 619-626.
5. Barabási, A.-L. and Albert, R., "Emergence of scaling in random graphs", Science 286 (1999), pp 509–512.
6. Bollobás, B., Riordan, O., Spencer, J. and Tusnády, G., "The degree sequence of a scale-free random graph process", Random Structures and Algorithms,Volume 18, Issue 3, 2001, pp 279–290.
7. Broder, A., Kumar, R., Maghoul, F., Raghavan, P., Rajagopalan, S., Stata, R., Tomikns, A. and Wiener, J., "Graph structure in the Web", Proc. 9th International World Wide Web Conference (WWW9)/Computer Networks, 33(1-6), 2000, pp. 309–320.
8. Chung, F.R.K. and Lu, L., "Connected components in random graphs with given degree sequences", Available at http://www.math.ucsd.edu/ fan.
9. Cooper, C. and Frieze, A., A general model for web graphs, *Proceedings of ESA*, 2001, pp 500–511.
10. Dill, S., Kumar, R., McCurley, K., Rajagopalan, S., Sivakumar, D. and Tomkins, A., "Self-similarity in the Web", In Proceedings of International Conference on Very Large Data Bases, Rome, 2001, pp. 69–78.
11. Dorogovtsev, S.N. and Mendes, J.F.F., "Evolution of Networks", *Advances in Physics*, to appear (2002). Available at http://www.fc.up.pt/fis/sdorogov.
12. Erdös. P. and Rényi, A., "On the Evolution of Random Graphs", *Publications of the Mathematical Institute of the Hungarian Academy of Science* 5, (1960), pp 17–61.
13. Fabrikant, A., Koutsoupias, E. and Papadimitriou, C.H. "Heuristically Optimized Tradeoffs", Available at http://www.cs.berkeley.edu/ christos.
14. Faloutsos, M., Faloutsos, P. and Faloutsos, C., "On Power-law Relationships of the Internet Topology", In Proceedings *Sigcomm* 1999, pp 251-262.
15. Farkas, I.J., Derényi, I., Barabási, A.L. and Vicsek, T., "Spectra of Real-World Graphs: Beyond the Semi-Circle Law", e-print cond-mat/0102335.
16. Goh, K.I., Kahng, B. and Kim, D., "Spectra and eigenvectors of scale-free networks", Physical Review E., Vol 64, 2001.
17. Husbands, P., Simon, H. and Ding, C., "On the use of the Singular Value Decomposition for Text Retrieval", 1st SIAM Computational Information Retrieval Workshop, October 2000, Raleigh, NC.

18. Jin, C., Chen, Q. and Jamin, S., "Inet: Internet Topology Generator", University of Michigan technical Report, CSE-TR-433-00. Available at http://irl.eecs.umich.edu/jamin.
19. Kleinberg, J., "Authoritative sources in a hyperlinked environment", Proc. 9th ACM-SIAM Symposium on Discrete Algorithms, 1998. Extended version in Journal of the ACM 46 (1999).
20. Kumar, R., Rajagopalan, S., Sivakumar, D. and Tomkins, A., "Trawling the web for emerging cyber-communities", *WWW8/Computer Networks*, Vol. 31, No 11-16, 1999, pp. 1481–1493.
21. Kumar, R., Raghavan, P., Rajagopalan, S., Sivakumar, D., Tomkins, A. and Upfal, E., "Stochastic models for the Web graph", *Proceedings of the 41st IEEE Symposium on Foundations of Computer Science*, (FOCS 2000), pp 57–65.
22. Lovász, L., *Combinatorial Problems and Exercises*, North-Holland Publishing Co., Amsterdam-New York, 1979.
23. Medina, A., Lakhina, A., Matta, I. and Byers, J., BRITE: Universal Topology Generation from a User's Perspective. Technical Report BUCS-TR2001 -003, Boston University, 2001. Available at http://www.cs.bu.edu/brite/publications.
24. Medina, A., Matta, I. and Byers, J., "On the origin of power laws in Internet topologies", *ACM Computer Communication Review*, vol. 30, no. 2, pp. 18–28, Apr. 2000.
25. Gkantsidis, C., Mihail, M. and Zegura, E., "Spectral Analysis of Internet Topologies", Georgia Institute of Technology, Technical Report GIT-CC-0710.
26. Palmer, C. and Steffan, J., "Generating network topologies that obey power laws", In Proceedings of *Globecom* 2000.
27. Papadimitriou, C.H., Raghavan, P., Tamaki, H. and Vempala, S., "Latent Semantic Indexing: A Probabilistic Analysis", *Journal of Computer and System Sciences*, 61, 2000, pp. 217–235.
28. Spencer, J., *Ten Lecture Notes on the Probabilistic Method*, SIAM Lecture Notes, Philadelphia, 1987.
29. Stewart, G. W. and Sun, J., *Matrix Perturbation Theory*, Academic Press, 1990.
30. Wilkinson, J. H., *The Algebraic Eigenvalue Problem*, Numerical Mathematics and Scientific Computation, Oxford University Press, 1965.

Classifying Special Interest Groups in Web Graphs

Colin Cooper

Department of Mathematical and Computing Sciences,
Goldsmiths College,
London SW14 6NW, UK.
`c.cooper@gold.ac.uk`

Abstract. We consider the problem of classifying special interest groups in web graphs. There is a secret society of blue vertices which link preferentially to each other. The other vertices, which are red, are unaware of the distinction between vertex colours and link to vertices arbitrarily. Each new vertex directs m edges towards the existing graph on joining it. The colour of the vertices is unknown. We give an algorithm which **whp** classifies all blue vertices, and all red vertices of high degree correctly. We also give an upper bound for the number of mis-classified red vertices.

1 Introduction

1.1 Definition of the Problem

We assume we have a graph which models the link structure of the world wide web. The vertices of this graph have a colour, either red or blue. The colour of a vertex is initially unknown to us. For example, this colour may indicate the response to some query, such as 'find all web pages which have property P'. Thus blue pages are those which have property P, and red pages are those which do not have this property.

In the *classification problem*, we attempt to determine which vertices are red and which vertices are blue. There are several versions of the classification problem. In some versions the colour can be decided by inspecting the page or the address of the page (eg. a vertex is blue if the url is `xxx.ac.uk`). In other cases a vertex would be blue if the page contained enough key words of the required type (eg. `graph`, `web`, `colour`).

The problem we consider here, is the problem of *secret societies*. The blue vertices form a special interest group, and link preferentially to each other. The red vertices are unaware of the colour distinction and can link to vertices of either colour. A blue vertex would never admit to being blue, ie. there is nothing on the page to allow a vertex colour to be discovered by inspection. The problem is to try to find the colours of the vertices using the structural properties of the graph.

J.D.P. Rolim and S. Vadhan (Eds.): RANDOM 2002, LNCS 2483, pp. 263–275, 2002.

Thus the problem is akin to the problem of finding a hidden vertex partition in a graph. Problem of this type have been studied in the context of standard random graphs (eg. $G_{n,p}$) by many authors. See for example [16] for a comprehensive discussion of this problem and an extensive bibliography. The analysis for web graphs differs from these cases, in that a web graph is not a standard random graph, web graphs are extremely sparse, continuously growing so that the partition is augmented at each step, and the edge probability of a vertex in a web graph is dependent on its age.

In a belief propagation model [8,9] the perceived colour (or probability of colour) of each vertex is iteratively updated in a Bayesian manner, based on the opinions of neighbours. Such an algorithm requires an initial classification of vertex colours in which each node of the graph is (independently) assigned classification (perceived colour) based on some rule. This estimate is then improved on the basis of the classification of neighbours, next neighbours and so on, until some sort of stable value is (hopefully) arrived at.

The algorithm (CLASSIFY) which we describe here, is a simple two pass algorithm which processes the vertices sorted by decreasing degree. The algorithm first makes an initial classification (based on vertex degree). This is followed by a restricted one pass belief propagation, in which the classification of some vertices is updated based on the perceived colour of in-neighbours.

We claim that the algorithm can work well for classifying vertices of high degree (hub-authority vertices) and quite well for vertices of low degree. For the simple web-graph model on which we evaluate the algorithm, a more precise statement of performance is given in Theorem 1.

The performance of the algorithm depends on a parameter η_A, $(A =$ red , blue). η_A is a simple function of the probability p that a vertex is red. We assume that p is known, or could be estimated. The maximum in-degree of the red (resp. blue) vertices is about t^{η_R} (resp. t^{η_B}). As $p \to 1$, $\eta_R \to 1/2$ and $\eta_B \to 2/3$. This difference in the degree sequences of the red and blue vertices is exploited by the algorithm.

A web graph is a sparse connected graph designed to capture some properties of the www. Studies of the graph structure of the www were made by [4] and [7] among others. There are many models of web graphs designed to capture the structure of the www found in the studies given above. For example see references [1], [2], [3], [5], [6], [10], [11], [12], [14] and [15] for various models.

In the simple model we consider, each new vertex directs m edges towards existing vertices according to the degree of existing vertices (copy model). With probability p the new vertex takes the colour red, and with probability $q = 1 - p$ it takes the colour blue.

The new vertex v_t selects m vertices of $G(t-1)$ as neighbours, and directs edges from itself to these vertices. We require that the colour of these neighbours depends on the colour of v_t as follows: In the model we consider, the blue vertices are *biased*, so a new blue vertex always chooses existing blue vertices as neighbours. The red vertices are *arbitrary* in their choice of neighbours, in the sense that they do not discriminate on vertex colour, or are even perhaps unaware of the distinction. We model this arbitrariness by allowing a red vertex to choose a mix of red and blue vertices according to a binomial process, $Bin(m,p)$.

At each step the decision to add a red vertex is made by an independent Bernoulii $Be(p)$ process, so that the expected number of red vertices added in steps $1, ..., t$ is pt. Thus, provided $t \to \infty$, $|R(t)| \sim pt$, so that $p \sim |R(t)|/|V(t)|$, and p good approximation for the probability that a red vertex is selected when a neighbour vertex is chosen arbitrarily.

Once the number of red and blue neighbours of the new vertex has been decided, then the actual neighbour vertices are selected. Suppose, for example, that at step t, a blue vertex is to be selected as the i-th neighbour $(1 \le i \le m)$ of v_t. Let $d(B, t-1)$ be the total degree in $G(t-1)$ of the set $B(t-1)$ of blue vertices existing at step $t-1$. Let $w \le t-1$ be a blue vertex of $G(t-1)$ of total degree $d(w, t-1)$, then

$$\mathbf{Pr}(v_t \text{ chooses } w) = \frac{d(w, t-1)}{d(B, t-1)}.$$

This sampling is repeated independently and with replacement for all edges directed out of v_t towards blue vertices. The sampling of red neighbours occurs similarly.

We suppose the initial graph $G(0)$ upon which $G(t)$ is built, is of constant size, is connected, and consists of some non-trivial mixture of red and blue vertices. We do not assume that the initial graph $G(0)$ is acyclic, but we require that all vertices of $G(0)$ have out-degree m.

Let $V(t)$ denote the vertex set of $G(t)$. Thus $V(t) = V(G(0)) \cup \{1, ..., t\}$. At time t, $G(t)$ has $\hat{t} = t + |G(0)|$ vertices. For simplicity we assume $t \to \infty$ so that $\hat{t}$ is well approximated by t.

We assume that we know the edges of $G(t)$, including their direction. The vertices are labelled in some arbitrary order which gives no information about their age. Each vertex v has exactly m out-edges, directed away from v, arising when v was added to G. Any other edges incident with v are directed towards v.

We suppose that do not know the colours of the vertices of $G(t)$. At first glance, there is not much about the structure of $G(t)$ which allows us to determine the colour of the vertices by inspecting the graph.

We describe a deterministic algorithm (algorithm CLASSIFY) which classifies the vertices of $G(t)$ into two sets $\boldsymbol{R}$, $\boldsymbol{B}$. The set $\boldsymbol{R}$ contains the vertices which we suppose to be red, and $\boldsymbol{B}$ those we suppose to be blue.

The algorithm classifies the vertices using the sorted degree sequence of the underlying graph. The only other information the algorithm uses is the directed adjacency structure of the vertices.

Let $\eta_R = \frac{p}{(1+p)}$, $\eta_B = \frac{(1+p)}{(2+p)}$. Let $\alpha = \eta_R^2/(\eta_R + m(1 - \eta_B)[(1 - \epsilon)^2 \eta_B - \eta_R])$ where $\epsilon > 0$ is a constant (eg $\epsilon = 0.01$). Let $d_0 = t^\alpha \log t$. Let λ be the smallest non-negative solution of $\lambda = (q + \lambda p)^m$.

Theorem 1. whp *algorithm* CLASSIFY

i)   *Correctly classifies all blue vertices.*
ii)   *Correctly classifies all red vertices of degree at least d_0.*
iii) *Provided $q > \lambda$, correctly classifies all red vertices of degree at least $K \log t$.*
iv)   *Mis-classifies at most a limiting proportion λ of red vertices.*

The time complexity of the algorithm is dominated by sorting the vertices of $G(t)$ into decreasing degree sequence, and building an adjacency list structure for $G(t)$.

A red vertex which chooses m blue vertices (an event of probability q^m) can be considered to be re-coloured *purple*. Red vertices which choose a mixture of blue and purple vertices also become purple. A major problem for the algorithm will be to correctly classify purple vertices.

We make the convention that $\omega = \omega(t)$ is a generic function of t which tends slowly to infinity.

2   Algorithm CLASSIFY-I

At the end of algorithm CLASSIFY-I we claim that **whp**

(i)   All blue vertices are correctly classified.
(ii)   All red vertices of degree at least d_0 are correctly classified.
(iii) All non-purple (distinctly red) vertices are correctly classified.

begin-algorithm-CLASSIFY-I

Let $\boldsymbol{R}$ be the set of vertices classified as red and let $\boldsymbol{B}$ be the set of vertices classified as blue. Initially $\boldsymbol{R}, \boldsymbol{B}$ are empty.
Let $U = V \setminus (\boldsymbol{R} \cup \boldsymbol{B})$ denote the set of unclassified vertices. Initially $U = V$.
Any vertex which is added to $\boldsymbol{R} \cup \boldsymbol{B}$ by the algorithm is simultaneously removed from U.

A1 Put all vertices of degree at least $\Delta_R = (t^{\frac{p}{(1+p)}} \omega \log t)$ into $\boldsymbol{B}$.
A2 If any vertex $v \in \boldsymbol{B}$ has an edge pointing to any w in U, when w is classified blue and added to $\boldsymbol{B}$, until this no longer occurs.
A3 Let d be the maximum degree of U.
    Let $\delta(d) = (d/(\omega \log t))^{(1-\epsilon)\eta_B/\eta_R}$.
    Let $K(d) = \{u \in \boldsymbol{B} : d(u) \geq \delta(d)\}$.
    Pick a vertex $v \in U$ of degree d.
    Let $F^+(v)$ be the fan-out of v within U.
    If $F^+(v)$ contains a directed cycle C, add the vertices of C to $\boldsymbol{R}$.
A4 For each $w \in F^+(v)$ we consider only the m edges pointing out of w.
    Choose $w \in F^+(v)$ such that all out-edges of w point to $\boldsymbol{R} \cup \boldsymbol{B}$.
    If any out-edge of any w points to $\boldsymbol{R}$ add w, to $\boldsymbol{R}$.
    If any w points to $K(d)$ with all m out-edges, add w to $\boldsymbol{R}$.
    Otherwise add w to $\boldsymbol{B}$.
A5 Repeat Steps A3-A4 until all vertices are classified.

end-algorithm-CLASSIFY-I

Analysis of Algorithm CLASSIFY-I

The main problem is that we would like to make our analysis using the time when a vertex was born, whereas the only information we have about the vertices is their degree. The distribution of degree $d(s,t)$ of any vertex s in $G(t)$ is a function of its age and colour. The following facts are established in the Appendix.

Lemma 1. *Let* $d_A(s,t)$ *be the degree after step* t *of the vertex born at step* s, *then*

i) *The expected degree at step* t *of an* A-coloured vertex $(A = B, R)$ *born at step* s *is*

$$\mu_A(s,t) = m\left(\frac{t}{s}\right)^{\eta_A}(1+o(1)),$$

where $\eta_R = \frac{p}{(1+p)}$, $\eta_B = \frac{(1+p)}{(2+p)}$. *We note that* $\eta_B > \eta_R$ *for all* $p \in [0,1]$.

ii)

$$\mathbf{Pr}(d(s,t) \le a) \le e^{-h\left(c\left(\frac{t}{s}\right)^{\eta(1-\epsilon)}-a\right)},$$

where $c, \epsilon > 0$ *constant and* $h = h(\epsilon)$, *constant.*

iii)

$$\mathbf{Pr}\left(d(s,t) \ge \left(\frac{t}{s}\right)^{\eta}\omega\log t\right) \le t^{-K},$$

where $\omega \to \infty$ *arbitrarily slowly and* K *is an arbitrary constant.*

Step A1 This initialization step classifies all vertices of degree at least $\Delta_R(t^{\frac{p}{(1+p)}}\omega\log t)$ as blue. This is correct **whp**, as by Lemma 1 (iii) no red vertex can have degree this large.

Step A2 This ensures that edges point from U to $\boldsymbol{B} \cup \boldsymbol{R}$. The algorithm maintains this at all future steps by processing the fan-out $F^+(v)$ of the selected vertex v.

Step A3 If $F^+(v)$ contains a directed cycle C then $C \subseteq G(0)$, as $G(t)$ is acyclic apart from (possibly) $G(0)$. Of necessity, all vertices of this cycle have the same colour, which must be red, as all the blue vertices of $G(0)$ have been moved into $\boldsymbol{B}$ during A1.

As the fan-out $F^+(v)$ is acyclic, we can always find a *fringe* vertex w with all m out-edges pointing towards classified vertices $\boldsymbol{R} \cup \boldsymbol{B}$. We reduce the fan-out from the fringe inwards to v.

There is a technical point here. The fan out of v, $F^+(v)$, may contain vertices w of degree $d(w,t)$ less than d. However, the correct classification of these vertices is assured. If w is blue, the analysis for A4 assures us that w cannot point only to $K(d)$. If w is red, then w was born before v, and can only point to red vertices of $\boldsymbol{R}$ or blue vertices of $K(d)$.

Step A4 Given a vertex v of degree d, which points exclusively to $\boldsymbol{B}$, how can we decide if it is red or blue?

The birth time of an A-coloured vertex (A = red, blue) of degree d lies in the interval $I_A(d) = (\tau_A, T_A)$ **whp** , where

$$I_A(d) = \left(t \left(\frac{c}{d} \right)^{1/((1-\epsilon)\eta_A)}, t \left(\frac{\omega \log t}{d} \right)^{1/\eta_A} \right).$$

At time t the lowest possible degree of a blue vertex born at or before time T_R is

$$\delta(d) = 2c \left(\frac{t}{T_R} \right)^{\eta_B(1-\epsilon)} = 2c \left(\frac{d}{\omega \log t} \right)^{\eta_B(1-\epsilon)/\eta_R}.$$

If v points exclusively to $\boldsymbol{B}$-vertices of degree at least δ (ie: to $K(d)$) we put v in $\boldsymbol{R}$. This would result in a mis-classification if a blue vertex of degree d pointed exclusively to blue vertices of degree at least δ.

The latest possible birth time of a blue vertex of degree $\delta(d)$ is $T' = T_B(\delta(d))$ where

$$T_B(\delta(d)) = t \left(\frac{\omega \log t}{\delta(d)} \right)^{1/\eta_B} = t \frac{(\omega \log t)^{1/\eta_B + (1-\epsilon)/\eta_R}}{c^{1/\eta_B} d^{(1-\epsilon)/\eta_R}}.$$

Let $B(T')$ denote blue vertices born at or before T'. Given v is blue, let $\pi(v)$ be the probability that v only chooses its neighbours from $B(T')$.

At time T' the degree of $B(T')$ was $\theta_B T'(1 + o(1))$, where $\theta_B = mq(2 + p)$. At time v the degree of $B(T')$ is at most

$$d(B(T'), v) = \theta_B T' \left(\frac{v}{T'} \right)^{\eta_B} \omega \log v.$$

Thus

$$\pi(v) = (1 + o(1)) \left(\frac{d(B(T'), v)}{\theta_B v} \right)^m \leq \left(\omega \log v \left(\frac{T'}{v} \right)^{1-\eta_B} \right)^m.$$

Now $d \geq c(t/v)^{(1-\epsilon)\eta_B}$, so

$$\pi(v) \leq \left((\omega \log t)^\xi \left(\frac{v}{t} \right)^\lambda \right)^m,$$

where $\xi = (1 + (1 - \eta_B)(1/\eta_B + 1/\eta_R))$ and $\lambda = (1 - \eta_B)((1 - \epsilon)^2 \eta_B / \eta_R - 1)$. We will choose v_0 such that $\sum_{v \leq v_0} \pi(v) = o(1)$. Now

$$\sum_{v \leq v_0} \pi(v) \leq \frac{(\omega \log t)^{\xi} v^{m\lambda + 1}}{t^{m\lambda}}.$$

Let

$$v_0 = \left(\frac{o(1)t^{m\lambda}}{(\omega \log t)^{\xi}} \right)^{\frac{1}{1+m\lambda}}, \qquad d_0 = \log t \left(\frac{t}{v_0} \right)^{\eta_R},$$

then v_0, d_0 are respectively the birth time and degree of the last red vertex which is **whp** guaranteed correctly classified by CLASSIFY-I. Below d_0, purple vertices may appear in $\boldsymbol{B}$.

3   Algorithm-CLASSIFY-II

The purpose of algorithm-CLASSIFY-II is to re-classify those purple vertices of degree between d_0 and $K \log t$ which have been mis-classified as blue. We regard all vertices of degree d_0 as correctly classified by algorithm-CLASSIFY-I.

We give an upper bound on the proportion of mis-classified red vertices of degree at most $K \log t$.

Let $\zeta_B = 1/(1+p), \zeta_R = \lambda/(q+\lambda p)$ where λ is the smallest non-negative solution of $\lambda = (q + \lambda p)^m$. We assume $\zeta_B > \zeta_R$.

begin-algorithm-CLASSIFY-II

Initially all vertices of $\boldsymbol{B}$ of degree at most d_0 are *unprocessed*.

Pick an unprocessed vertex v of maximum degree d. Mark v as processed.

If $d_B^-(v) > \frac{1}{2}(\zeta_B + \zeta_R)(d(v) - m)$ return v to $\boldsymbol{B}$, else place v into $\boldsymbol{R}$.

Any vertex which points to a red vertex is classified red.

end-algorithm-CLASSIFY-II

Some care is needed in defining what we mean by a purple vertex in the context of the algorithm. It may well be that every red vertex of $G(0)$ points only to blue vertices, so that every red vertex of $G(t)$ is (in this sense) purple. However we are assured **whp** no red vertex of degree at least d_0 is in $\boldsymbol{B}$. Thus a purple vertex is a red vertex of degree at most d_0 mis-classified into $\boldsymbol{B}$.

As an upper bound on the probability a vertex is purple, we can consider the extinction probability β of a binomial branching process $Bin(m, p)$. The process becomes extinct if all branches eventually have zero progeny (eventually only blue vertices are reached). The value of β is the smallest non-negative root of $g(x) = (q+px)^m - x$. The value of β is an upper bound on the limiting proportion $\lambda(t)$ of purple vertices in $G(t)$ for several reasons.

i) $G(t)$ is finite so $\beta(t) \leq \beta$. This follows because extinction probabilities are monotone non-decreasing as the number of levels in the branching increases.

ii) If the branching process hits a red vertex of degree at least $d_0(t)$, it cannot become extinct.

iii) Red neighbours are selected based on degree, so the process is biased towards red vertices of degree at least d_0.

We can imagine our branching process modified as follows. After the number, j say, of red descendants of vertex $t+1$ is sampled, there is a second sampling process, based on degree, to determine the labels of the descendants. If the descendant τ has degree $d(\tau) \geq d_0$ then the process is halted without extinction. When a label τ is selected then further branching can only occur to labels less than τ. The second sampling process occurs with replacement, so there is a possibility that a vertex may be hit more than once during a branching. However this is also biased towards vertices of larger degree.

Let $t \geq \tau$. Let $d_P(t)$ denote the total degree in $G(t)$ of purple vertices, then

$$d_P(t+1) = d_P(t) + 1\{c(t+1) = R\} \times$$
$$\sum \binom{m}{j} q^{m-j} p^j \left(m \left(\frac{d_P(t)}{d_R(t)} \right)^j + \sum \binom{j}{k} k \left(\frac{d_P(t)}{d_R(t)} \right)^k \left(1 - \frac{d_P(t)}{d_R(t)} \right)^{j-k} \right).$$

Now $\mathbf{E}\, d_P(t) \geq 2mpq^m(t - \tau)$ and $\mathbf{E}\, d_P(t) < \mathbf{E}\, d_R(t)$ as there is a positive probability $(1-\beta)$ of a red vertex surviving in the worst case binomial branching. Thus $\mathbf{E}\, d_P(t) = c(t)t$ for $t \gg \tau$.

The values $d_P(t), d_R(t)$ are sharply concentrated within $O(\sqrt{t}\log t)$ of their expected values **whp** (see [10]), so for $1 \leq j \leq m$, $\mathbf{E}\,(d_P(t)/d_R(t))^j$ is well approximated by $f(t) = (\mathbf{E}\, d_P(t)/\mathbf{E}\, d_R(t))^j$. Thus

$$\theta f(t+1) \sim \theta f(t) + p \sum \binom{m}{j} q^{m-j} p^j (m(f(t))^j + j f(t)).$$

Noting that $\theta_R = mp(1 + p)$ we obtain

$$(1+p)f(t+1) = (1+p)f(t) + (q + f(t)p)^m + pf(t).$$

If $f(t)$ ever increases to $f(t) = \lambda t$ where λ is the smallest non-negative root of

$$\lambda = (q + \lambda p)^m. \tag{1}$$

We should in principle be able to discriminate between a blue vertex v and a purple vertex w of the same degree d, based on the number of edges directed into v (resp. w) from $\mathbf{B}$ (vertices classified blue/purple) and from $\mathbf{R}$ (vertices classified red).

Lemma 2. *Let* $\zeta_B = 1/(1 + p)$, $\zeta_P = \lambda/(q + \lambda p)$. *Provided* $\zeta_B > \zeta_P$ **whp** *algorithm* CLASSIFY-II *can correctly re-classify all purple vertices in* $\mathbf{B}$ *of degree at least* $K \log t$ *into* $\mathbf{R}$.

Proof.

Let $d_A^-(s,t)$ denote the in-degree after step t of an A-coloured vertex s, from A-coloured vertices τ, $(s < \tau \leq t)$.

$$\mathbf{E}\, d_B^-(s,t) \mid d(s,t), c(s) = B = \frac{1}{1+p}(d(s,t) - m)(1 + o(1))$$

$$\mathbf{E}\, d_P^-(s,t) \mid d(s,t), c(s) = P = \frac{\lambda}{q + \lambda p}(d(s,t) - m)(1 + o(1)).$$

The random variable $d_A^-(s,t)$ is the sum of $l = d(s,t) - m$ independent $\{0,1\}$-indicator variables.

We first consider the case where s is blue. What is the probability that s will be chosen by vertex τ?

$$\mathbf{Pr}(\tau \text{ chooses } s, c(\tau) = A \mid c(s) = B) = \begin{cases} pmq\frac{d(s,\tau)}{d_B(\tau)}(1+o(1)) \ c(\tau) = R \\ mq\frac{d(s,\tau)}{d_B(\tau)}(1+o(1)) \ \ c(\tau) = B \end{cases}.$$

Thus

$$\mathbf{Pr}(c(\tau) = B \mid \tau \text{ chooses } s, c(s) = B) = (1+o(1))\frac{1}{1+p},$$

and

$$\mathbf{E}\, d_{\overline{B}}(s,t) \mid d(s,t), c(s) = B = (1+o(1))\frac{1}{1+p}(d(s,t) - m).$$

We next consider the case where the colour of s is purple (P).

$$\mathbf{Pr}(\tau \text{ chooses } s \mid c(s) = P) = mp^2\frac{d(s,\tau)}{d_R(\tau)}(1+o(1)).$$

$$\mathbf{Pr}(\tau \text{ chooses } s, \ c(\tau) = P) = p\sum \binom{m}{j}p^j\left(\frac{d_P(\tau)}{d_R(\tau)}\right)^{j-1} q^{m-j}j\frac{d(s,\tau)}{d_R(\tau)}(1+o(1))$$

$$= (1+o(1))mp^2(q+\lambda p)^{m-1}\frac{d(s,\tau)}{d_R(\tau)}.$$

Thus

$$\mathbf{Pr}(c(\tau)=P \mid \tau \text{ chooses } s, c(s)=P) = (q+\lambda p)^{m-1}(1+o(1))=\frac{\lambda}{q+\lambda p}(1+o(1)).$$

By the Hoeffding inequality [13]

$$\mathbf{Pr}(d_{\overline{A}}(s,t) \notin [(1-\epsilon)\mathbf{E}\, d^-{}_A(s,t), (1+\epsilon)\mathbf{E}\, d^-{}_A(s,t)]) \leq 2\exp\left(-\frac{\epsilon^2}{3}\zeta_A d(s,t)\right).$$

If v is an un-processed vertex of $\mathbf{B}$ of degree $d(v)$ the algorithm classifies v as blue if

$$d_{\overline{B}}(v) > \frac{1}{2}(\zeta_B + \zeta_P)(d(v) - m),$$

which corresponds to the value of $\epsilon_B = \frac{1}{2}(\zeta_B - \zeta_P)/\zeta_B$, and $\epsilon_P = \frac{1}{2}(\zeta_B - \zeta_P)/\zeta_P$.
    Thus the probability that a vertex of degree ω is mis-classified is $O(e^{-c\omega})$. Choosing $\omega = K\log t$ where K is a large constant, Theorem 1(ii) follows.

References

1. D. Achlioptas, A. Fiat, A.R. Karlin and F. McSherry, Web search via hub synthesis, *Proceedings of the 42nd Annual IEEE Symposium on Foundations of Computer Science* (2001) 500-509.
2. M. Adler and M. Mitzenmacher, *Toward Compressing Web Graphs*, To appear in the 2001 Data Compression Conference.
3. W. Aiello, F. Chung and L. Lu, Random evolution in massive graphs, *Proceedings of the 42nd Annual IEEE Symposium on Foundations of Computer Science* (2001) 510-519.
4. R. Albert, A. Barabasi and H. Jeong. *Diameter of the world wide web.* Nature 401:103-131 (1999) see also http://xxx.lanl.gov/abs/cond-mat/9907038
5. B. Bollobás, O. Riordan and J. Spencer, *The degree sequence of a scale free random graph process*, to appear.
6. B. Bollobás and O. Riordan, *The diameter of a scale free random graph*, to appear.
7. A. Broder, R. Kumar, F.Maghoul, P. Raghavan, S. Rajagopalan, R. Stata, A. Tomkins and J. Wiener. *Graph structure in the web.*
   http://gatekeeper.dec.com/pub/DEC/SRC /publications/stata/www9.htm
8. A. Broder, R. Krauthgamer and M. Mitzenmacher. *Improved classification via connectivity information.*
9. S. Chakrabarti, B. Dom and P. Indyk. *Enhanced hypertext categorization using hyperlinks.* Proceedings of ACM SIGMOD Conference on management of Data (1998) p307-318.
10. C. Cooper and A.M. Frieze, A general model of web graphs, *Proceedings of ESA 2001*, 500-511.
11. E. Drinea, M. Enachescu and M. Mitzenmacher, *Variations on random graph models for the web.*
12. M.R. Henzinger, A. Heydon, M. Mitzenmacher and M. Najork, Measuring Index Quality Using Random Walks on the Web, *WWW8 / Computer Networks* 31 (1999) 1291-1303.
13. W. Hoeffding, Probability inequalities for sums of bounded random variables, *Journal of the American Statistical Association* 58 (1963) 13-30.
14. R. Kumar, P. Raghavan, S. Rajagopalan, D. Sivakumar, A. Tomkins and E. Upfal. *The web as a graph.* www.almaden.ibm.com
15. R. Kumar, P. Raghavan, S. Rajagopalan, D. Sivakumar, A. Tomkins and E. Upfal. *Stochastic models for the web graph.* www.almaden.ibm.com
16. F. McSherry. *Spectral partitioning of random graphs* FOCS 2001

4   Appendix: Distribution of Vertex Degree

Let A can mean either red or blue, and B is the opposite colour (local change of notation). Let $N(t)$ be the number of edges directed out of vertex t which are incident with A-coloured vertices other than t. Let p_A be the probability that vertex t is A-coloured, and let $q_A = 1 - p_A$ be the probability that vertex t is B-coloured. Let $\widehat{p}$ be the probability of an A-coloured vertex choosing an A-coloured neighbour, and $\widetilde{q}$ be the probability of a B-coloured vertex choosing an A-coloured neighbour. The distribution of $N(t)$ is given by

$$N(t) \sim 1_A Bin(m, \widehat{p}) + 1_B Bin(m, \widetilde{q}), \tag{2}$$

where $Bin(m, p)$ is a Binomial random variable with parameters m, p and $1_A = 1\{c(t) = A\}$ is the indicator for the event that vertex t is A-coloured.

In the notation of the main body of this paper, when A denotes red, then $p_A = p$, $q_A = q = 1 - p$, $\widehat{p} = p$, $\widetilde{q} = 0$, and when A denotes blue, then $p_A = q$, $\widehat{p} = 1$, $\widetilde{q} = q$.

Let $d(A, t)$ be the total degree of the A-coloured vertices of $G(t)$. Then

$$\mathbf{E}\, d(A, t) = t(pm + pm\widehat{p} + qm\widetilde{q}) = \theta_A t,$$

say. Thus in red, blue notation $\theta_R = mp(1 + p)$ and $\theta_B = mq(2 + p)$.

The value of $d(A, t)$ is the sum of t independent $\{0, ..., 2m\}$-random variables and is sharply concentrated in t. The inequality

$$d(A, t) = \theta_A t \left(1 + O\left(\sqrt{\tfrac{K \log t}{\theta t}} \right) \right) \tag{3}$$

holds simultaneously with probability $1 - t_0^{-K+1}$ for all $t \geq t_0$ and $t_0 > 0$.

Let $X_t = X_t(v)$ denote $d(v, t)$ the degree of vertex v at time t.

Expected value of vertex degree.

Let $\eta_A = m(p_A\widehat{p} + q_A\widetilde{q})/\theta_A$. Let v be A-coloured. For any $v \leq \tau \leq t$ let $d(v, \tau)$ be given, then **whp**

$$\mathbf{E}\, d(v, t) = d(v, \tau) \left(\frac{t}{\tau} \right)^{\eta_A} \left(1 + O\left(\sqrt{\tfrac{\log \tau}{\tau}} \right) \right). \tag{4}$$

From (2) $X_{t+1} = X_t + Y_{t+1}$ where

$$Y_{t+1} = \sum_j Bin\left(j, \frac{X_t}{d(A, t)} \right) \left(1_A \binom{m}{j} \widehat{p}^j\, \widehat{q}^{m-j} + 1_B \binom{m}{j} \widetilde{q}^j\, \widetilde{p}^{m-j} \right).$$

Thus

$$\mathbf{E}\, X_{t+1} = X_t \left(1 + \frac{m(p\widehat{p} + q\widetilde{q})}{d(A, t)} \right).$$

The result (4) follows from iterating the expectation of this identity and applying (3).

Lower bound on vertex degree.

Let $X_t = X_t(s)$. The lower bound will follow from

$$\mathbf{Pr}(X_t \leq a) \leq e^{ha}\mathbf{E}\left(e^{-hX_t} \right).$$

As before, let $X_t = X_{t-1} + Y_t$. Let $Z_t = Z_{t-1} + U_t$ where U_t is a Bernoulii random variable $Be(\eta' Z_{t-1}/(t-1))$ and $\eta' = \eta_t = \eta(1 - O(\sqrt{\log t/t}))$. Thus Y_t is a Binomial random variable with the same parameter as U_t. Hence X_t stochastically dominates Z_t, and thus

$$\mathbf{E}\left(e^{-hX_t} \right) \leq \mathbf{E}\left(e^{-hZ_t} \right).$$

We now prove that

$$\mathbf{E}\left(e^{-hZ_t}\right) \le e^{-hc_s m\left(\frac{t}{s}\right)^{(1-\epsilon)\eta}},$$

where $c_s \ge 1$ is a constant depending on s and $\epsilon = \epsilon(h)$ is such that $1 - e^{-h} \ge (1 - \epsilon)h$.

$$\mathbf{E}\left(e^{-hU_{t+1}} \mid Z_t\right) = 1 - \frac{\eta' Z_t}{t}(1 - e^{-h})$$

$$\le 1 - \frac{\eta' Z_t}{t}(1 - \epsilon)h$$

$$\le \exp\left(-hZ_t\frac{\eta'(1 - \epsilon)}{t}\right).$$

Thus

$$\mathbf{E}\left(e^{-hZ_t}\right) \le \mathbf{E}\ \exp\left(-h\left(1 + \frac{\eta'(1-\epsilon)}{t-1}\right)Z_{t-1}\right)$$

$$\le \exp\left(-hm\prod_{\tau=s...t-1}\left(1 + \frac{\eta'(1-\epsilon)}{\tau-1}\right)\right)$$

$$\le \exp\left(-hm\left(\frac{t}{s}\right)^{(1-\epsilon)\eta}\left(1 + O\left(\sqrt{\frac{\log s}{s}}\right)\right)\right).$$

The result follows on choosing $c_s = \left(1 + O\left(\sqrt{\frac{\log s}{s}}\right)\right)$.

Upper bound on vertex degree.

This will follow from

$$\mathbf{Pr}(X_t \ge a) \le e^{-ha}\mathbf{E}\left(e^{hX_t}\right),$$

on choosing

$$h = \frac{1}{\omega}\left(\frac{s}{t}\right)^\eta \qquad a = \left(\frac{t}{s}\right)^\eta \omega K \log t.$$

Let $\lambda = X_t/d(A,t)$ then from (2)

$$\mathbf{E}\left(e^{hY_{t+1}} \mid X_t\right) = p(1 + \widehat{p}\lambda(e^h - 1))^m + q(1 + \widetilde{q}\lambda(e^h - 1))^m.$$

Now, for $h \le 1$, $e^h - 1 \le h + h^2$ and $\lambda h = O(h)$ as $h \to 0$. Thus

$$\mathbf{E}\left(e^{hY_{t+1}} \mid X_t\right) \le 1 + m(p\widehat{p} + q\widetilde{q})\lambda(h + h^2) + O((\lambda h)^2)$$

$$= 1 + hX_t\frac{\eta}{t}(1 + O(h))\left(1 + O(\sqrt{(\log t)/t})\right).$$

Thus

$$\mathbf{E}\ e^{h_t X_t} \le \mathbf{E}\ e^{h_{t-1}X_{t-1}},$$

where

$$h_{t-1} = h_t\left(1 + \frac{\eta}{t-1}(1 + O(h_t))\left(1 + O\left(\sqrt{\frac{\log t-1}{t-1}}\right)\right)\right).$$

Let $h = (1/\omega)(s/t)^\eta$, and let $h_t = h$. We claim that, for $\tau \le t$

$$h_\tau = c_\tau h \left(\frac{t}{\tau}\right)^\eta, \qquad c_\tau = \Omega(1).$$

We note first that

$$h_{\tau-1} = c_\tau h \left(\frac{t}{\tau}\right)^\eta \left(1 + \frac{\eta}{\tau-1} + \frac{O(h_\tau)}{\tau-1} + \frac{O((1+h_\tau)\log(\tau-1))^{1/2}}{(\tau-1)^{3/2}}\right)$$

$$= h \left(\frac{t}{\tau-1}\right)^\eta c_\tau \left(1 + \frac{\eta}{\tau(\tau-1)} + \frac{O(1)}{\tau^2} + \frac{O(h_\tau)}{\tau-1} + \frac{O((1+c_\tau/\omega)\log(\tau-1))^{1/2}}{(\tau-1)^{3/2}}\right),$$

where the second line follows from expanding $(\tau-1)/\tau)^\eta$ and multiplying out.
Now, certainly $c_t = 1$ and suppose $c_t, ..., c_\tau$ are $\Omega(1)$. Let $c^* = \max(c_t, ..., c_\tau)$.

$$c_{\tau-1} = \prod_{r=\tau...t} \left(1 + \frac{O(1)}{r^2} + \frac{O(c^* h t^\eta)}{r^{1+\eta}} + \frac{O(\log r)^{1/2}}{r^{3/2}}\right)$$

$$\le \exp\left(O(1)\left(\sum \frac{1}{r^2} + hc^*\left(\frac{t}{\tau}\right)^\eta + \frac{O(1)\sqrt{\log\tau}}{\sqrt{\tau}}\right)\right)$$

$$\le \exp\left(O\left(1 + \sqrt{\frac{\log s}{s}}\right)\right).$$

Author Index

Lecture Notes in Computer Science

For information about Vols. 1–2371
please contact your bookseller or Springer-Verlag